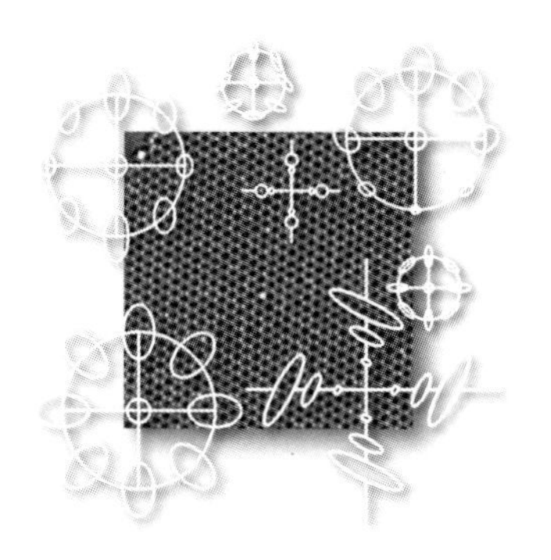

# Aberration-Corrected Imaging in Transmission Electron Microscopy

## An Introduction

Rolf Erni

Swiss Federal Laboratories for Materials Science and Technology (Empa),
Switzerland

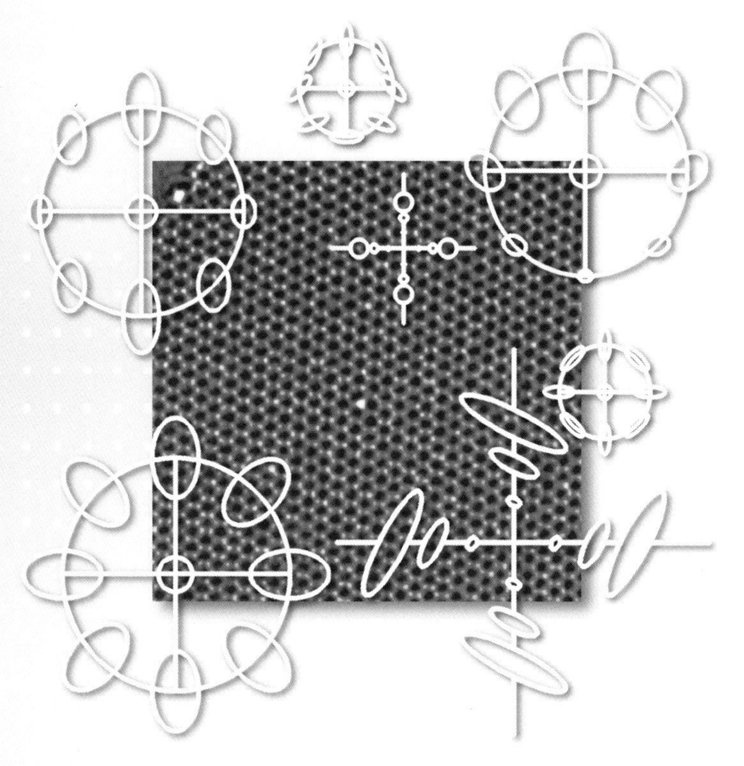

# Aberration-Corrected Imaging in Transmission Electron Microscopy

## An Introduction

ICP

Imperial College Press

*Published by*

Imperial College Press
57 Shelton Street
Covent Garden
London WC2H 9HE

*Distributed by*

World Scientific Publishing Co. Pte. Ltd.
5 Toh Tuck Link, Singapore 596224
*USA office:* 27 Warren Street, Suite 401-402, Hackensack, NJ 07601
*UK office:* 57 Shelton Street, Covent Garden, London WC2H 9HE

**British Library Cataloguing-in-Publication Data**
A catalogue record for this book is available from the British Library.

**ABERRATION-CORRECTED IMAGING IN TRANSMISSION ELECTRON MICROSCOPY**



ISBN-13 978-1-84816-536-6
ISBN-10 1-84816-536-6

Printed in Singapore by Mainland Press Pte Ltd.

# Preface

*The tendency of people to*
*take small things as important*
*has resulted in many great things.*

G.C. Lichtenberg

Aberration-corrected electron microscopy is a key term in the headline of numerous research reports which involve state-of-the-art electron microscopy equipment. The implementation of spherical aberration correctors in transmission electron microscopes has directly and indirectly enabled revolutionary applications which a little more than ten years ago would not have been feasible or even thinkable. Nonetheless, aberration correction is not the only advancement which enriches the repertoire of experiments that can be carried out in electron microscopes. Besides aberration correction, electron microscopes have been brought to application which, for instance, provide temporal resolution to study dynamics on the nanosecond scale. Another branch of instrument development focuses on *in situ* measurements. Devices have been realized which enable the study of materials and their evolution in liquid media or in controlled gaseous atmospheres, or which make it feasible to monitor the very basic deformation mechanism of materials. This involves the controlled deformation of nanometer-sized objects. Furthermore, many of the analytical capabilities of electron microscopes can nowadays be considered to be comparable to dedicated analytical instruments, which, however, do not provide the unique spatial resolution that is inherent for electron microscopes. Yet, spatial resolution is often indispensable for systematically analyzing complex nanostructured materials. Of course, there are many more directions in the development of state-of-the-art electron microscopy equipment. Nevertheless, many of these developments can potentially benefit from aberration-corrected electron optics. As such, the correction of the resolution-limiting aberration of conventional electron microscopy has to be considered as one of the great inventions which has advanced materials science and our experimental understanding of matter. There is no reason why this trend should not be continued.

Although aberration-corrected electron microscopy has many facets, the primary advantage of overcoming the fundamental geometrical limitation of round electron lenses is the enhanced resolution in atomic-resolution imaging. This book is about spherical aberration-corrected electron microscopy, focusing on this particular aspect: on the application of spherical aberration correctors for atomic-resolution imaging in the broad-beam transmission mode, as well as in the scanning transmission mode. This book does not provide a review of recent experimental achievements that have become possible through the application of top-notch microscopes. This work shall serve as an introductory text which explains some of the fundamental concepts in aberration-corrected microscopy. The text also addresses practical aspects. For instance, it outlines strategies on how to optimize the electron optical setup in order to improve the data that can be collected with an aberration-corrected microscope. The central topic is spherical aberration correction, yet, some of the very recent developments, which go beyond spherical aberration correction, are also elucidated in the text.

The bulk of this book was written in 2009, i.e., roughly ten years after the successful implementation of the first spherical aberration correctors in (scanning) transmission electron microscopes. This book was written with the intention of providing a text which bridges between application-oriented reviews about aberration-corrected electron microscopy and advanced physics textbooks which focus on the electron optical concepts underlying this key technology in electron microscopy imaging.

During the time I have worked with advanced and, in particular, aberration-corrected electron microscopes, I have had the privilege of being guided by various very knowledgable persons to whom I owe many thoughts that went into this text. Without exceptions, my friends, colleagues and collaborators have been my teachers. I would especially like to acknowledge Drs. Peter Hartel, Heiko Müller and Maximilian Haider from CEOS GmbH, who introduced me to the field of aberration-corrected electron microscopy and who have always been very supportive in any matter that arose in the practical handling and optimization of aberration-corrected instruments. I would like to acknowledge Drs. Ondrej Krivanek and Niklas Dellby from NION Company for open conversations, instructive explanations and for sharing their expertise with me. I further wish to thank my former colleagues at FEI Company, in particular Drs. Stephan Kujawa, Bert Freitag, Peter Tiemeijer, Sergei Lopatin, Maarten Bischoff, Michiel van der Stam and Hans van Lin. I would like to acknowledge former colleagues at the National Center for Electron Microscopy at the Lawrence Berkeley National Laboratory and collaborators of the TEAM project, namely Drs. Quentin Ramasse, Marta D. Rossell, Prof. Masashi Watanabe, Christian Kisielowski, Prof. Andrew Minor, Peter Denes, Ulrich Dahmen, and Drs. Andrew Lupini, Juan-Carlos Idrobo and Steve Pennycook from Oak Ridge National Laboratory. I wish to thank Prof. Nigel Browning who to a great extent paved the way for me to work with high-end electron-optical instruments. In addition, I would like to thank the following persons with whom I have collaborated on research projects that helped me to understand topics addressed in this book: Dr. Sandra Van Aert from Antwerp University and Prof. Joanne Etheridge, Drs. Christian Dwyer and Philip Nakashima from Monash University. I would like to express my gratitude to Dr. Helge Heinrich and Prof. Gernot Kostorz, who introduced me to various fields in materials science and in particular to electron microscopy, and taught me the scientific way of addressing research topics during my Ph.D. studies at ETH Zurich. Special thanks go to the staff of the library at ETH Zurich, whom I numerous times challenged, deliberately, but not unintentionally, in retrieving source texts which were of extraordinary value to me. Finally, I would like to thank Marta D. Rossell and my parents Gaby and Peter Erni for their patience, understanding and continuous support.

*Rolf Erni*

January, 2010

Empa, Dübendorf, Switzerland

Swiss Federal Laboratories for Materials Science and Technology

# Contents

# Chapter 1

# Introduction

## 1.1 Where are the Atoms?

Uncovering objects that are too small to be seen by eye: this is the primary purpose of microscopy. The human eye enables us to resolve objects that are as small as a few tens of micrometers. Many objects and structures that affect and control our daily life are orders of magnitudes smaller than what is observable by eye. In order to explore and understand objects like individual cells or viruses, the grain structure of crystalline materials or the self-organization of nanocrystals, microscopes are the tools that make it possible to surpass our natural physical limits. They enable us to observe the very details of our environment. Microscopes are our keys to the micro- and nano-universe. Keys that are good for specific purposes but which are neither perfect nor universal.

There is no microscope that uncovers the platonic essence of the object being imaged. Every microscope elucidates a very specific characteristic of the object. Indeed, there is no need for a microscope to unravel the whole nature of an object. In most cases we know quite a lot about the object which is analyzed in the microscope. We might know its basic chemical composition, what type of cell it is, or which micro- or crystal structure it is supposed to exhibit. Typically, there is a whole set of *a priori* information available, and there are very specific questions which need to be addressed in an investigation.

We expect that a microscope enables us to form a magnified image of a given object that answers one of the questions we have. An image collected in a microscope is supposed to reflect a certain characteristics of the object, like its structure, its constitution or, for example, its shape. The microscope provides us with an image of details that we cannot see by eye. Yet an image is never equivalent to the object. An image is an interpretation of the object. This is true for a painting, a photograph and a micrograph. In microscopy, the interpretation strongly depends on the probe or the radiation which is used to explore the object. Different techniques and different microscopes might have different resolutions, but the information in the micrograph, regardless of whether a certain feature is resolved, is still supposed to reflect true object information. What should be prevented is that the imaging characteristics of the microscope alter the object information in such a way that the relation between image and object becomes obscure. The microscope must not complicate the relation between object and image. Hence, the interpretation of the object shall be kept simple.

The resolving power of a conventional optical microscope is of the order of the wavelength of the light that is used for imaging. By utilizing an optical microscope we can thus expand our observations down to objects that are a bit smaller than a micrometer. Even though this is substantially better than what is doable without a microscope, it is still by far insufficient to resolve the atomic structure of materials; i.e. the basic skeleton of materials.

In order to be able to derive an understanding about what fundamentally constitutes a material and, in particular, how the structure of a material is related to its physical properties, it is often necessary to measure information about the material's atomic framework. Of course, diffraction techniques such as X-ray diffraction or neutron diffraction are extremely powerful in indirectly accessing atomic structure information. But in general, a diffraction pattern obtained from X-rays or neutrons does not contain local information. For a perfect crystal or for a material whose properties can be described from the statistical average of the structure over a large volume fraction, local information is not necessarily needed, and global information is even more valuable. Hence, it is not always necessary to access atomic-scale information to understand a particular property of a material and, indeed, a micro- or even a macroscopic model can provide sufficient depth to explain a particular property. Such micro- or macroscopic approaches are often based on a continuous model where the atomic structure and the atoms' individual contributions to the overall properties are averaged; i.e. where the atomic discontinuity is neglected or levelled out over small sub-volumes. Nevertheless, local information is of crucial importance for the study of individual small objects, like nanoparticles, but also for investigating defects such as dislocations, grain boundaries or individual precipitates. Indeed, it is often the case that local deviations from perfect crystallinity, like defects or boundaries, define the real functionality of a material.

Knowing where which species of atoms does what, i.e. observing the location of each individual atom, measuring which element it is and how it is connected to the neighboring atoms: this is the fundamental level at which materials science starts. As a matter of fact, the properties of a material are determined by the atomic constitution, by its skeleton and, in particular, by the configuration and density of defects. Eventually, it is the agglomerate of *discrete* atoms that constitutes the material and defines its physical properties.

The smaller a system, the higher the importance of the building blocks that constitute the system. This is true for any kind of network, and it is especially applicable to nanomaterials. Nanomaterials are in principle ordinary materials, except for the fact that their physical extension in at least one direction is of the order of nanometers to tens of nanometers. Nanomaterials are small systems. Therefore, when we deal with nanomaterials, such as clusters of atoms that consist merely of hundreds or thousands of atoms, it is undeniable that atomic-scale information is of the highest importance. Unlike bulk materials whose intrinsic physical properties do not depend on the size of the piece of material under consideration, for nanomaterials this can be fundamentally different.

A bulk semiconductor has a characteristic bandgap energy. This is an intrinsic physical property of the semiconductor which does not depend on the volume or mass of the material. The type of atoms forming the bulk material and the way they are arranged define the physical properties. But for a semiconductor nanocrystal this can be different. Below a critical size of the nanocrystal, say for instance 10 nm, the bandgap energy can become a function of the size of the crystal. The bandgap essentially transforms into an extrinsic physical property which of course still depends on the type of atoms constituting the nanocrystal and how these atoms are arranged, *but* also on the number of atoms that are in the system. In order to relate the property of the material to its structure, it is mandatory to know how many atoms of which type form the nanocrystal. Of course, this is not limited to the case of quantum confinement effects of nanocrystals. Nanomaterials can in general exhibit physical properties which depend strongly on the size, shape or, for instance, on the surface configuration of the atoms. Similarly, the core structure of a dislocation, the surface structure of solids or the arrangement of precipitates in an alloy — for all these cases it can be of crucial importance to know how the atoms are arranged on the skeleton of the solid. Hence, there are many cases where we need to know where which atomic species does what. It is for these cases that atomic-resolution imaging is an indispensable tool in materials science.

Electron microscopes enable us to observe objects that are smaller than what is accessible by optical microscopes. Instead of using optical light for imaging, transmission electron microscopes and scanning transmission electron microscopes employ electrons with energies in the range of about 50 to 1,000 keV. The wavelength of such electrons is between 5 and 1 pm. While in light optics it is feasible to access structural information that is of the order of the wavelength of the optical light that is used for imaging, this is not the case in electron optics. Though the wavelength of the electrons is in the range of a few picometers, the resolution of conventional high-resolution microscopes is roughly two orders of magnitude larger. Since the spacing between atoms arranged on a crystal lattice is in the range of the order of 200 pm, the resolution is sufficient to access some kind of atomic-resolution information. Indeed both modes, i.e. the broad-beam transmission mode and the scanning transmission mode, have been employed to resolve the atomic structure of crystals. Even the detection of individual heavy atoms on thin support films has been possible.

Since electron microscopy enables atomic-resolution imaging, one is tempted to say that by having a small enough wavelength the problem of atomic-resolution imaging is solved. This conclusion is indeed partially true. The higher the energy of the electrons used for imaging, the better the resolution. However, especially when working with small materials whose many surface atoms are removable quite easily, the damage the electrons can cause to the material is significant. Hence, one cannot simply increase the electron energy to see more details of the specimen — one also needs to make sure that the object is not severely modified by probing it with electrons.

Furthermore, the resolution problem in conventional electron microscopy is not directly related to the electron wavelength. There are numerous factors which can affect the resolution of a microscope, regardless of whether this is a light or an electron microscope. One of these factors is the effect of aberrations, i.e. the non-ideal imaging characteristics of lenses. An aberration describes an imaging characteristic which causes the image to deviate from the object. The image is no longer a simple interpretation of the object. Aberrations lead to false and truncated images.

With respect to aberrations, light and electron optics are not equivalent; the imaging characteristics of electron lenses is not perfect. Of course, neither is any real-light optical lens perfect. But the difference between light and electron optics lies in the fact that in light optics the dominant lens aberrations can be compensated by, for example, serially arranging convex and concave electron lenses. In electron optics on the other hand, this is not the case. There are no concave electron lenses. The inherent lens errors of electron lenses cannot be corrected simply by adding a complementary lens of opposite error. Instead, the errors add up. Electron lenses, and in particular round electron lenses, i.e. lenses which are rotationally symmetric about the optical axis, suffer from aberrations which cannot be compensated simply by a second lens. Though affected by aberrations, this does not imply that electron micrographs necessarily show false object information. By strongly narrowing the path of the electron beam through the microscope, it is possible to approach a nearly perfect imaging characteristic. False information is minimized and the micrographs show true object information. Minimizing the impact of the aberrations on the image comes, however, at the expense of a very limited resolution.

The resolution limitation imposed by the aberrations does not mean that there is no object information beyond the resolution limit contained in a micrograph. The resolution limit is a quantity that describes the smallest object detail that can be interpreted directly in a micrograph. This is not necessarily the smallest object detail that contributes to the image. Indeed, modern, but conventional, electron microscopes transfer object information to the image which is beyond the actual resolution limit. However, since this high-resolution information is not free of aberrations, we cannot read it directly from a micrograph. Hence, although atomic-resolution imaging is feasible with conventional (scanning) transmission electron microscopes, lens aberrations can cause image artefacts which obscure the object information contained in a micrograph. Whether an atom is imaged as a white or as dark spot in the image is a question of the imaging characteristics of the microscope, which can be dominated by the lens aberrations. But it is not only the image contrast. Object details are imaged into areas of the micrograph such that a simple relation between image and object is lost. Under these circumstances the micrographs do not tell us where the atoms are.

Aberration correction is about correcting the intrinsic aberration(s) of round electromagnetic electron lenses which are typically used in conventional electron microscopes. Aberration-corrected imaging is about the formation of micrographs which no longer suffer from unwanted aberrations, which thus show an enhanced resolution and which contain information that is a true and simple interpretation of the object.

The improved resolution and the simplified image interpretation are only two aspects of the benefits of aberration-corrected electron microscopy. The smaller electron probes, which can be formed in aberration-corrected probe-forming instruments, enable a higher probe current. This enhances the signal-to-noise ratio in analytical measurements. Furthermore, since the enhanced resolution reduces the instrument-specific blurring of the object information contained in a micrograph, the object information can be imaged with a higher signal-to-noise ratio. This simplifies quantification of micrographs, as for instance the derivation of quantitative chemical information or distortion measurements on grounds of atomic-resolution micrographs. Hence, there is a whole set of side effects which comes along with the enhanced resolution of aberration-corrected electron microscopes.

## 1.2 Brief Historical Overview

Compared to other fields in physics, electron optics is a rather young discipline. It might not be wrong to state that its launch happened around 1924 when Louis de Broglie postulated the wave characteristics of the electron in his doctoral thesis. Shortly thereafter, Hans Busch (1926) described the focusing characteristics of electromagnetic lenses on pencils of electrons. Both findings were crucial for electron microscopy; while de Broglie showed that electrons possess a characteristic which is similar to the wave characteristic of optical light, Busch (1926) revealed that the trajectories of electrons can be controlled by electromagnetic fields in a way similar to which light rays can be controlled by glass lenses. What occurred during the following five to ten years was a technological and scientific breakthrough that is undoubtedly astonishing. In less than a decade, the fundamental principles of electron optics were developed. Devices were constructed that generate electrons of high flux and instruments were developed that modify the trajectories of electrons in a controlled manner. The development of electron sources and electron lenses was crucial for the invention of the electron microscope in 1931 by Ernst Ruska and Max Knoll (see, e.g. Ruska 1987). At that time, Ruska worked as a doctoral student in a group headed by Knoll in Berlin. In parallel with the technical development of the electron optical instruments, the development of electron optics, i.e. the development of the theoretical tools needed to understand the path of electrons exposed to well characterized electromagnetic fields, was carried out.

In 1932, Knoll and Ruska had already predicted that the theoretical resolution limit of the electron microscope is of the order of 2 Ångström (Knoll and Ruska, 1932). The first electron microscopes that were built in the 1930s did not reach a resolution that even came close to the Ångström scale though. However, it was clear that similar to light optics, the inherent imaging properties of the electron lenses need to be fully understood in order to be able to gather the fundamental characteristics of the imaging process in a multi-lens electron optical system and to reveal the resolution limiting factors. A first theoretical derivation of the inherent aberrations of a purely electric, i.e. non-magnetic, electron lens was carried out by Otto Scherzer (1933). Shortly after that, Scherzer (1936b) showed that for stationary round electron lenses which are free of charges, the constant of spherical aberration and the constant of chromatic aberration are finite positive. This result is known as the *Scherzer theorem*. At the time Otto Scherzer came to this conclusion, the electron microscopes were not limited by lens aberrations but by other factors, such as the brightness of the electron sources, the (mechanical) stability of the systems or by the lack of axial symmetry of the round lenses, which arose due to mechanical and material-specific imperfections.

Roughly ten years after Scherzer's theoretical study of the aberration characteristic of round electron lenses and their impact on the resolution limit (Scherzer, 1939), the first report on approaching the theoretical resolution of an electron microscope was published by James Hillier (1946). By developing a new electron source that enabled an electron 'intensity' that was a factor of twenty better than previous sources, as well as by improving the axial symmetry of the magnetic electron lenses, Hillier (1946) reported a spatial resolution of about 10 Å. This was the first report of an electron microscope that approached its theoretical resolution determined by the lens aberrations, which, according to Scherzer (1933), were unavoidable. From then on, the performance of electron microscopes had continuously improved, but still their ultimate resolution performance had always been limited by the unavoidable aberrations of round electron lenses.

As pointed out by Scherzer (1936b), the aberrations of round electron lenses impose a practical but not a fundamental barrier. In 1947, Scherzer laid out strategies to correct for the chromatic and the spherical aberration of a round electron lens. One of these strategies involved non-round optical elements, i.e. optical lenses which are not rotationally symmetric about the optical axis but which produce an electromagnetic field of reduced azimuthal symmetry. The application of such multi-pole lenses turned out to be the most promising way of correcting lens aberrations. Scherzer's work (1947) was the starting point of a 50-year period of various attempts to design and build electron optical units which would correct the lens aberrations and thus enable for an enhanced resolution. Indeed, a first corrector was built by Scherzer's student Robert Seeliger (1949, 1951, 1953), who essentially showed that the spherical aberration can be decreased by employing three so-called *Korrekturstücke*, i.e. correction units, which consist of electrostatic octupole fields. This corrector was then further developed by Möllenstedt (1956). Even though this corrector was functional, it could not improve the resolution. There were other attempts, for instance the one from Deltrap (1964), who built a corrector based on a design suggested earlier by Archard (1955), or a corrector based on a fundamentally different principle which was suggested by Beck (1979) and further advanced by Crewe and Kopf (1980), Crewe and Salzman (1982), Crewe (1984) and Rose (1981). The variety of approaches was so large that it was even necessary to mention *How Not To Correct An Electron Lens* — as the title of a short note by Scherzer (1982) suggests — which was a response to a less fruitful idea.

The first aberration correctors that proved to increase the resolution of electron microscopes were built in the 1990s. The very first of these workable aberration correctors was installed in a scanning electron microscope (Zach and Haider, 1995). Shortly after that, aberration correctors for transmission electron microscopes were brought to application: one for a dedicated scanning transmission electron microscope (Krivanek *et al.*, 1997) and one for a broad-beam illumination transmission electron microscope (Haider *et al.*, 1998). These pioneering inventions were the starting point of a new era in electron microscopy, an era that has led to electron microscopes which are largely indispensible in the research of nanomaterials.

This very brief introductory overview is certainly not complete and more snapshots about the chronological evolution of today's working aberration correctors are elucidated in Chapter 8. For now it shall suffice to point out that the development of working aberration correctors has a long history. A long history of attempts, which were not all as successful as hoped for, yet all these practical attempts and the theoretical knowledge that was developed in parallel (see, e.g. Rose (1971b)) Rose, 1971a, 1971b) helped to build electron microscopes which are no longer limited by the intrinsic aberrations of round electron lenses (Zach and Haider, 1995; Krivanek *et al.*, 1997; Haider *et al.*, 1998). Indeed, from the beginning of electron microscopy in 1932 until the first working aberration correctors were built, a period passed which is identical to the time that passed between the first flying motorized manned aircraft built by the Wright brothers in 1903 and the moment the first man put a foot on the moon. Of course, these two evolutionary processes are not comparable, neither on the relevant scale of length nor in the money that was needed to warrant a certain degree of progress. The point, however, is that building a working aberration corrector needed the results of technological advancements that span more than half a century. As in many areas of science and technology, the basic principles are often quite straightforward. The difficulty lies in realizing them. Indeed, the complexity of the correctors is not just defined by the ingenious optical elements employed but also arises from the fact that once a potential corrector is available, it needs to be set up such that the corrector does what it is supposed to do without doing things it should not do.

## 1.3 Scope of the Book

The scope of this book is to provide an introduction to aberration-corrected atomic-resolution transmission electron microscopy. The book covers both the scanning transmission mode and the broad-beam transmission mode. As such, it is not mode-specific. Preferably, the reader is familiar with the concepts of electron diffraction and conventional electron microscopy, and has some practical experience with electron microscopes. The text does not focus on the mathematical derivation of how an aberration corrector works. It is a book intended, but not exclusively, for experimentalists and students who want to know how this black box on an aberration-corrected microscope works and who want to know the guidelines on how (and why) to optimize the experimental conditions on aberration-corrected microscopes. It aims at making the reader familiar with concepts, strategies and terms which are used in the daily handling of aberration-corrected microscopes, as well as in the specific literature about it. Although the book does not aim at treating the concepts about electron optics, aberrations and aberration correction from a strictly mathematical point of view, it shall make the reader familiar with some of the fundamental ideas in electron optics which are applied in conventional as well as in aberration-corrected microscopy. The text can also serve as a guide for readers who want to start with atomic-resolution imaging on conventional electron microscopes.

This work shall bridge the gap between texts focusing on electron and charged-particle optics on the one hand and application-oriented books on the other hand, whose focus is on materials science. It shall motivate the reader to dig deeper into the literature about aberration-corrected electron microscopy and electron optics. It is an introductory text — it is a primer.

This book provides an 'image' of aberration-corrected electron microscopy from a more application-oriented rather than a purely theoretical point of view. As any image, it is an interpretation of the subject.

## 1.4 General Outline

The text is structured in three main parts, each comprising two to three chapters. Part I, entitled Fundamentals, introduces the reader to the fundamental concepts of the two most widely used atomic-resolution electron microscopy techniques: phase contrast imaging in the broad-beam illumination mode (Chapter 2) and scanning transmission electron microscopy (Chapter 3). The focus lies on conventional electron microscopes which are not equipped with aberration correctors. Chapter 4 illustrates the limits of conventional electron microscopes.

The title of the second part is Electron Optics. It is split into two chapters. Chapter 5 introduces the reader to some of the fundamental concepts of electron optics, discussing topics like the Hamiltonian analogy, the point eikonal, the refractive index of electrons and geometrical wave surfaces. Chapter 6 deals with the Gaussian dioptrics of a round electron lens with a straight optical axis. The path equations of a round electromagnetic lens are derived, the theorem of optical imaging is discussed and cardinal elements, as well as some of the fundamental rays, are defined and illustrated. This chapter also elucidates the sources of aberrations in conventional electron microscopy.

The third part of the book comprises three chapters. Chapter 7 introduces the concepts of image and wave aberrations, and discusses axial and off-axial as well as geometrical and, more briefly, chromatic aberrations. It deals with the isoplanatic approximation and defines the axial aberration function. Chapter 8 deals with the basic concepts of spherical aberration correctors. It first illustrates the effect of multi-pole lenses and then introduces the reader to the basic concepts of the quadrupole–octupole corrector and the hexapole corrector. The chapter also contains a more detailed historical overview and it provides a brief outlook to the next generation of aberration correctors. Chapter 9 focuses on practical aspects when working with aberration-corrected electron microscopes. It describes the effect of aberration correction on the imaging characteristics. It also provides practical guidelines on how to work with an aberration-corrected microscope and how the imaging performance can be optimized.

Though the three parts of the book provide a comprehensive introduction to aberration-corrected electron microscopy, the individual parts of the book can in principle be read independently from each other. Each part of the book is largely self-contained.

# PART I
# Fundamentals

Chapter 2

# High-Resolution Transmission Electron Microscopy

The most fundamental imaging mode of a transmission electron microscope is realized by illuminating an electron-transparent specimen with a broad electron beam followed by recording an image of the magnified electron beam after it transmitted the specimen. What we naturally expect from this operation mode is that the image produced by the microscope, i.e. the micrograph, reflects in a distinct way a projection of the specimen along the main beam direction.

There are many different contrast mechanisms that can be used to form an image in an electron microscope (see, e.g. Williams and Carter, 1996). Each technique elucidates a certain aspect of the specimen and thus can be used to address a specific question about the specimen, such as whether there are dislocations present, which type they are and how they are arranged. The choice of the contrast mechanism is based on the object feature that shall be imaged in the micrograph.

Here we restrict the discussion of imaging techniques in transmission electron microscopy to the two fundamental and most common techniques that are used for atomic-resolution imaging. In the present chapter we discuss phase contrast imaging, and in Chapter 3 we discuss the basics of scanning transmission electron microscopy (STEM). Though the contrast mechanisms and acquisition processes are fundamentally different, both the phase contrast method and STEM are most suitable for imaging the arrangement of the atoms in the specimen, i.e. for uncovering the basic skeleton of a material.

The goal of this part of the book is firstly to introduce the atomic-resolution imaging techniques and secondly to elucidate the limiting factors in conventional electron microscopy[1]. We do not draw our attention to the electron-specimen and electron-atom interactions. We restrict the treatment of the imaging process to purely elastic scattering and therefore neglect inelastic interactions and absorption. This implies that the atomic potentials as well as the scattering factors are real scalar functions. For detailed discussions about inelastic electron scattering and the image formation using inelastically scattered electrons we refer to specific literature (see, e.g. Rose, 1976a, 1976b; Kohl and Rose, 1985; Wang, 1995; Egerton, 1996).

[1] *Conventional* electron microscopy shall refer to instruments which are not equipped with aberration correctors.

## 2.1 Overview

The basic picture that is commonly used to explain the imaging process in high-resolution phase contrast imaging can be summarized as follows. The specimen is illuminated with a broad beam that ideally is described by a plane electron wave whose wave vector is parallel to the optical axis of the microscope. On transmitting a zone-axis oriented crystalline specimen, the incoming electron beam undergoes diffraction. Determined by the structure factor of the crystal, each diffracted beam experiences a specific phase shift. The diffracted beams leave the specimen directed to specific angles determined by Bragg's law. As illustrated in Fig. 2.1, the objective lens focuses these partial beams in the back focal plane where a diffraction pattern is formed. Each diffraction spot $\boldsymbol{g}$ reflects a partial beam of wave vector $\boldsymbol{k_g}$. The partial beams are then transferred to the image plane where they are brought to interference and form the image.

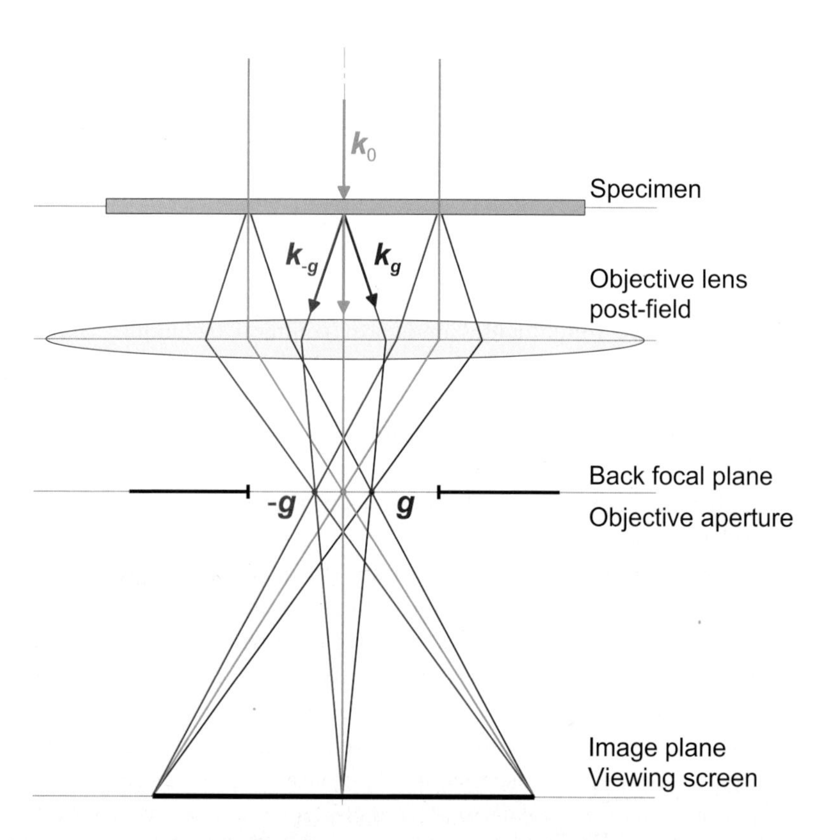


Fig. 2.1 HRTEM imaging. Transfer of the exit-plane wave by the post-field of the objective lens to the viewing screen. For simplicity, the projection system of the microscope, which is necessary to further magnify the image, is omitted.

For our discussion of high-resolution transmission electron microscopy, we simplify the formation of the image to the steps illustrated in Fig. 2.2. We assume that there is an electron wave at the exit plane of the specimen, called the *exit-plane wave*, which contains the structural information of the specimen. We do not focus in detail on how this exit-plane wave is formed. The exit-plane wave, which is a complex function, is transferred by the objective lens to the image plane where an image is formed. This image reflects structural information which is contained in the exit-plane wave.

Without too much loss of generality, we assume that we deal with a phase object. A phase object is a specimen which changes the phase of the transmitting electron wave but not its amplitude. This implies that the exit-plane wave shows a phase modulation, but no amplitude modulation. However, since the phase information of a wave is lost on deriving its intensity, the phase modulation of the electron wave is not *per se* detectable as an intensity modulation in the image. In order to form an image of finite contrast, we need to translate the phase modulation into an amplitude modulation. This technique, i.e. the translation from phase to amplitude modulation, is known as the *phase contrast method*. It finds its origin in light optics and was developed by Frits Zernike (see, e.g. Born and Wolf, 2001).

While in light optics the phase contrast is primarily obtained by employing a physical phase plate, Fig. 2.2 illustrates that in transmission electron microscopy the phase contrast transfer is normally determined by the characteristic of the objective lens' post-field.

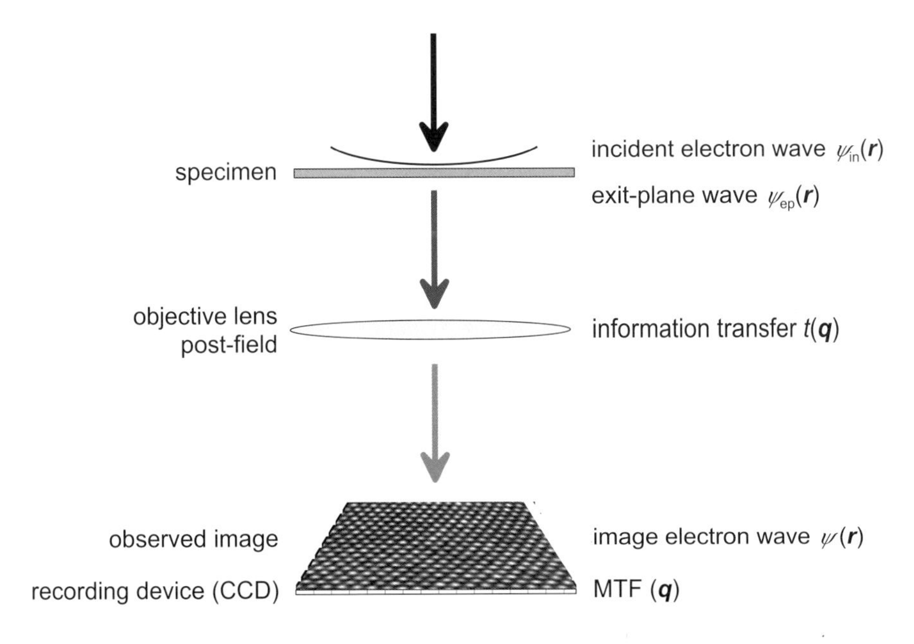


Fig. 2.2 The information transfer in the electron microscope. The case of HRTEM imaging.

The phase contrast transfer function determines the information content of a phase contrast micrograph, and in particular whether and how atomic-resolution phase contrast information is contained in it. Hence, the figure of merit of phase contrast imaging is the phase contrast transfer function. What defines it? To answer this question we picture the image formation process split into different steps as depicted in Fig. 2.2. We employ the idea that HRTEM can be regarded as an information transfer process and that the electron microscope acts as an information channel (De Jong and Van Dyck, 1992). The factors that need to be considered in the evaluation of the phase contrast transfer function can then be grouped into (i) optical effects, which are related to the imaging characteristics of the objective lens, including the limited coherence of the electron beam; (ii) stability issues such as mechanical instabilities of the specimen and drift; and (iii) an incoherent contribution stemming from the actual recording process (De Jong and Van Dyck, 1993; Van Dyck *et al.*, 2003). The latter effect essentially leads to a blurring of the information which is due to the point spread function of the pixelated recording device (De Ruijter, 1995).

In the foregoing text, we refer to the phase contrast imaging mode simply as high-resolution transmission electron microscopy (HRTEM). With TEM imaging we simply mean a broad-beam illumination technique which also covers HRTEM. For a comprehensive treatise of various aspects of HRTEM imaging see, for example, Spence (1981); Buseck *et al.* (1992).

## 2.2 The Exit-Plane Wave

### 2.2.1 *The weak phase object approximation*

Dynamic scattering theory is in general indispensable in describing elastic electron scattering. However, without too much loss of generality we shall base the present discussion on the case of kinematic scattering. We start with a thin specimen which is illuminated by a plane electron wave. The incident electron wave is given by $\psi_{\text{in}}(\boldsymbol{r}) = 1$, where $\boldsymbol{r} = (x, y)$ is a two-dimensional vector in the object plane.

On transmitting a thin specimen which alters the phase but not the amplitude of the electron wave, the incident electron wave experiences a phase modulation which can be expressed by $\exp\{-\mathrm{i}\phi(\boldsymbol{r})\}$. The phase modulation is determined by the object function $\phi(\boldsymbol{r})$, which is a real function that reflects the phase shifts induced by the specimen as a function of the position $\boldsymbol{r}$. In fact, it can be shown that $\phi(\boldsymbol{r})$ is proportional to the projected electrostatic crystal potential $V_t(\boldsymbol{r})$ by

$$\phi(\boldsymbol{r}) = \frac{\pi}{\lambda U} \int_0^t V(\boldsymbol{r}, z) dz = \frac{\pi}{\lambda U} V_t(\boldsymbol{r}) = \sigma V_t(\boldsymbol{r}) \,, \tag{2.1}$$

with $U$ the acceleration voltage of the microscope, the electrostatic crystal potential $V(\boldsymbol{r}, z)$, $z$ the cartesian coordinate along the optical axis, the specimen thickness $t$

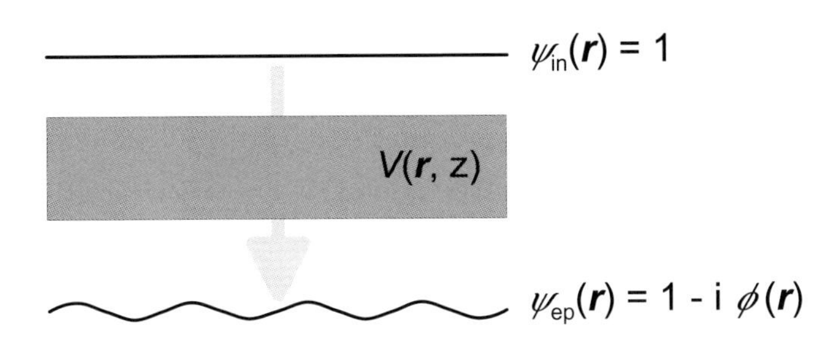


Fig. 2.3 The weak phase object approximation. A plane wave transmits a phase object and experiences a position dependent phase modulation.

and $\lambda$ the relativistic wavelength of the electron, written as

$$\lambda = 2\pi\hbar \left( 2m_0\, eU \left[ 1 + \frac{eU}{2\, m_0\, c^2} \right] \right)^{-\frac{1}{2}}. \tag{2.2}$$

The elementary charge is denoted by $e$, $m_0$ is the rest mass of the electron and $c$ is the speed of light in a vacuum (see Appendix A.1).

For a weak phase object, i.e. $\phi \ll 1$, the exponential containing $\phi(\boldsymbol{r})$ can be expanded by $\exp\{-\mathrm{i}\phi(\boldsymbol{r})\} \approx 1 - \mathrm{i}\phi(\boldsymbol{r})$. Given the incident plane wave $\psi_{\mathrm{in}} = 1$, the exit-plane wave can be written as the product of the incident electron wave $\psi_{\mathrm{in}}(\boldsymbol{r})$ and the induced phase shift. This yields

$$\psi_{\mathrm{ep}}(\boldsymbol{r}) = 1 - \mathrm{i}\,\phi(\boldsymbol{r}). \tag{2.3}$$

This relation, which is illustrated in Fig. 2.3, is known as the *weak phase object approximation* (see, e.g. Williams and Carter, 1996). In order to include absorption of a weak, i.e. thin specimen, the object function $\phi(\boldsymbol{r})$ can be replaced by a complex function whose negative imaginary part represents absorption or simply a loss of electrons due to scattering outside of the aperture opening in the back focal plane (see Fig. 2.1) (Hawkes and Kasper, 1994). For the case that $\phi$ has a small finite imaginary part, one speaks of a weak object rather than a weak phase object (Born and Wolf, 2001).

It has to be emphasized that the validity of the weak phase object approximation is strongly limited. However, it reveals the essential characteristic of the exit-plane wave. Equation (2.3) shows that the imaginary of the exit-plane wave contains the structural information of the object. The real part of the exit-plane wave is unaffected by the presence of a weak phase object.

An alternative way of illustrating the phase modulation of the exit-plane wave induced by the projected crystal potential is based on the assumption that the incident electron wave consists of individual electrons which are all in phase. An electron that traverses the specimen at a position $\boldsymbol{r}_1$ experiences the crystal potential $V_t(\boldsymbol{r}_1)$. This electron gains the potential energy $eV_t(\boldsymbol{r}_1)$ while it is within the crystal. For the time the electron is within the crystal, the gain in energy reduces the electron's wavelength from $\lambda$ to $\lambda_1$, i.e. $\lambda_1 < \lambda$. Another electron, which shall be in phase with the first electron when entering the crystal, traverses the specimen at position $\boldsymbol{r}_2$ and thus experiences the crystal potential $V_t(\boldsymbol{r}_2)$. Hence, for the time this second electron is within the crystal, it has a wavelength, $\lambda_2$, which is different from $\lambda_1$ provided that $V_t(\boldsymbol{r}_1) \neq V_t(\boldsymbol{r}_2)$. Comparing the phases of the two electrons at the exit plane of the specimen shows that they have a specific phase difference which reflects that $\lambda_1 \neq \lambda_2$. The variation of the projected crystal potential thus defines the relative phase shifts which electrons experience on traversing the specimen at their specific $(x, y)$ positions. The entity of electrons at the exit plane can again be regarded as one wave. The phase modulation of the electron wave at the exit plane of the specimen thus contains the structural information about the specimen. Therefore, it is the phase modulation reflecting the projected crystal potential that we would like to have imaged in order to learn about the atomic structure of the specimen. The concept of individual electrons transmitting a specimen and being affected by the local crystal potential forms the basis of the channelling theory, which, however, also considers dynamic scattering effects (see, e.g. Van Dyck, 1999).

### 2.2.2 *Dynamic scattering theories*

The derivation of an exit-plane wave based on the idea of a weak phase object neglects for instance the angular distribution of the electron scattering. Indeed, the weak phase object approximation must be regarded as a very basic and rather pictorial approach which is based on a kinematic scattering model. There are alternative, more elaborate approaches which lead to the concept of an exit-plane wave. We do not intend to discuss them in detail but we think that it is of importance to mention them in the present context.

A real material is not just a simple, even non-uniform potential well. The propagation of the incident electron wave through the crystal is determined by the scattering of the electrons by atoms in the material. It is the Coulomb interaction between atoms and electrons that leads to electron scattering in the specimen. For elastic (coherent) scattering[2], the electron wave is said to be diffracted by the crystal. Hence, the incident electron wave described by the wave vector $\boldsymbol{k}_0$ is diffracted by the crystal, leading to Bragg diffracted partial beams $\boldsymbol{k}_{\boldsymbol{g}}$, where $\boldsymbol{g}$ is a vector of the reciprocal lattice of the crystal. Each diffraction spot $\boldsymbol{g}$ in the diffraction pattern, which is contained in the back focal plane, reflects a Bragg diffracted beam (Fig. 2.1). Since the Coulomb interaction between electrons and atoms is strong, the elastic scattering cross-section for electrons is large. For this, the probability

[2]Elastic electron scattering is always coherent, regardless of whether this is scattering of the electron at the electron cloud of the atom or Rutherford scattering, i.e. scattering at the nucleus. There is no incoherent elastic scattering cross-section for electrons. Neutrons, on the other hand, have a coherent and an incoherent elastic scattering cross-section (Schwartz and Cohen, 1987).

that an electron is scattered more than once on transmitting a specimen of finite thickness is not negligible. Multiple scattering means that with increasing thickness, the dynamic exchange of intensity between different Bragg diffracted beams alters the intensity distribution in the back focal plane. This effect increases with increasing specimen thickness and increasing atomic number of the elements in the specimen. As a consequence, the presence of multiple scattering makes it necessary to describe electron diffraction by considering a dynamic scattering theory.

A dynamic scattering theory, which goes back to Hans Bethe, is based on the so-called Bloch wave approach. It describes electron scattering within a crystal as the excitation of Bloch waves, i.e. solutions of the stationary Schrödinger equation for a periodic crystal potential (Bethe, 1928; Metherell, 1975). The Bethe theory solves the problem of dynamic electron scattering by a crystalline specimen under the assumption of a perfectly symmetrical crystal, employing suitable boundary conditions. The periodicity of the crystal potential is directly reflected in the Bloch waves. Each diffracted beam $\boldsymbol{k_g}$ can then be described as a superposition of a set of Bloch waves. There are expansions of the basic Bloch wave theory that incorporate the effect of inelastic scattering and, for example, thermal diffuse scattering (Wang, 1995). Another dynamic scattering theory is the multi-slice approach, which is widely used for electron microscopy simulations (Cowley and Moodie, 1957, 1959a, 1959b). A particular advantage of the multi-slice approach is that it can be applied to non-periodic objects.

Dynamic scattering theories are certainly closer to the physical reality and are more adequate to describe the elastic scattering of electrons in a material of finite thickness. However, even though the derivation of the exit-plane wave based on the kinematic weak phase object approximation is very basic, it still covers the essential aspects of phase contrast imaging that are of relevance for our purpose.

## 2.3 Image Formation

In the previous section we saw that the exit-plane wave contains the structural information about the specimen. The complex exit-plane wave $\psi_{\mathrm{ep}}(\boldsymbol{r})$ is a function of the (two-dimensional) real space coordinate $\boldsymbol{r} = (x, y)$. If we consider the exit-plane wave to constitute of a set of Bragg diffracted partial beams $\boldsymbol{k_g}$, we see from Fig. 2.1 that each partial beam $\boldsymbol{k_g}$ is focused by the objective lens to a single spot in the back focal plane. The location of a given spot depends on the Bragg diffraction angle, which is specific for its particular $\boldsymbol{g}$-vector. Each spot can be labelled with a (three-dimensional) reciprocal lattice vector $\boldsymbol{g}$ of the corresponding crystal lattice. The partial beams brought to focus in the back focal plane form a diffraction pattern. Hence, similarly to the case where the observation point is at infinity, the back focal plane contains a Fraunhofer diffraction pattern. The wave function in the back focal plane $\psi_{\mathrm{bfp}}(\boldsymbol{q})$ can thus be written as the Fourier transform of the wave function at the exit plane of the specimen (Wang, 1995)

$$\psi_{\mathrm{bfp}}(\boldsymbol{q}) = \mathcal{F}[\psi_{\mathrm{ep}}(\boldsymbol{r})] = \psi_{\mathrm{ep}}(\boldsymbol{q}). \tag{2.4}$$

The Fourier transform is denoted by $\mathcal{F}$, and $\boldsymbol{q}$ is a two-dimensional vector of the spatial frequency spectrum, i.e. a spatial frequency component which is the transform variable of $\boldsymbol{r}$. While $\boldsymbol{q}$ is a general two-dimensional reciprocal space vector that allows for navigating in the back-focal plane, the vectors $\boldsymbol{g}$ are reciprocal lattice vectors which are specific for the crystal lattice. The intensity $I_{\mathrm{d}}$ of the diffraction pattern is given by

$$I_{\mathrm{d}}(\boldsymbol{q}) = |\psi_{\mathrm{ep}}(\boldsymbol{q})|^2 = \psi_{\mathrm{ep}}(\boldsymbol{q})\overline{\psi}_{\mathrm{ep}}(\boldsymbol{q}), \tag{2.5}$$

where the bar denotes the complex conjugate. On propagating the electron wave in the back focal plane $\psi_{\mathrm{ep}}(\boldsymbol{q})$ to the image plane, one might expect that because of symmetry the image wave function $\psi(\boldsymbol{r})$ is the inverse Fourier transform of the wave function at the back focal plane (see Fig. 2.1). However, it is in this particular step where the impact of the transfer characteristic of the imaging system becomes apparent. The wave function in the image plane is not simply given by the inverse Fourier transform of $\psi_{\mathrm{ep}}(\boldsymbol{q})$ but by

$$\psi(\boldsymbol{r}) = \mathcal{F}^{-1}[\psi_{\mathrm{ep}}(\boldsymbol{q})t(\boldsymbol{q})]. \tag{2.6}$$

The symbol $\mathcal{F}^{-1}$ denotes the inverse Fourier transform and $t(\boldsymbol{q})$ is a two-dimensional complex transfer function describing the imaging characteristics of the microscope. The transfer function $t(\boldsymbol{r})$ consists of a real $\Re[t(\boldsymbol{r})]$ and an imaginary part $\Im[t(\boldsymbol{r})]$. Multiplication of two functions in Fourier space corresponds to the convolution of the functions in real space. Hence, the electron wave function in the image plane is given by the convolution of the exit-plane wave $\psi_{\mathrm{ep}}(\boldsymbol{r})$ with the complex transfer function $t(\boldsymbol{r})$

$$\psi(\boldsymbol{r}) = \psi_{\mathrm{ep}}(\boldsymbol{r}) \otimes t(\boldsymbol{r}), \tag{2.7}$$

with $\otimes$ referring to the convolution. Equation (2.7) is a description of the coherent imaging model. Furthermore, Eq. (2.7) is independent of the approximation that are made to derive the exit-plane wave $\psi_{\mathrm{ep}}(\boldsymbol{r})$. However, substituting in Eq. (2.7) the exit-plane wave of the weak phase object approximation given in Eq. (2.3) yields the complex wave function $\psi(\boldsymbol{r})$ of a weak phase object in the image plane

$$\psi(\boldsymbol{r}) = 1 + \phi(\boldsymbol{r}) \otimes \Im[t(\boldsymbol{r})] \; - \mathrm{i}\phi(\boldsymbol{r}) \otimes \Re[t(\boldsymbol{r})]. \tag{2.8}$$

Neglecting terms that are quadratic in $\phi(\boldsymbol{r})$, the image intensity $I(\boldsymbol{r})$ of a weak phase object is then

$$I(\boldsymbol{r}) = |\psi(\boldsymbol{r})|^2 = 1 + 2(\phi(\boldsymbol{r}) \otimes \Im[t(\boldsymbol{r})]). \tag{2.9}$$

Equation (2.9) describes how a (weak) phase modulation of the exit-plane wave is translated into an image intensity. The Fourier transform of $\Im[t(\boldsymbol{r})]$, i.e. $\mathcal{F}[\Im[t(\boldsymbol{r})]] = \Im[t(\boldsymbol{q})]$, is the phase contrast transfer function, which determines how each spatial frequency component $\boldsymbol{q}$ of the exit-plane wave $\psi_{\mathrm{ep}}(\boldsymbol{r})$ is transferred to the image plane as an intensity modulation of frequency $\boldsymbol{q}$.

As mentioned above, in case of absorption or scattering beyond the aperture opening, i.e. if $\Im[\phi(\boldsymbol{r})] \neq 0$, the exit-plane wave shows in addition to the phase modulation an amplitude modulation. Similarly to the case discussed above, it can be shown that the real part of $t(\boldsymbol{r})$ translates the amplitude modulation of the exit-plane wave into an amplitude modulation which contributes to the recorded image intensity (Hawkes and Kasper, 1994). This is known as amplitude contrast.

Comparing the intensity of the diffraction pattern given in Eq. (2.5) with the intensity of the image in Eq. (2.9) reveals that the image is affected by the microscope's transfer function $t(\boldsymbol{r})$, while the intensity of the diffraction pattern is not. Generally, the image frequency spectrum $\mathcal{F}[I(\boldsymbol{r})] = I(\boldsymbol{q})$, which can be derived from a numerical (fast) Fourier transformation of the image, does not correspond to the diffraction pattern $I_{\mathrm{d}}(\boldsymbol{q})$. While the information content of the diffraction pattern is independent of the transfer characteristics of the microscope, a HRTEM micrograph reflects the microscope's transfer characteristics.

## 2.4 The Phase Contrast Transfer Function

Now we draw our attention to the the transfer function $t(\boldsymbol{q})$. An ideal information transfer is described by a frequency-independent transfer function of constant value. For an ideal phase contrast transfer, an additional requirement needs to be considered; the phase contrast should be the sole contribution to the image.

As explained in Sec. 2.3, the second requirement can be fulfilled only if the imaginary part of the object function $\phi(\boldsymbol{r})$ is vanishingly small. Otherwise, $\Re[t(\boldsymbol{r})]$ transfers the amplitude modulation of the exit-plane wave into an amplitude modulation in the image plane. The amplitude contrast would then, of course, contribute to the image intensity, which, however, might still be dominated by the phase contrast.

In order to discuss the first requirement, we derive the frequency spectrum of the image intensity. According to Eq. (2.9), the spatial frequency spectrum of the image intensity of a weak phase object is given by

$$I(\boldsymbol{q}) = \mathcal{F}[I(\boldsymbol{r})] = \delta(\boldsymbol{0}) + 2\phi(\boldsymbol{q})\Im[t(\boldsymbol{q})], \tag{2.10}$$

where $\delta(\boldsymbol{0})$ is the Dirac delta function. Equation (2.10), which is the Fourier transform of Eq. (2.9), reveals how the phase contrast transfer function $\Im[t(\boldsymbol{q})]$ linearly translates the frequency spectrum of the exit-plane wave $\phi(\boldsymbol{q})$ into the amplitude modulation of the image plane. For an ideal phase contrast, the relative intensities of the frequency components of the phase modulation should be preserved in the amplitude modulation of the image wave function. Hence, in order to ideally translate the phase modulation of the exit-plane wave to the image plane, the phase contrast transfer function $\Im[t(\boldsymbol{q})]$ would need to have a constant finite value.

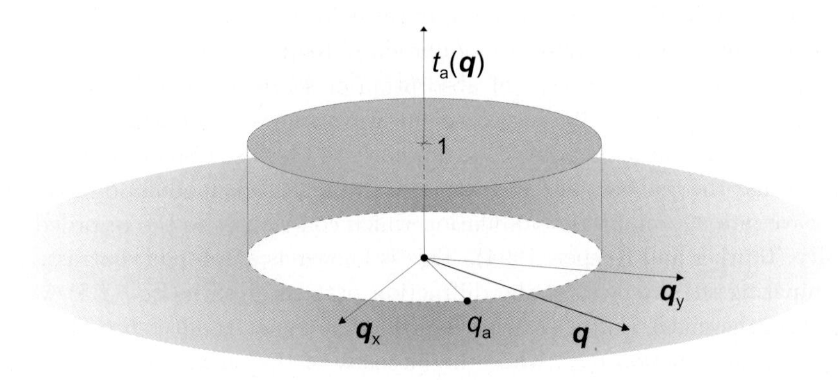


Fig. 2.4 Definition of the vector $\boldsymbol{q}$ and the top-hat aperture function according to Eq. (2.11).

A transmission electron microscope, however, is a non-ideal information channel (De Jong and Van Dyck, 1992). The individual spatial frequencies of the phase of the exit-plane wave $\psi_{\mathrm{ep}}(\boldsymbol{r})$ are unequally transferred to the image plane; that is, $\Im[t(\boldsymbol{q})]$ is in general a function which is not a constant. In the following we discuss the individual contributions that define the characteristics of the phase contrast transfer function $\Im[t(\boldsymbol{q})]$ of HRTEM imaging.

### 2.4.1 *The objective aperture*

The objective aperture is located in the back focal plane of the objective lens, i.e. in a plane which contains a diffraction pattern (see Fig. 2.1). The wave function in the back focal plane is described by Eq. (2.4). Hence, a circular aperture centered on the optical axis in the back focal plane acts as a low-pass frequency filter that removes components of high spatial frequencies. Alternatively, it can be understood as a filter for electrons that are scattered to high angles. Hence, the aperture removes electrons from the image formation which carry high resolution information. The effect of the aperture on the wave function in the back focal plane can be described by a top-hat function $t_{\mathrm{a}}(\boldsymbol{q})$

$$t_{\mathrm{a}}(\boldsymbol{q}) = \begin{cases} 1 \text{ if } |\boldsymbol{q}| < |\boldsymbol{q}_{\mathrm{a}}| \\ 0 \text{ otherwise} \end{cases}, \tag{2.11}$$

where $|\boldsymbol{q}_{\mathrm{a}}| = q_{\mathrm{a}}$ is the radius of the aperture in the back focal plane, i.e. the cut-off frequency. Usually diffraction at the edge of the aperture is not considered. Instead of using a top-hat function, we can employ the Fermi function to approximately describe the aperture function in an alternative way:

$$t_{\mathrm{a}}(\boldsymbol{q}) = \frac{1}{1 + \exp\left\{\dfrac{|\boldsymbol{q}|^2 - |\boldsymbol{q}_{\mathrm{a}}|^2}{\delta_{\mathrm{a}}^2}\right\}}. \tag{2.12}$$

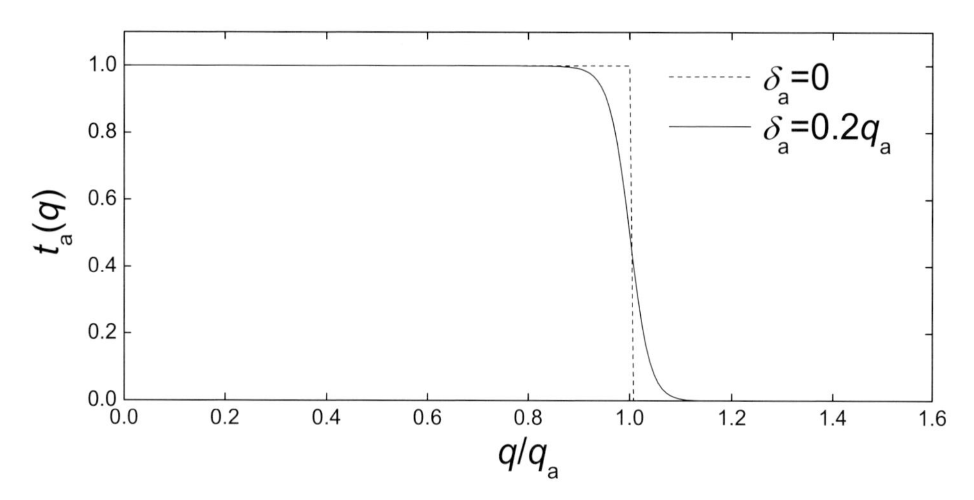


Fig. 2.5 The Fermi function. The Fermi function according to Eq. (2.12) is drawn for $\delta_a \to 0$ (dotted line) and for $\delta_a = 0.2q_a$ (solid line).

Here, $\delta_a$ is a small fraction of the aperture radius $|\boldsymbol{q}_a|$. For $\delta_a \to 0$, Eq. (2.12) is equivalent to Eq. (2.11), and for arbitrarily small but finite values of $\delta_a$, the edge of the top-hat function defined in Eq. (2.11) is rounded off. This is depicted in Fig. 2.5. Equation (2.12) is particularly useful for calculations where a closed function can be of some advantage. Furthermore, a non-sharp edge decreases the risk of introducing sampling artifacts in numerical methods (see, e.g. Kirkland *et al.*, 1987). Figure 2.5 shows the Fermi function according to Eq. (2.12) for two distinctly different values of $\delta_a$.

Though the usage of an objective aperture reduces the impact of aberrations whose contributions increases with increasing scattering angle, many HRTEM applications are carried out without employing an objective aperture. The transfer at high spatial frequencies is then essentially determined by the information limit of the microscope. However, for cases where an objective aperture is employed to limit the transfer of high spatial frequencies, $t_a(\boldsymbol{q})$ has to be considered in the evaluation of the phase contrast transfer function.

### 2.4.2 *Lens effects*

The post-field of the objective lens, which transfers the exit-plane wave to the image plane, is not the ideal lens. Indeed, by shifting the relative phases of the individual spatial frequency components of the electron wave, the characteristic of the lens defines the fundamental transfer characteristics of the microscope. It is this phase change mechanism induced by the errors of the lens which is essential for the phase contrast. The aim of doing phase contrast imaging is indeed to control the phase shifts in such a way that the phase contrast micrographs directly reflect the structure of the specimen under consideration. One could say that the phase shifts induced by the objective lens correspond to the effect of a physical phase plate, which is used to control the phase contrast in light optics.

The effect of the lens errors on the contrast transfer is expressed by a transfer function $t_{\mathrm{L}}(\boldsymbol{q})$, which is given by

$$t_{\mathrm{L}}(\boldsymbol{q}) = \exp\left\{-\frac{2\pi\mathrm{i}}{\lambda}\chi(\boldsymbol{q})\right\}. \tag{2.13}$$

The function $\chi(\boldsymbol{q})$ is called the lens or wave aberration function, which describes the phase shifts $\gamma(\boldsymbol{q})$ induced by the post-field of the objective lens

$$\gamma(\boldsymbol{q}) = \frac{2\pi}{\lambda}\chi(\boldsymbol{q}). \tag{2.14}$$

According to Eq. (2.9), it is the imaginary part of $t_{\mathrm{L}}(\boldsymbol{q})$ that is of importance for phase contrast imaging. The phase plate of coherent phase contrast imaging is thus described by $\Im[t_{\mathrm{L}}(\boldsymbol{q})]$, which we denote by $t_{\mathrm{c}}(\boldsymbol{q}) \equiv \Im[t_{\mathrm{L}}(\boldsymbol{q})]$.

The general form of the wave aberration function $\chi(\boldsymbol{q})$ is discussed in Chapter 7. For now and for the case of conventional electron microscopes, which are not equipped with an aberration corrector, it is the third-order spherical aberration $C_3$ of the objective lens and the defocus $C_1$ that determine the wave aberration function[3]. Because of the dominant effect of the third-order spherical aberration $C_3$, other contributions, particularly anisotropic contributions that might arise due to a slight misalignment, can in general be neglected. The wave aberration function can thus be written as

$$\chi(\boldsymbol{q}) = \chi(q) = \frac{1}{2}q^2\lambda^2 C_1 + \frac{1}{4}q^4\lambda^4 C_3. \tag{2.15}$$

Both defocus $C_1$ and the third-order spherical aberration $C_3$ are isotropic aberrations, i.e. they do not depend on the azimuth angle. Therefore, $\chi$ is isotropic and without loss of generality we can substitute $\boldsymbol{q}$ with $q = |\boldsymbol{q}|$. Furthermore, on introducing the scattering angle[4] $\theta \approx q\lambda$ we can substitute $q$ with the scattering angle $\theta$. This yields for the aberration function

$$\chi(\theta) = \frac{1}{2}\theta^2 C_1 + \frac{1}{4}\theta^4 C_3. \tag{2.16}$$

Substituting Eq. (2.15) in Eq. (2.13) yields the coherent phase contrast transfer function $t_{\mathrm{c}}$

[3] In other textbooks, the defocus is often denoted as $\Delta f$ and the third-order spherical aberration is often denoted as $C_{\mathrm{S}}$. We employ the notation $C_1$ for defocus and $C_3$ for the constant of third-order spherical aberration for consistency with subsequent chapters.

[4] This approximation only holds for small $\theta$, as commonly fulfilled in electron microscopy.

$$
\begin{aligned}
t_c(q) &= \Im\left[\exp\left\{-\mathrm{i}\pi q^2\lambda C_1 - \frac{\mathrm{i}}{2}\pi q^4\lambda^3 C_3\right\}\right] \\
t_c(\theta) &= \Im\left[\exp\left\{-\frac{\mathrm{i}\pi}{\lambda}\left(\theta^2 C_1 + \frac{1}{2}\theta^4 C_3\right)\right\}\right] \\
&= \sin\left\{-\frac{\pi}{\lambda}\left(\theta^2 C_1 + \frac{1}{2}\theta^4 C_3\right)\right\}.
\end{aligned}
\tag{2.17}
$$

Equation (2.17) has the familiar form of the coherent phase contrast transfer function for non-aberration-corrected transmission electron microscopes of isotropic information transfer (see, e.g. Williams and Carter, 1996). For aberration-corrected instruments, Eq. (2.13) is still valid, however additional aberrations need to be considered in the wave aberration function $\chi(\boldsymbol{q})$ (see Chapter 7).

Equation (2.17) reveals that if both defocus $C_1$ and the third-order spherical aberration $C_3$ are equal to zero, $t_c = 0$. Hence, in the absence of both $C_1$ and $C_3$ there is no phase contrast in the image. As mentioned above, it is the characteristic of the objective lens which determines the phase contrast. In order to optimize the phase contrast transfer function in a way such that structure information contained in the exit-plane wave is translated into a directly interpretable HRTEM micrograph, it is crucial to control the phase shifts induced by the objective lens. According to the criteria mentioned above, there is no ideal phase contrast transfer function in HRTEM imaging, but the parameters that define the phase shifts

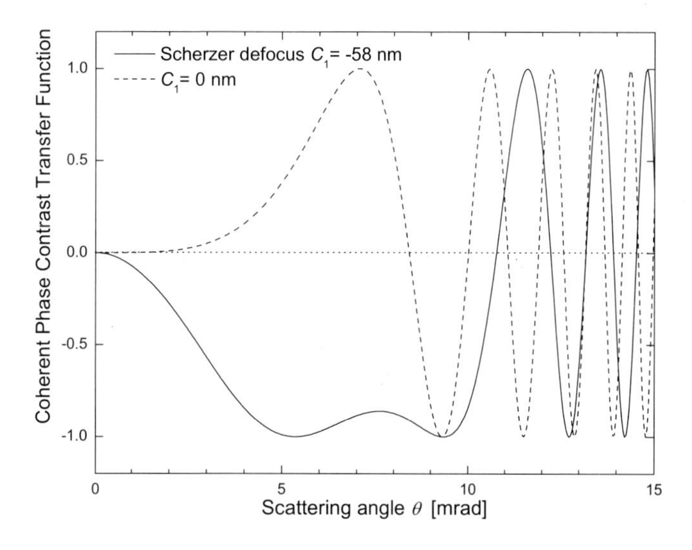


Fig. 2.6 Coherent phase contrast transfer function $t_c(\theta)$ according to Eq. (2.17) for 200 kV and $C_3 = 1$ mm. The dashed line shows $t_c(\theta)$ for zero defocus $C_1 = 0$ and the full line shows $t_c(\theta)$ for $C_{1\,\mathrm{Scherzer}} = -58$ mm, which corresponds to the Scherzer focus (see next subsection).

induced by the objective lens can in principle be set such that an approximately optimized phase contrast transfer function is obtained. However, Eq. (2.17) shows that $t_c$ is essentially a sine-function. Hence, one peculiarity of the phase contrast transfer function is that it is always zero for $q \rightarrow 0$ or $\theta \rightarrow 0$. Object features described by large spatial frequency components do not contribute to the phase contrast. Unless a physical phase plate is employed, this characteristic of the phase contrast transfer function cannot be changed (see, e.g. Majorovits *et al.* 2007, Gamm *et al.* 2008, Danev *et al.*, 2009). Optimizing the phase contrast transfer function means forming a broad passband which enables the transfer of a large frequency range under similar phase shift.

If only isotropic aberrations are present, like defocus $C_1$ and spherical aberration $C_3$, the description of the coherent phase contrast transfer function according to Eq. (2.17) is sufficient. However, one has to keep in mind that the transfer function affects a plane. The plane which is affected by the transfer function is the back focal plane. To navigate in the back focal plane we use either the two-dimensional reciprocal vector $\boldsymbol{q}$ or the scattering angle $\theta$. However, the scattering angle as a scalar quantity does not allow for describing an arbitrary point in the back focal plane. We therefore need to relate the scattering angle $\theta$ to either cartesian or cylindrical coordinates. What turns out to be useful for the navigation in the back focal plane is the usage of complex cartesian coordinates. The complex notation allows for handling two-dimensional quantities by practically handling a single quantity. This prevents more laborious vectorial notation. Figure 2.7 illustrates the idea: we introduce a cartesian coordinate system with a real and an imaginary axis, which intersect in the optical axis denoted by $z$. Both axes refer to the scattering angle. A point $P$ in the plane is then described by a pair of values $\theta_x$ and $\theta_y$, which is summarized in one complex quantity $\omega = \theta_x + \mathrm{i}\theta_y$. Hence, $\omega$ is the complex scattering angle and $\overline{\omega} = \theta_x - \mathrm{i}\theta_y$ is its complex conjugate.

Having introduced the complex scattering angle $\omega$, we can rewrite the wave aberration function as

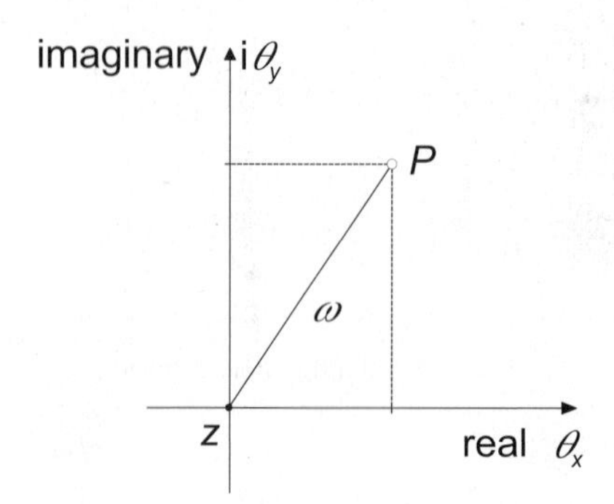


Fig. 2.7 The complex angular coordinate $\omega = \theta_x + \mathrm{i}\theta_y$.

$$\chi(\omega) = \Re\left(\frac{1}{2}\omega\overline{\omega}C_1 + \frac{1}{4}\left(\omega\overline{\omega}\right)^2 C_3\right), \tag{2.18}$$

which is still a real scalar function. The coherent phase contrast transfer function can then be written as

$$t_c(\omega) = \Im\left[\exp\left\{-\frac{i\pi}{\lambda}\Re\left(\omega\overline{\omega}C_1 + \frac{1}{2}\left(\omega\overline{\omega}\right)^2 C_3\right)\right\}\right]. \tag{2.19}$$

This is essentially the same expression as in Eq. (2.17), however while Eq. (2.17) corresponds to a one-dimensional description of the transfer function, Eq. (2.19) can be used to derive the two-dimensional characteristics of the transfer, which is indispensable for considering the effect of anisotropic aberrations.

#### 2.4.2.1 *Scherzer focus*

The coherent phase contrast transfer function as given in Eq. (2.19) is determined by two parameters: the defocus $C_1$ and the constant of spherical aberration $C_3$ of the objective lens. For a given electron microscope, the constant of spherical aberration is determined by the design of the lens. Typical objective lenses of conventional high-resolution microscopes have an inherent $C_3$ of 0.5 to about 1.5 mm. Since this value is given by the design of the lens, it cannot be changed, at least not significantly. However, what can be changed is the defocus $C_1$. The phase contrast transfer function as given in Eq. (2.19) can be optimized by adjusting $C_1$ such that it balances the effect of $C_3$ in a way that $t_c(\omega)$ approaches the characteristics of an ideal transfer function. This leads us to the well-known treatise of Otto Scherzer (1949), who analyzed the characteristics of the phase contrast transfer function and derived criteria to optimize the transfer. For an optimized phase contrast, Scherzer

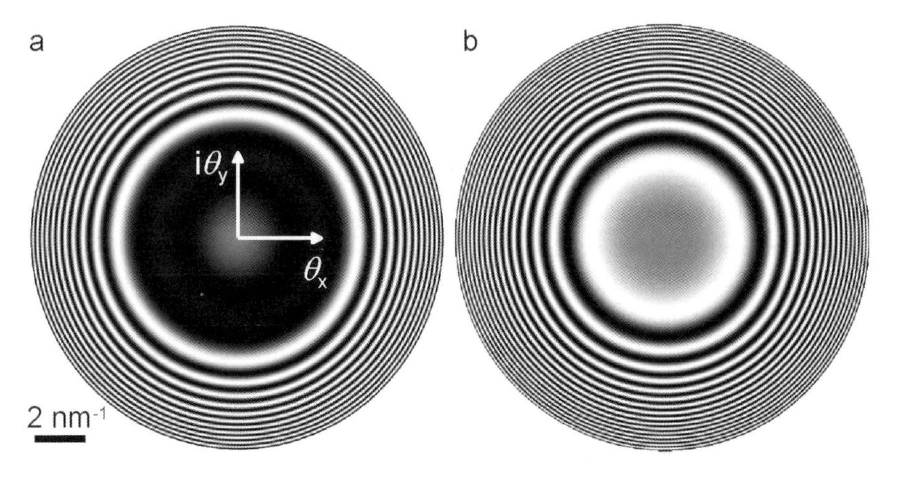


Fig. 2.8 Coherent phase contrast transfer functions $t_c(\omega)$ according to Eq. (2.19) for 200 kV and $C_3 = 1$ mm; (**a**) for Scherzer focus $C_{1\,\mathrm{Scherzer}} = -58$ mm and (**b**) for zero defocus $C_1 = 0$.

derives that the defocus should be chosen as

$$C_{1\,\mathrm{Scherzer}} = -\sqrt{\frac{\tau^2}{2\pi}}\,\sqrt{\lambda C_3}\,, \tag{2.20}$$

with the constant $\tau$ between 2.5 and 3.0. Equation (2.20) defines the *Scherzer focus* for coherent phase contrast imaging. An alternative expression that can be derived straightforwardly (see, e.g. Williams and Carter, 1996) and is often found in modern textbooks defines the Scherzer focus as

$$C_{1\,\mathrm{Scherzer}} = -\sqrt{\frac{4}{3}\lambda C_3}\,, \tag{2.21}$$

which implies that $\tau = 2.89$ in Eq. (2.20). According to Eq. (2.2) the wavelength of a 200 keV electron is 2.508 pm, and with $C_3 = 1\,\mathrm{mm}$ we obtain from Eq. (2.21) a Scherzer focus of $-58$ nm. The corresponding transfer function $t_c(\theta)$ is plotted in Figs. 2.6 and 2.8. Comparing the coherent phase contrast transfer function for vanishing defocus with the one derived for Scherzer focus, Figs. 2.6 and 2.8 reveals that for Scherzer focus there is broad and flat passband present where the transfer function is negative. For image frequencies that lie within this passband there is a direct relation between the object, i.e. the phase modulation of the exit-plane wave, and the image. However, for image frequencies that are outside the passband and fall into the frequency range where the transfer function oscillates, the interpretation is more difficult. However, as will be discussed in the next subsection, whether and how image components of high frequencies are transferred to the image depends on additional factors.

### 2.4.3 *Partial spatial and partial temporal coherence*

The information transfer in the transmission electron microscope not only depends on the objective aperture and the phase shifts, which are induced by the post-field of the objective lens, but also on the degree of coherence of the electron beam that illuminates the specimen.

Electrons, which either originate from different sources or from two distant points of a large source, are said to be incoherent. Incoherence means that there is no phase relation between the electrons. If an image is formed by electrons belonging to two partial beams which are incoherent, the image is given by the superposition of the intensity of the partial beams. The optical phenomena that dominate our daily life are mostly incoherent in nature.

On the other hand, if electrons originate from a small source, their phases can be correlated. Depending on whether the correlation is complete or partial, the beams are said to be coherent or partially coherent, respectively. The superposition of coherent or partially coherent beams leads to interference effects. An interference pattern reflects the phase relation between the individual waves that are brought to superposition.

The coherence not only depends on the point of origin of the electrons but also on their energy. While the first effect refers to spatial coherence or spatial incoherence, the second effect refers to temporal coherence or temporal incoherence. Hence, electrons which originate from the same point but have different energies are temporally incoherent.

In transmission electron microscopy, an electron beam generally originates from a small electron source. An ideal electron source is point-like and emits electrons of equal energy. A point-like monochromatic electron source generates electrons which are fully coherent and thus form a coherent electron beam which can be transferred by the condenser lens system to the specimen. Due to diffraction in a crystalline specimen, the incoming beam $\boldsymbol{k}_0$ is split into different components $\boldsymbol{k}_g$. Although diffraction leads to characteristic phase shifts of the Bragg diffracted beams, a phase relation between the different beams is maintained, i.e. elastic electron scattering is a coherent scattering process (Schwartz and Cohen, 1987). Below the specimen, the post-field of the objective lens transfers the diffracted beams to the image plane, where they are brought to superposition. We have seen in Sec. 2.4.2 that the (coherent) lens characteristics lead to an additional phase shift which does not affect the phase correlation between the partial beams either. Hence, neither diffraction in the specimen nor the transfer of the beams through the objective lens alters the coherence between the different partial beams. In every step, each partial beam experiences a defined phase shift, but the coherence between the individual partial beams is maintained. Therefore, one can say that the superposition of the Bragg diffracted beams in the image plane forms a interference pattern which we identify with a HRTEM micrograph (see Fig. 2.1). One could say that coherence is a quantity similar to entropy in thermodynamics. However, while the entropy in a closed system can only increase, the coherence of a beam can only decrease in an optical system. Coherence can become lost by inelastic scattering or for instance by a noisy lens system.

For an *ideal* electron source, the image formation in phase contrast microscopy is thus a fully coherent imaging process. This coherent transfer function is given by Eqs. (2.17) and (2.19). However, a *real* electron source has a finite size, it is never a single point and it produces electrons of slightly varying energies. Therefore, the incident electron beam $\boldsymbol{k}_0$ is not fully coherent but partially coherent. The finite size of the source leads to partial spatial coherence, and the finite energy spread of the beam leads to partial temporal coherence of the beam. Therefore, phase contrast microscopy is not a coherent but a partially coherent imaging technique. The imaging process can be described by a fully coherent imaging model as given in Eq. (2.7), whereas, however, partial coherence is taken into account by damping envelope functions that essentially reduce and limit the information transfer at high spatial frequencies (Wade and Frank, 1977).

#### 2.4.3.1 *Partial temporal coherence*

The damping envelope function due to partial temporal coherence $E_t(\omega)$ is essentially caused by the chromatic aberration of the lens and the finite energy spread of the electrons that are emitted by the electron source. It can be described by

$$E_t(\omega) = \exp\left\{-\frac{2\pi^2 \Delta C_1^2}{\lambda^2}\left(\frac{\partial\chi}{\partial C_1}\right)^2\right\}, \tag{2.22}$$

with the chromatic defocus spread $\Delta C_1$ given by

$$\Delta C_1 = C_C\sqrt{\left(\frac{\Delta U}{U_0}\right)^2 + 4\left(\frac{\Delta I}{I_0}\right)^2 + \left(\frac{\Delta E_{\text{rms}}}{E_0}\right)^2}. \tag{2.23}$$

In Eq. (2.23), $C_C$ is the constant of chromatic aberration, $\Delta U/U$ is the instability of the high tension $U$, called high tension ripple, $\Delta I/I$ is the instability of the lens current $I$, $E_0$ is the primary electron energy given by $eU$, and $\Delta E_{\text{rms}}$ is the root-mean-square energy spread of the electron beam, which is related to the full width at half maximum (FWHM) of the energy spread $\Delta E$ by $\Delta E = 2\sqrt{2\ln 2}\,\Delta E_{\text{rms}} \approx 2.355\Delta E_{\text{rms}}$. Equation (2.23) reveals that the chromatic defocus spread $\Delta C_1$ is determined by the high tension ripple, fluctuations of the lens current and the inherent energy spread of the electron source. If the current of the electron beam is high, an additional factor needs to be considered which describes the interactions between individual electrons in the beam: this contribution is referred to as the Boersch effect (Wade and Frank, 1977).

Equation (2.22) is valid for any wave aberration function $\chi$. However, for the special isotropic case with $\chi$ given according to Eq. (2.18), the damping envelope function due to partial temporal coherence $E_t(\theta)$ becomes

$$E_t(\omega) = \exp\left\{-\frac{1}{2}\frac{\pi^2}{\lambda^2}\Delta C_1^2\,(\omega\overline{\omega})^2\right\}, \tag{2.24}$$

where we employed the complex scattering angle $\omega$ (and its complex conjugate $\overline{\omega}$). Figure 2.9 shows $E_t$ for a 200 kV electron microscope with an energy spread of $\Delta E = 1$ eV (FWHM) and a constant of chromatic aberration $C_C = 1$ mm assuming that lens current fluctuation and the high tension ripple are negligible.

Though Eq. (2.24) is valuable for the practical evaluation of $E_t$, Eq. (2.22) reveals the nature of the damping envelope function. The damping envelope function due to partial temporal coherence is caused by the defocus dependence of the wave aberration function $\chi$. Hence, the energy spread of the electron beam, including high tension ripple and lens current fluctuations, leads to a blurring of the defocus in the image. The focus spread depends on the constant of chromatic aberration $C_C$ of the objective lens. It is the incoherent superposition of images, each formed with a slightly different defocus, which leads to the dampening of high spatial frequencies in the resulting micrograph.

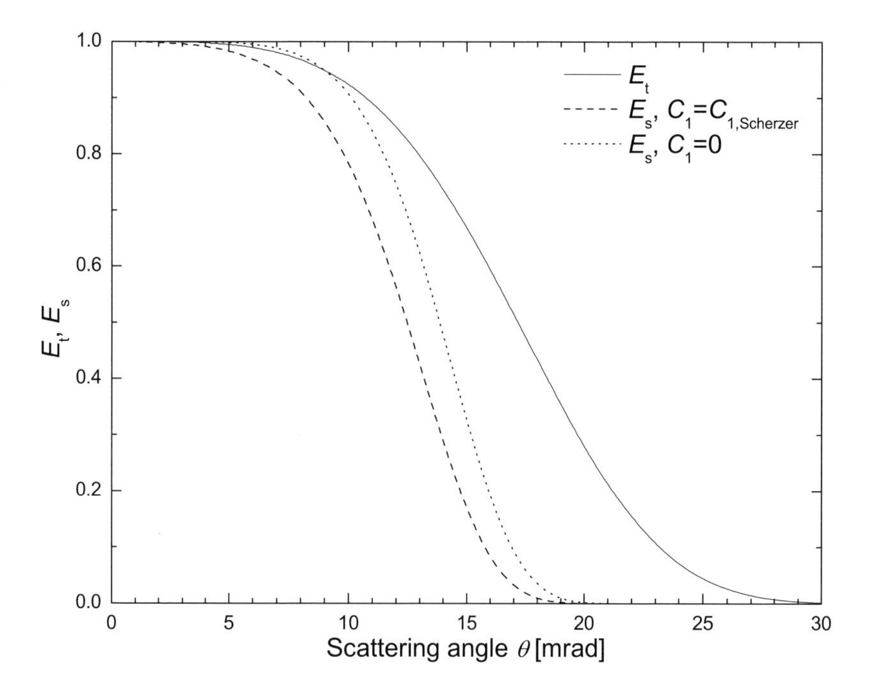


Fig. 2.9 Damping envelope functions plotted as a function of the scattering angle $\theta = |\omega|$. The damping envelope function due to partial temporal coherence $E_{\mathrm{t}}$ according to Eqs. (2.24) and (2.23) for 200 kV, $\Delta E = 1$ eV, $C_{\mathrm{C}} = 1$ mm, and $\Delta U = \Delta I = 0$ (full line), and the damping envelope function due to partial spatial coherence $E_{\mathrm{s}}$ according to Eq. (2.27) for 200 kV, $C_3 = 1$ mm, $\theta_{\mathrm{s}} = 0.2$ mrad and $C_1 = 0$ nm (dotted line) and $C_1 = -58$ nm (dashed line), respectively.

### 2.4.3.2 *Partial spatial coherence*

So far, we assumed that the illumination of the specimen can be described by a plane wave whose wave vector is parallel to the optical axis of the microscope. For a parallel illumination, the illumination angle is equal zero. Though a parallel illumination is in principle feasible, due to the finite brightness of the electron source the electron beam is often condensed in order to increase the dose on the specimen. Furthermore, even if the illumination is fully parallel, we have to consider that the size of the electron source is finite. This means that each point of the electron source illuminates the specimen independently from the other source points. On the other hand, if we consider a point on the specimen, it is illuminated by a multitude of individual point sources. A point on the specimen is thus illuminated by a cone of electrons, or as Wade and Frank (1977) brought it to the point; the object is illuminated by electrons emitted randomly from all points of the source. The image intensities due to each source element combine incoherently in the image plane to yield the complete image.

The smaller the source, the smaller the angle of the illumination cone under which each object point is illuminated. Hence, it is not of crucial importance whether the actual illumination on the specimen is parallel, but the important parameter is the divergence semi-angle $\theta_s$ under which an object point sees the electron source.

In principle, the source can be demagnified by the condenser lenses. For a finite brightness $B$ of the source, however, it is not possible to reduce its size significantly without severe loss of beam current. The brightness can be expressed as

$$B = \frac{I_s}{\Omega A_s}. \tag{2.25}$$

The brightness $B$ is emission current $I_s$ per solid angle $\Omega$ and emission area $A_s$. Hence, the beam current is proportional to the source size. If the source size is theoretically demagnified to a single point, i.e. if $A_s \to 0$, the beam current goes to zero as well.

Hence, because the brightness is finite and because point sources do not exist, under experimental conditions we always have to deal with a finite beam divergence semi-angle $\theta_s$. Of course, the smaller the source and the higher its brightness, the smaller $\theta_s$ can be set to still achieve a given beam current on the specimen. The source size and the brightness are thus the crucial factors which affect the damping envelope function due to partial spatial coherence, and eventually contribute to the limited information transfer in TEM imaging.

Because of the finite size of the electron source, we have to consider the illumination consisting of partial waves of varying wave vectors which form a cone whose opening angle is given by the beam divergence semi-angle $\theta_s$. Each partial wave within the illumination cone thus produces a partial image. Similar to the focus spread, it is the incoherent superposition of images each formed from a slightly different illumination angle that leads to the dampening of high spatial frequencies in the resulting micrograph. This loss of coherent image information is described by the damping envelope function $E_s(\boldsymbol{q})$, which accounts for the finite size of the electron source, i.e. for the partial spatial coherence. It can be written as

$$E_s(\omega) = \exp\left\{-\frac{\pi^2\theta_s^2}{\lambda^2}\left[\nabla\chi(\omega)\right]^2\right\}, \tag{2.26}$$

where $\nabla$ is the nabla operator referring to the gradient of $\chi$ in respect to $\omega$. We follow strictly the description of J. Frank (1976)[5], where the beam divergence semi-angle $\theta_s$ is related to the radius $r_s$ of the source distribution function by $\theta_s = \lambda/r_s$. Assuming a Gaussian source distribution function (see Chapter 3), the full width at half maximum (FWHM) of the source distribution function is then given by $\delta_{geo} = r_s/\ln 2$. Equation (2.26) is valid regardless of the form of $\chi(\omega)$. However, if

[5]For a discussion on how the beam divergence semi-angle $\theta_s$ is measured and weighted to describe the damping due to the finite source size see, e.g. De Jong and Van Dyck (1992); Malm and O'Keefe (1993); O'Keefe *et al.* (2001a).

we make use of Eq. (2.18), we obtain

$$E_{\mathrm{s}}(\omega) = \exp\left\{-\frac{\pi^2\theta_{\mathrm{s}}^2}{\lambda^2}\left(\omega\overline{\omega}C_1^2 + 2\left(\omega\overline{\omega}\right)^2 C_1C_3 + \left(\omega\overline{\omega}\right)^3 C_3^2\right)\right\}. \tag{2.27}$$

With the damping envelope functions due to partial temporal coherence in Eq. (2.24) and partial spatial coherence in Eq. (2.27), all generally relevant optical effects that contribute to the information transfer are defined. We can thus summarize the total phase contrast transfer function in the next subsection.

#### 2.4.3.3 *The phase contrast transfer function for partial coherent illumination*

Considering the effects of the objective aperture $t_{\mathrm{a}}$, the lens errors $t_{\mathrm{L}}$ and the effects due to partial spatial $E_{\mathrm{s}}$ and partial temporal coherence $E_{\mathrm{t}}$ discussed in Eqs. (2.11), (2.13), (2.22) and (2.26), the frequency spectrum of the complex transfer function $t(r)$ in Eq. (2.7) can be written as the product of the individual terms

$$t(\omega) = t_{\mathrm{a}}(\omega)\, t_{\mathrm{L}}(\omega)\, E_{\mathrm{s}}(\omega)\, E_{\mathrm{t}}(\omega). \tag{2.28}$$

Equation (2.11) needs to be adjusted to express $t_{\mathrm{a}}$ as a function of the complex coordinate $\omega$. Since $t_{\mathrm{a}}$, $E_{\mathrm{s}}$ and $E_{\mathrm{t}}$ are real scalar functions, while only $t_{\mathrm{L}}$ is a

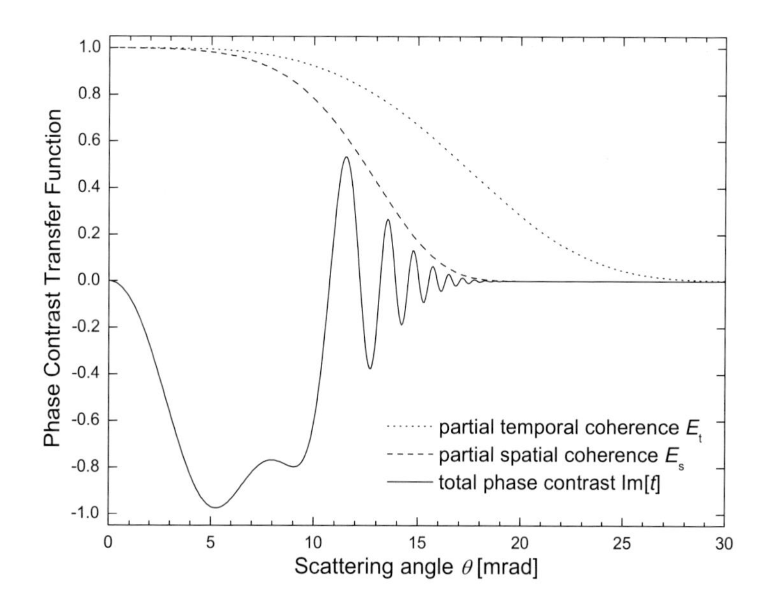


Fig. 2.10 Phase contrast transfer functions as a function of the scattering angle $\theta = |\omega|$. The total phase contrast transfer function according to Eq. (2.29) drawn for a 200 kV electron microscope at Scherzer focus (full line). Parameters are the same as in Figs. 2.8 and 2.9. The damping envelope functions due to partial temporal coherence (dotted line) and due to partial spatial coherence (dashed line) are plotted separately. It is assumed that $t_{\mathrm{a}} = 1$, i.e. there is no objective aperture.

complex function, we can write according to Eq. (2.17) that the phase contrast transfer function for HRTEM imaging is

$$\Im[t(\omega)] = t_a(\omega)\, t_c(\omega)\, E_s(\omega)\, E_t(\omega). \tag{2.29}$$

The phase contrast transfer function for partial coherent imaging as given in Eq. (2.29) is thus a real scalar function. Equation (2.29) is the fundamental relation of phase contrast electron microscopy. The phase contrast transfer function of a 200 kV electron microscope is shown in Fig. 2.10, employing the same parameters used in Figs. 2.8 and 2.9.

### 2.4.4 *Point resolution and information limit*

With the phase contrast transfer function for partial coherent imaging given in Eq. (2.29) we can characterize the information transfer in terms of point resolution and information limit.

#### 2.4.4.1 *Point resolution*

The point resolution in phase contrast electron microscopy reflects the highest spatial frequency and the corresponding spatial distance, which can be transferred from the object to the image such that a direct correlation between object and image is feasible. Equation (2.17) shows that for a conventional electron microscope, the phase contrast transfer function can be modified by adjusting the defocus. Under optimized conditions the applied defocus corresponds to the Scherzer focus, which is given in Eq. (2.21). The phase contrast transfer function for Scherzer focus shows a broad passband. Object information contained in the image can directly be interpreted up to the first zero crossing of the phase contrast transfer function for Scherzer focus. This reasons the definition of the point resolution for phase contrast imaging; the first zero crossing of the the phase contrast transfer function at Scherzer focus reflects the point resolution of the microscope. For a conventional electron microscope, this value is determined by the electron wavelength and the constant of spherical aberration.

In order to derive the point resolution we simply have to insert the Scherzer focus from Eq. (2.20) into Eq. (2.17) and set $t_c(\theta) = 0$ for $\theta \neq 0$. From this we obtain the characteristic scattering angle $\theta_r$, corresponding to the point resolution

$$\theta_r = \left(\frac{2\tau^2\lambda}{\pi C_3}\right)^{\frac{1}{4}}. \tag{2.30}$$

Furthermore, since $\theta \approx \lambda q$, we obtain the spatial frequency $q_r$ of the point resolution and its reciprocal value, i.e. the point resolution $\rho_r$

$$\rho_r = \frac{1}{q_r} = \left(\frac{\pi C_3 \lambda^3}{2\tau^2}\right)^{\frac{1}{4}}. \tag{2.31}$$

With $\tau = 2.89$ in Eq. (2.21), the point resolution for HRTEM becomes (Williams and Carter, 1996)

$$\boxed{\rho_{\rm r} \approx 0.66 \left(C_3 \lambda^3\right)^{\frac{1}{4}}\,.} \tag{2.32}$$

As pointed out above, the point resolution reflects the smallest distance in a micrograph which under optimized conditions can directly be interpreted. It is independent of the effects of the damping envelope functions due to partial coherence. Of course, if the impact of the damping envelope functions is too large, the image intensity of components of high spatial frequency is low. Hence, whether there is detectable intensity transferred at the point resolution depends on the damping envelope functions. On the other hand, if the effect of the damping envelope functions is small at $q_{\rm r}$, object information will be transferred to the image, which exceeds the point resolution. Because the transfer function is oscillating beyond $q_{\rm r}$, this information is, in general, not directly interpretable (see, e.g. Otten and Coene, 1993). The situation where the information transfer goes beyond the point resolution is illustrated in Fig. 2.10.

#### 2.4.4.2 *Information limit*

The above discussion reveals that there can be a gap between the point resolution and the actual information which contributes to a given micrograph. Hence, while the point resolution is a measure which reflects the shape of the optimized phase contrast transfer function, the information limit provides a measure for the highest spatial frequency component that significantly contributes to the image. In the absence of the damping envelope functions, the coherent phase contrast transfer function oscillates between $-1$ and $+1$. The damping envelope functions due to partial coherence determine the maximum amplitude of the phase contrast transfer function. Hence, it is the characteristics of the damping envelope functions which determine the highest spatial frequency component that significantly contributes to the image. Of course, the damping envelope functions never reach a value equal to zero, but above a certain limit, which is called the information limit, the transfer is too weak to cause a significant contribution to the image intensity. The information limit is defined as the distance for which either $E_{\rm t}$ or $E_{\rm s}$ reaches a value of $1/e^2$.

To derive the information limit $q_{\rm t}$, due to partial temporal coherence and the corresponding spatial distance $\rho_{\rm t}$, we simply set $E_{\rm t}$ from Eq. (2.27) equal to $1/e^2$. This yields for the information limit due to partial temporal coherence (see, e.g. Zandbergen and Van Dyck, 2000)

$$\rho_{\rm t} = \frac{1}{q_{\rm t}} = \left(\frac{\pi \Delta C_1 \lambda}{2}\right)^{\frac{1}{2}}\,. \tag{2.33}$$

Similarly, one can deduce the information limit $q_{\rm s}$ due to partial spatial coherence and the corresponding spatial distance $\rho_{\rm s}$ by setting $E_{\rm s} = 1/e^2$. However, in this case the envelope function depends on the applied focus $C_1$. For $C_1 = 0$, we obtain from Eq. (2.27) the information limit due to partial spatial coherence

$$\rho_s = \frac{1}{q_s} = \left(\frac{\pi^2 \theta_s^2 \lambda^4 C_3^2}{2}\right)^{\frac{1}{6}} \approx \left(\pi \theta_s \lambda^2 C_3\right)^{\frac{1}{3}}. \tag{2.34}$$

For dedicated high-resolution transmission electron microscopes, which are operated at 200–300 kV and which are typically equipped with field-emission electron sources and dedicated high-resolution pole-piece objective lenses with $C_3$ as small as 0.5 mm, it is often the information limit due to partial temporal coherence given in Eq. (2.33) that is limiting the transfer at high spatial frequencies (see, e.g. Van Dyck *et al.*, 1996).

For the transfer function depicted in Fig. 2.10, the information limit due to partial spatial coherence is $\rho_s = 0.17$ nm and the information limit due to partial temporal coherence is $\rho_t = 0.14$ nm. Hence, for this case it is the partial spatial coherence that limits the information transfer. However, if $C_3$ is changed from 1 mm to 0.5 mm, one obtains for the same set of microscope parameters $\rho_s = 0.13$ nm and $\rho_t = 0.14$ nm. Hence, in the second case it is the information limit due to partial temporal coherence which is limiting the information transfer. As will be shown in the last part of the book, this is particularly true for aberration-corrected instruments with $C_3 \rightarrow 0$.

The information limit of transmission electron microscopes is commonly measured by the so-called Young's fringes method (Frank, 1976; Zemlin and Schiske, 2000). However, recently it has been argued that this method is not adequate to measure the information limit (see, e.g. O'Keefe *et al.*, 2008) and for this reason alternative methods have been suggested (Barthel and Thust, 2008). This trend reflects the difficulty in precisely assessing the information limit of a microscope.

### 2.4.5 *Other factors*

Apart from the partial temporal and partial spatial coherence, there are additional factors which can lead to an incoherent blurring of high resolution micrographs (see, e.g. De Jong and Van Dyck, 1992; Van Dyck *et al.*, 2003). These contributions, some which of are discussed below, are not directly related to the optics of the microscope but concern mechanical aspects and detector properties as well as physical limitations due to the limited scattering power of the elements.

#### 2.4.5.1 *Non-ideal recording devices*

A major contribution which can severely influence the observable resolution of a micrograph is due to the non-ideality of the recording process. Employing any kind of detector consisting of a two-dimensional array of pixels implies that the electron wave in the image plane, which in fact is a continuous function, is sampled according to the chosen magnification and the pixel size of the detector. Hence, the resolution of the micrograph cannot be better than the pixel size. For instance, if a detector is employed which has a physical pixel size of $a = 20$ $\mu$m and an effective magnification of $M = 10^5$ is chosen, the resolution cannot be better than $a/M = 0.2$ nm. This type of resolution limitation only holds in the case of an ideal pixelated detector, but real detectors are not even ideal pixelated detectors.

The non-ideality of a detector causes a blurring of information over several pixels, even if in principle the area of only one pixel is excited. Though slow-scan CCD cameras have many advantages (Krivanek and Mooney, 1993), this blurring is particularly pronounced for this type of electron detector (De Ruijter, 1995). While a point-like electron impacts an area which in principle is smaller than a pixel, the translation of the electron into a cloud of photons registered as individual counts causes the localized signal of the electron to spread over a finite area covering several pixels. This blurring is described by a point spread function (see Fig. 2.11a). The point spread function is a characteristic of the detector and does not depend on the magnification. It describes the area of the detector which contributes to the image if one pixel is hit by an incoming electron. Ideally, the only pixel that contributes to the image intensity is the one which is hit by the electron. However, for non-ideal recording devices the point spread function is finite, i.e. information of one pixel is spread over a certain area. In order to minimize the impact of the point spread function on the resolution of a micrograph, one has to choose a higher magnification than would be required by the actual pixel size. Hence, one has to over-sample the image to balance the blurring caused by the point spread function (see, e.g. Campbell *et al.*, 1997).

The point spread function is the real-space measure which describes the blurring of information over a certain area of the detector. However, the blurring can also be described in the spatial frequency space. For this, one employs the modulation transfer function (MTF), which essentially represents the Fourier transform of the point spread function. While the characteristic length of the point spread function is the pixel size, for the modulation transfer function it is the Nyquist frequency, which is defined as the reciprocal value of twice the pixel size, i.e. $f_{\mathrm{N}} = 1/2a$. Figure 2.11b shows the modulation transfer function of a slow-scan CCD camera operated with 300 keV electrons. The MTF is plotted from zero to the Nyquist frequency. In order to compare easily the modulation transfer function of different detectors, one typically compares the value of the modulation transfer function at the Nyquist frequency. For the MTF shown in Fig. 2.11, the value MTF($f_{\mathrm{N}}$) is 0.19, which is a typical value for state-of-the-art CCDs employed for 300 keV electrons and comparable effective pixel size (Meyer *et al.*, 2000). In general, the smaller the pixel size and the higher the electron energy, the larger the impact of the MTF on the acquisition process.

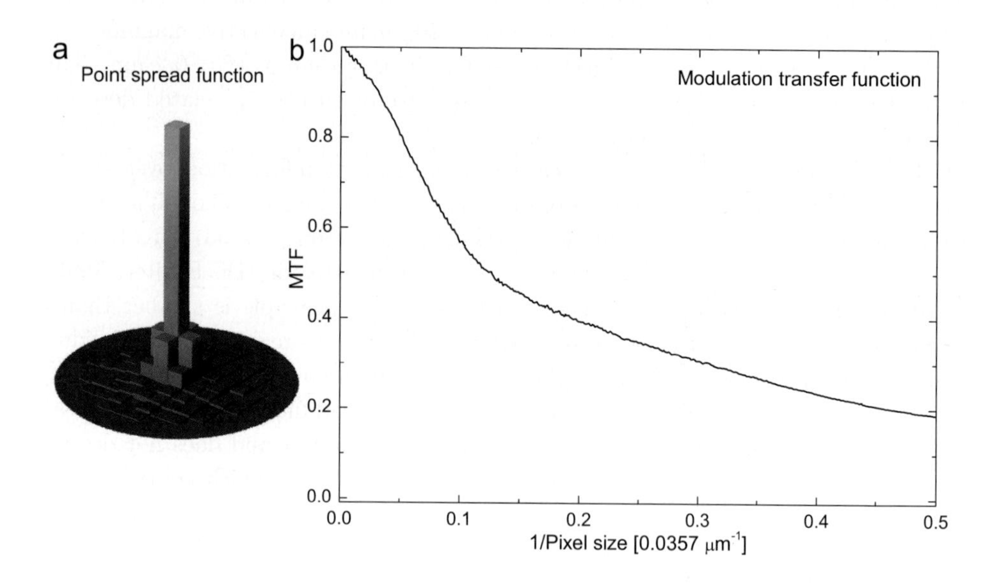


Fig. 2.11 Point spread function and modulation transfer function. Experimentally measured modulation transfer function of a slow-scan CCD camera for 300 keV electrons (Erni *et al.*, 2010). The MTF was measured according to the method described by Nakashima and Johnson (2003). The CCD has $1024^2$ effective pixels with a size of 28 $\mu$m. (**a**) shows the point spread function — each rectangular box reflects a pixel. (**b**) shows the corresponding modulation transfer function along the spatial frequency axis from zero to the Nyquist frequency $f_{\mathrm{N}}$.

### 2.4.5.2 *Stability issues*

Another contribution which can affect or limit the information transfer of a microscope is its overall mechanical stability. This applies to the stability of the specimen in the holder, including drift and vibrations in lateral and vertical directions.

For the case in which a specimen drifts with a constant rate of let us say, 0.5 nm/s in one direction, the best resolution that can be achieved for the drift direction is 0.5 nm if an exposure time of 1 s is chosen. With a distinct drift direction, the information transfer can become anisotropic. Similar restrictions can be imposed by mechanical vibrations where, however, the amplitude and the frequency of the vibrations are of importance.

Furthermore, the stability of the high tension and the stability of the electric currents of the lenses, stigmators and deflectors are also crucial for the information transfer. However, these two aspects are considered in the damping envelope function due to partial temporal coherence (see Eq. (2.23)).

Conceptually, atomic vibrations are also of importance and can in certain cases indeed limit the resolution. However, this is not meant when talking about the stability of the specimen. This effect can be considered in the description of the elastic scattering amplitude at finite temperature (see below).

#### 2.4.5.3 *The scattering amplitude*

Mentioning the elastic scattering factor raises the question about the scattering power of a single atom and whether this could affect the resolution. Provided that none of the above factors limit the information transfer, the information transfer reflects the scattering power of the atoms in the specimen. The elastic electron scattering factor, as for instance described by the Mott formula (see, e.g. Schwartz and Cohen, 1987), is a function which rapidly decays with increasing scattering angle. Furthermore, the smaller the atomic number of the element, the smaller its scattering factor. Hence, if the scattering factor is small, the atom cannot transfer sufficient electrons to high spatial frequencies which contains the high resolution information. In such cases, the resolution can be limited by the object. This, in fact, is a desirable situation. Whenever the object limits the resolution, the optical and mechanical performance of the microscope are no longer critical and one hits a real physical limitation which cannot be overcome.

Although the characteristics of the scattering factor do not impose a real limitation to the actual transfer of the microscope, one has to consider the scattering power of the elements if experiments are performed which aim at revealing the optical information limit of the instrument. Since under the assumption of single scattering events the intensity scattered to high spatial frequencies is directly related to the scattering factor of the element under consideration, light elements of weak scattering power might not allow for revealing the information limit of the instrument. For this reason, resolution tests are typically performed with specimens which contain heavy elements like gold or tantalum, and thus enable sufficient electron scattering to high scattering angles (see, e.g. Barthel and Thust, 2008).

The picture of having individual scatterers in the specimen might be an oversimplification. The arrangement of the atoms in the crystal can in general not be neglected when discussing the effect of the individual atoms on the achievable resolution. Dynamic scattering and particularly channelling of the electrons along the atomic columns can complicate the situation.

Atomic vibrations caused by phonons are normally considered by including a Debye–Waller factor in the description of the elastic scattering factor. The Debye–Waller factor attenuates the scattering factor. Hence, the thermal motion of the atoms can lead to a reduction of contrast at high spatial frequencies.

#### 2.4.5.4 *The finite field of view*

Though generally negligible, a further factor that can be of importance for the limitation of the information transfer in an electron microscope is the variation of the transfer function across the field of view. Commonly, it is assumed that for a given field of view, like the object area which is imaged on the recording device, the transfer function is identical for all object and image points, respectively. However, this approximation is only valid if the field of view is small enough. This condition is mostly valid for high-resolution imaging where the field of view is merely tens of nanometers. The assumption that the entire field of view is affected by the same transfer function is called the isoplanatic approximation. What happens if we consider a particular deviation from the isoplanatic approximation?

Let us picture a case where some optical effect causes a focus change across the field of view of a perfectly flat object with its foil normal parallel to the optical axis. If we analyze the information transfer locally on the image, we would not be able to see a difference in the information transfer compared to the case where the focus is constant across the field of view. However, if we consider the information transfer summed up across the field of view, there would be an additional blurring of information due to the focus variation. This is an additional focus blur which essentially has the same effect as the focus blur due to the chromatic aberration. However, compared to the focus variation caused by the chromatic aberration, which is local and affects each image point equally, the focus variation across the field of view is not local and its impact on the information transfer can only be revealed if a finite area is analyzed. A focus variation across the field of view leads to a transfer reduction along the gradient of the focus change. As will be discussed in Chapter 7, such effects can be caused by aberrations which do not affect object points in the immediate neighborhood of the optical axis but which become apparent for object points in a finite distance from the optical axis. Such aberrations are called off-axial aberrations.

For conventional electron microscopes, the invariance of the transfer function across the field of view is an issue if a large field of view of identical resolution is required, such as in biological applications where, for instance, a large number of molecules dispersed on a support film should be imaged with equal information transfer. For this particular case, the homogeneity, or isoplanaticity of the illumination is of crucial importance.

## 2.5 Non-Linear Imaging

In the previous sections about HRTEM imaging we dealt with a linear imaging model. Each spatial frequency component $\boldsymbol{q}$ of the exit-plane wave $\psi_{\mathrm{ep}}(\boldsymbol{r})$ has a corresponding component $\boldsymbol{q}$ in the image wave $\psi(\boldsymbol{r})$. The relation between a given frequency component $\boldsymbol{q}$ in the exit-plane wave and the corresponding component in the image wave is given by the (linear) transfer function $t(\boldsymbol{q})$ in Eq. (2.29). Equation (2.7) describes this linear relation between $\psi_{\mathrm{ep}}(\boldsymbol{r})$ and $\psi(\boldsymbol{r})$.

Under which assumption can we talk about linear imaging and when is this assumption violated? Let us assume we have a crystalline specimen free of any defects such that the spatial frequency components of the exit-plane wave $\psi_{\mathrm{ep}}(\boldsymbol{r})$ are fully determined by a discrete set of $\boldsymbol{q}$ vectors which correspond to the Bragg diffracted beams $\boldsymbol{k_g}$ for a particular crystal orientation (see Fig. 2.1). The entire frequency spectrum $\boldsymbol{q}$ of the exit-plane wave $\psi_{\mathrm{ep}}(\boldsymbol{r})$ is then transferred via objective lens to the image plane, where the individual spatial frequency components interfere and form a HRTEM micrograph.

It is in the step of forming the interference pattern where we make the assumption of linear imaging. The linear imaging model only accounts for interference between a frequency $\boldsymbol{q} \neq \boldsymbol{0}$ and the forward scattered beam described by the spatial frequency $\boldsymbol{q} = \boldsymbol{0}$. Yet, no interference takes place between different $\boldsymbol{q} \neq \boldsymbol{0}$ vectors (Saxton, 1978).

If the intensity of the forward scattered beam, i.e. $\boldsymbol{q} = \boldsymbol{0}$, is much larger than the intensity of the diffracted beams $\boldsymbol{q} \neq \boldsymbol{0}$, the linear imaging model can be considered to be a good approximation to describe the imaging process. This condition is usually met if the specimen is very thin, i.e. if it is for instance a weak (phase) object which does not diffract too much intensity away from the forward scattered beam. For a thin specimen, the contribution to the total image intensity stemming from the interference between two diffracted but weak beams can be neglected. However, in general and in particular for specimens of finite thickness, the contributions of interference terms between two different Bragg diffracted beams and thus any two different spatial frequency components of the exit-plane wave $\boldsymbol{q}''$ and $\boldsymbol{q}'$ cannot be neglected[6]. Hence, for specimens of finite thickness, non-linear image contributions arise from the interference between two spatial frequency components $\boldsymbol{q} \neq \boldsymbol{0}$.

In order to describe non-linear imaging in HRTEM, the *linear* transfer function in Eq. (2.7) is replaced by a transfer function that takes into account the interference between diffracted beams. This *non-linear* transfer function is called transmission cross-coefficient (Wade and Frank, 1977) and shall be denoted by $t_{\mathrm{tcc}}$. Since the transmission cross coefficient describes interference between diffracted beams, it is a function that depends on two $\boldsymbol{q}$ vectors, i.e., $t_{\mathrm{tcc}} = t_{\mathrm{tcc}}(\boldsymbol{q}', \boldsymbol{q}'')$.

The transmission cross-coefficient specifies how the interference between two diffracted beams is affected by the lens errors described by $\chi$, the contrast dampening due to partial spatial and partial temporal coherence, and likewise by the objective aperture. For the case that one of the $\boldsymbol{q}$ vectors is equal $\boldsymbol{0}$, the transmission cross coefficient is equivalent to the transfer function of Eq. (2.7) and $\Im[t_{\mathrm{tcc}}(\boldsymbol{g}', \boldsymbol{0})]$ is the phase contrast transfer function for linear imaging given in Eq. (2.29).

In case of a crystalline specimen illuminated with a parallel beam, the frequency spectrum of the electron wave is dominated by a few discrete $\boldsymbol{q}$ values, which are determined by the Bragg diffracted beams $\boldsymbol{k_g}$. In general, however, the entire spatial frequency spectrum $\boldsymbol{q}$ of the wave function has to be taken into account. Any arbitrary spatial frequency $\boldsymbol{q}$ can interfere with a frequency $\boldsymbol{q}'$. According to Ishizuka (1980), the frequency spectrum of the image intensity $I(\boldsymbol{q})$ can then be written as

$$I(\boldsymbol{q}) = \int t_{\mathrm{tcc}}(\boldsymbol{q}' + \boldsymbol{q}, \boldsymbol{q}')\, \psi_{\mathrm{ep}}(\boldsymbol{q}' + \boldsymbol{q})\, \bar{\psi}_{\mathrm{ep}}(\boldsymbol{q}')\; d\boldsymbol{q}', \tag{2.35}$$

[6]Within this section, the prime in $\boldsymbol{q}'$ and $\boldsymbol{q}''$ shall indicate the difference between two $\boldsymbol{q}$ vectors, while in upcoming chapters it has the meaning of a derivative. Its actual meaning should be clear from the context.

where the transmission cross-coefficient $t_{\mathrm{tcc}}(\boldsymbol{q}'', \boldsymbol{q}')$ is given by

$$t_{\mathrm{tcc}}(\boldsymbol{q}'', \boldsymbol{q}') = t_{\mathrm{a}}(\boldsymbol{q}'')t_{\mathrm{a}}(\boldsymbol{q}') \exp\left\{-\frac{2\pi \mathrm{i}}{\lambda}\left(\chi(\boldsymbol{q}'') - \chi(\boldsymbol{q}')\right)\right\} E_{\mathrm{t}}(\boldsymbol{q}'', \boldsymbol{q}')\, E_{\mathrm{s}}(\boldsymbol{q}'', \boldsymbol{q}'). \tag{2.36}$$

The function $t_{\mathrm{a}}(\boldsymbol{q})$ is the aperture function given in Eq. (2.11), and the wave aberration function $\chi(\boldsymbol{q})$ is given in Eq. (2.15). The damping envelope functions $E_{\mathrm{t}}$ and $E_{\mathrm{s}}$ describing the effect of partial temporal and partial spatial coherence need to be rewritten in order to take into account cross-terms between different $\boldsymbol{q}$ vectors (Coene *et al.*, 1996):

$$E_{\mathrm{t}}(\boldsymbol{q}'', \boldsymbol{q}') = \exp\left\{-\frac{2\pi^2 \Delta C_1^2}{\lambda^2}\left(\frac{\partial\chi(\boldsymbol{q}'')}{\partial C_1} - \frac{\partial\chi(\boldsymbol{q}')}{\partial C_1}\right)^2\right\}, \tag{2.37}$$

$$E_{\mathrm{s}}(\boldsymbol{q}'', \boldsymbol{q}') = \exp\left\{-\frac{\pi^2\theta_{\mathrm{s}}^2}{\lambda^2}\left(\nabla\chi(\boldsymbol{q}'') - \nabla\chi(\boldsymbol{q}')\right)^2\right\}. \tag{2.38}$$

The source-size-dependent envelope function $E_{\mathrm{s}}$ (Ishizuka, 1980) depends on the beam divergence semi-angle $\theta_{\mathrm{s}}$; $\nabla\chi(\boldsymbol{q})$ describes the gradient of $\chi(\boldsymbol{q})$, given in Eq. (2.15), with respect to $\boldsymbol{q}$. One easily sees that if $\boldsymbol{q}'$ is set equal to zero, the above equations for non-linear imaging are equivalent to the case of linear imaging. The damping envelope functions for non-linear imaging have essentially the same effect as the damping envelope functions for linear imaging given in Eqs. (2.27) and (2.24), except that in Eqs. (2.37) and (2.38) cross terms between different $\boldsymbol{q}$ vectors are taken into account.

The non-linear imaging theory can for instance explain the presence of spatial frequency components which can be observed in the image, but which do not have real structural counterparts in the object. But even the non-linear imaging theory on the basis of the transmission cross coefficient given in Eq. (2.36) is only approximately valid. Hence, while the classical non-linear imaging theory described above and essentially introduced by Ishizuka (1980) can be considered as a first-order approximation, more elaborate non-linear imaging contrast transfer theories under slightly different approximations have been developed by Pulvermacher (1981) and Bonevich and Marks (1988). As shown by Bonevich and Marks (1988), consideration of a full non-linear description of the contrast transfer can lead to results that are in contradiction with the linear and even with the classical non-linear imaging theory. This full non-linear imaging theory can for instance explain information transfer to very high spatial frequencies in the presence of beam illumination angles which, according to the classical theory, would prevent such transfer (see, e.g. Eqs. (2.26) and (2.38)).

It is important to note that the non-linear image formation is not related to the dynamic scattering of electrons in the specimen. These are two different effects. In dynamic scattering interactions between different Bragg diffracted beams are considered and not just the loss of intensity of the forward scattered beam, on account of the increase of intensity of the diffracted beams as this is done in the kinematic description. In the non-linear imaging theory on the other hand, interactions between any different image frequencies $\boldsymbol{q}'$ and $\boldsymbol{q}''$ are considered compared to the case of linear imaging, where only interference terms between $\boldsymbol{q} \neq \mathbf{0}$ and $\boldsymbol{q} = \mathbf{0}$ are taken into account. Dynamic scattering occurs in a specimen of finite thickness and it leads to the dynamic exchange of scattering intensity between different Bragg diffracted beams, including, of course, the forward scattered beam. But this occurs within the specimen.

On the other hand, the theory of non-linear imaging considers deviations from the linear image formation, which become important if the intensity of diffracted beams becomes more intense such that their interference in the image plane can no longer be ignored. Obviously, the diffracted intensity is more intense in case of thick specimens, which then requires the dynamic scattering theory. Of course, it is for these situations, i.e. that the diffracted intensity becomes comparable to the intensity of the forward scattered beam, that the linear image formation theory needs to be replaced with a non-linear image formation theory. However, one has to be aware that dynamic scattering in the specimen and the non-linear image formation are two separate processes which both can complicate the interpretation of HRTEM micrographs. The dynamic scattering can lead to the formation of an exit-plane wave which does not correspond directly to the projected structure of the specimen. Hence, the object which is imaged, i.e. the exit-plane wave, does not simply reflect the structure of the specimen. On the other hand, due to the interference of image frequencies $\boldsymbol{q}'$ and $\boldsymbol{q}''$, non-linear imaging can give rise to the occurrence of image frequencies which are not present in the dynamically or even kinematically formed exit-plane wave.

With this short look into the non-linear imaging theory of high-resolution imaging, we close the part about conventional high-resolution phase contrast imaging. In forthcoming chapters we will make use of the concepts introduced here and discuss the image formation for aberration-corrected instrument under the linear approximation.

## 2.6 Summary

In this chapter, we described from a fundamental point of view how the exit-plane wave is formed and how the structural information contained in the exit-plane wave is translated into a phase-contrast micrograph. The discussion of this topic is based on the following keywords: weak phase object; exit-plane wave; phase contrast transfer function; isotropic aberration function; temporal and spatial partial coherence; Scherzer focus; point resolution; and information limit.

# Chapter 3

# Scanning Transmission Electron Microscopy

Shortly after the invention of the broad-beam illumination transmission electron microscope by Knoll and Ruska (1932), Manfred von Ardenne (1938a), a German physicist, noticed the great potential of using a focused probe rather than a broad beam to study microscopic objects by electrons. Von Ardenne, who was awarded for the invention of the table-top electron microscope by the former Soviet Union, was active as an inventor in various fields of physics, including research in communication and radar technology as well as in medical physics. After the Second World War, he conducted research in nuclear and plasma physics for the Soviet Union, whereupon he returned to Germany in 1953. Without trying to elucidate the historical details of the early years of electron microscopy, it can be summarized that from the invention of the electron microscope in 1932 and von Ardenne's first electron optical instrument, which made use of a focused electron probe that was scanned across a specimen (von Ardenne, 1938b), it took roughly 30 years to the realization of a dedicated scanning transmission electron microscope which was capable of producing results of similar quality to the broad-beam equivalent. The pioneering work of Albert V. Crewe (1966) marks the beginning of practical scanning transmission electron microscopy (STEM). From then on, the scanning probe mode was developed as a complementary technique to the broad-beam illumination mode. Although the actual realization of STEM has a handicap of roughly 30 years, its fast development has certainly benefited from the electron optical know-how derived from the broad-beam illumination mode. Nowadays, STEM can be regarded as a powerful operation mode which, on many state-of-the-art electron microscopes, i.e. STEM/TEM instruments, provides a wealth of complementary information that elucidates the properties of a material from a slightly different point of view.

## 3.1 Overview

In STEM, information about the specimen is collected in a serial acquisition mode. The specimen is illuminated with a convergent electron beam which is focused to a small spot at the height of the specimen (see Fig. 3.1). To record an image, the electron probe is scanned within a rectangular frame on the specimen. On each scan position, the electron probe is propagated through the specimen. As a consequence of electron scattering within the specimen, part of the electrons are scattered away from their initial trajectories. The scattering distribution of the electrons in the far field behind the specimen corresponds to a diffraction pattern. Since the electron probe is convergent, the diffraction pattern is a convergent electron diffraction pattern. If the crystal spacing is large enough, or if the convergence angle of the electron beam is sufficiently large, the diffraction disks in the diffraction pattern partially overlap. Indeed, as will be shown below, the coherent partial overlap of diffraction disks is a requirement for resolving a given crystal spacing in an atomic-resolution scanning transmission electron micrograph.

Each scan position produces a site-specific diffraction pattern. If the electron beam is positioned on an area that contains strong scatterers, like heavy atoms, the intensity at high scattering angles is enhanced, whereas for the case that the electron beam is positioned on an area of weak scatterers or on a thin area, the scattering intensity at high angles is low and the intensity in forward direction and at small scattering angles remains high. This is a very pictorial explanation, and generally the propagation of the focused electron beam through the specimen and the intensity distribution in the diffraction pattern are complex matters which require consideration of dynamical scattering, channelling effects as well as quasi-elastic scattering, such as thermal diffuse scattering. Nonetheless, for our present purpose, where we do not focus on the electron-specimen interaction, this simple explanation shall suffice. For an in-depth discussion of the image formation in STEM and its dependence on the specimen we refer to more specific literature (Rose, 1975; Fertig and Rose, 1981; Nellist and Pennycook, 2000).

A two dimensional STEM micrograph corresponds to a two-dimensional array of data points. Each of these data points reflects the detector signal collected during the dwell time, i.e. during the time the electron probe was stationed on a given scan position. The information contained in a STEM micrograph depends on the position and size of the detector in the diffraction plane. Let us assume we have an infinitely large detector which detects all electrons in the diffraction plane behind the specimen. Neglecting back-scattered electrons, the detector would thus produce a constant signal and we would not learn anything about the specimen. Hence, while scanning the focused electron beam across the specimen, only a certain part of the intensity of the diffraction pattern is recorded as a function of the beam position. The choice of the area of the diffraction pattern that is positioned on the detector determines the image contrast.

Positioning a circular electron detector on the forward scattered beam yields a bright-field (BF) scanning transmission electron micrograph (see Fig. 3.1). For the case where there is no scatterer in the path of the beam, the BF signal reflects the total beam current, whereas for the case that there are scatterers in the path of the beam, the BF signal corresponds to the total beam current minus the integrated

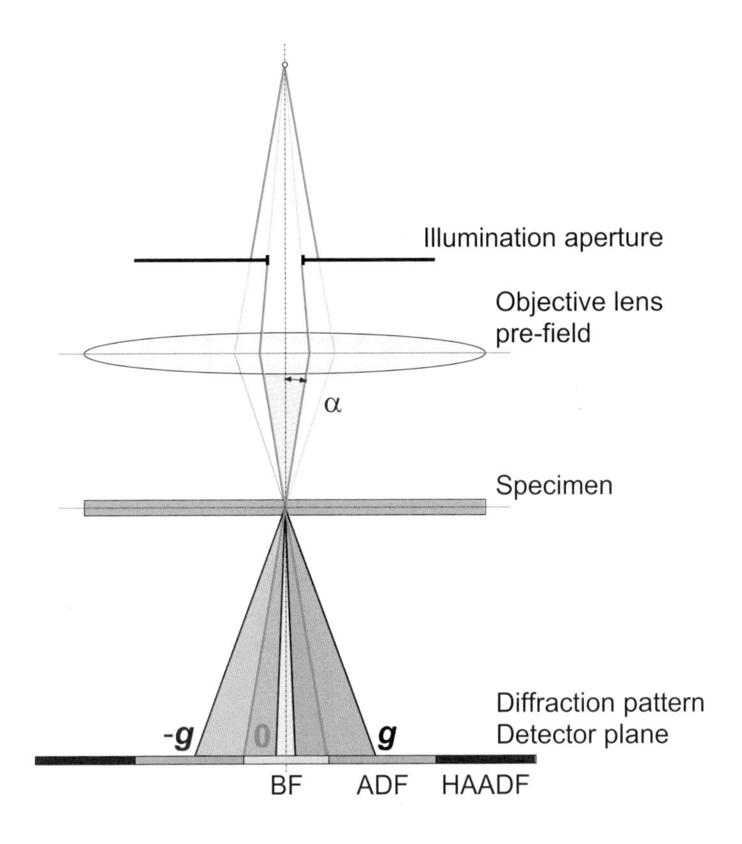


Fig. 3.1 STEM setup.

intensity that is scattered to angles beyond the area of the bright-field detector. Alternatively, an annular detector can be used which, instead of detecting the forward scattered beam, records an annular dark-field (ADF) signal. The BF and the ADF signals are in a qualitative way complementary to each other. If there is no scatterer in the path of the beam, the ADF intensity is zero, and for the case that there is scattering, the ADF signal reflects the scattering power — for the selected angular range — of the object that is in the path of the beam. Hence, the ADF signal increases with the scattering factor of the elements in the specimen as well as with the thickness of the specimen. The angular range of the annular detector, which in general is a fixed detector of a given size, can be adjusted by changing the camera length, i.e. the magnification by which the diffraction pattern is projected onto the detector.

A widely applied STEM imaging mode concerns the case for which the ADF detector is setup such that it collects over a large angular area electrons scattered to high angles. The integration of the high-angle scattering over a large area warrants that coherence effects between the diffracted beams are averaged out.

Hence, the corresponding high-angle annular dark-field (HAADF) STEM micrograph essentially reflects an incoherent signal (Rose, 1975; Hartel *et al.*, 1996; Nellist and Pennycook, 1999). Though (incoherent) thermal diffuse scattering contributes to the high-angle scattering, it is not fundamental in explaining the incoherence of the detected signal (Loane *et al.*, 1992; Hartel *et al.*, 1996; Nellist and Pennycook, 1999; Muller *et al.*, 2001). The crucial point that enables a largely incoherent signal is the size of the detector[1]. The incoherence of the HAADF STEM signal makes the specimen appear self-luminous. This simplifies image interpretation. Moreover, since the high-angle electron scattering is dominated by Rutherford scattering, the scattered intensity scales with the atomic number $Z$ of the elements in the sample. For pure Rutherford scattering, one expects a $Z^2$ dependence of the signal (Schwartz and Cohen, 1987). Experiments and calculations reveal that the actual exponent is around 1.6–1.8 instead of 2 (Hillyard and Silcox, 1995; Rafferty *et al.*, 2001; Erni *et al.*, 2003b). This difference can be explained by the fact that the electron cloud surrounding the nucleus screens the Coulomb potential of the nucleus, which is of relevance for Rutherford scattering (Hartel *et al.*, 1996).

For a specimen of constant thickness, a HAADF STEM micrograph maps the atomic number of the elements in the specimen. Due to its favorable atomic-number dependence, HAADF STEM is usually referred to as $Z$-contrast imaging (Nellist and Pennycook, 2000).

Apart from the common BF, ADF and HAADF detector settings, special detector setups have been discussed in the literature which, for instance, are suitable for enhancing the contrast of light atoms (Cowley *et al.*, 1996) or can be used for phase contrast imaging in STEM (Rose, 1974).

However, independent of the detector, the critical part of the scanning transmission mode is the characteristics of the focused electron beam. If the electron beam can be focused to a probe that is of the size of the atomic spacing of a zone-axis oriented crystal, a STEM micrograph reveals modulations which correspond directly to the atomic spacing of the crystal. Hence, it is the electron probe which is decisive for the resolution in STEM; the smaller the electron probe, the better the lateral resolution. Furthermore, similar to HRTEM, it is the characteristics of the objective lens that are of fundamental importance to achieve a small electron probe. However, as can be seen from Fig. 3.1, it is not the post-field that is relevant for the electron probe, but the pre-field of the objective lens.

In the following sections we draw our attention to the central point of STEM imaging which is the formation of the electron probe. Similar to Chapter 2, the

[1]The formation of an incoherent image in HAADF STEM can be explained by employing the *principle of reciprocity*. Consider the following situation: a source is placed in point A which emits a wave I. The wave is scattered at point P and arrives at point B. The principle of reciprocity states that the amplitude of wave I in point B is equal to the amplitude of a wave II in point A if the source is placed in B (Pogany and Turner, 1968). On the basis of the principle of reciprocity, it can be shown that a large, i.e. spatially incoherent, electron source in TEM is equivalent to a large detector in STEM (Cowley, 1969). Both the large electron source and the large detector provide an incoherent image.

electron–specimen interaction is not discussed in detail, i.e. the propagation of the electron probe through the sample and the detection of the scattered intensity to form a scanning transmission electron micrograph are not discussed in this context.

## 3.2 Geometrical Considerations

In the previous chapter on HRTEM imaging, we saw that the information transfer in phase contrast imaging is determined by the aperture function, the characteristics of the objective lens and by the limited degree of coherence of the electron beam, namely by the partial temporal coherence and the partial spatial coherence. In the following we will see that these four factors equivalently determine the characteristics of the electron probe and thus the information transfer in STEM imaging. We start discussing these effects from a geometrical point of view. The geometrical treatment of the individual contributions is particularly useful to understand their impact on certain microscope parameters (Crewe, 1987, 1997). However, in order to describe the combined effect of these contributions, it is essential to switch to a wave optical description of the electron probe. This will be done in the subsequent section.

### 3.2.1 *The diffraction limit*

In a twin-type objective lens, the specimen is immersed in the magnetic field formed by both the pre- and post-field of the objective lens. In analogy to the treatment of TEM imaging (see Chapter 2), we can simplify this situation by treating the pre- and post-field of the objective lens separately. With this simplification, the formation of an electron probe is determined on how the pre-field of the objective lens focuses the electron beam onto the specimen plane. The focused electron beam is the electron probe. As illustrated in Fig. 3.1, the electron probe and the specimen can be considered to be located in the back focal plane of the objective lens' pre-field.

In the broad-beam TEM mode, the post-field of the objective lens produces a diffraction pattern in the back focal plane of the objective lens' post-field (see Fig. 2.1). The back focal plane containing the diffraction pattern is conjugate to the plane of the electron source and a diffraction spot can be regarded as an image of the source. For STEM, the electron probe is located in the back focal plane of the objective lens' pre-field and, similar to TEM, the electron probe represents a demagnified image of the electron source. This image is not perfect. One of the factors explaining why there is no stigmatic image of the source is that there is an aperture present which limits the angular range of the illuminating electrons that form the electron probe.

Figure 3.1 shows that the effect of the illumination aperture is to control the illumination (or convergence) angle of the electron probe. As long as the aperture defines the illumination angle of the focused electron beam on the height of the object, its location along the optical axis is not critical. If the aperture is approximately illuminated by a parallel beam, the electron probe at the object plane is an Airy pattern (see, e.g. Born and Wolf, 2001). In the presence of aberrations, an illumination aperture of finite size is needed to optimize the size of the electron probe. Therefore, an Airy-pattern-type electron probe is practically unavoidable.

The Airy pattern is dominated by a central maximum surrounded by concentrical side lobes of distinctly lower intensity (see Fig. 3.2a). In order to relate the characteristics of the Airy pattern to the size of the electron probe, we can choose the first zero of the Airy pattern as the radius $\delta_D$ of the diffraction-limited electron probe. This can be written as

$$\boxed{\delta_D = 0.61\frac{\lambda}{\alpha}}, \tag{3.1}$$

where $\lambda$ is the electron wavelength given in Eq. (2.2) and $\alpha$ is the illumination (or convergence) semi-angle defined by the aperture opening (see Fig. 3.1). The value $\delta_D$ expresses the size of an electron probe, which is solely determined by

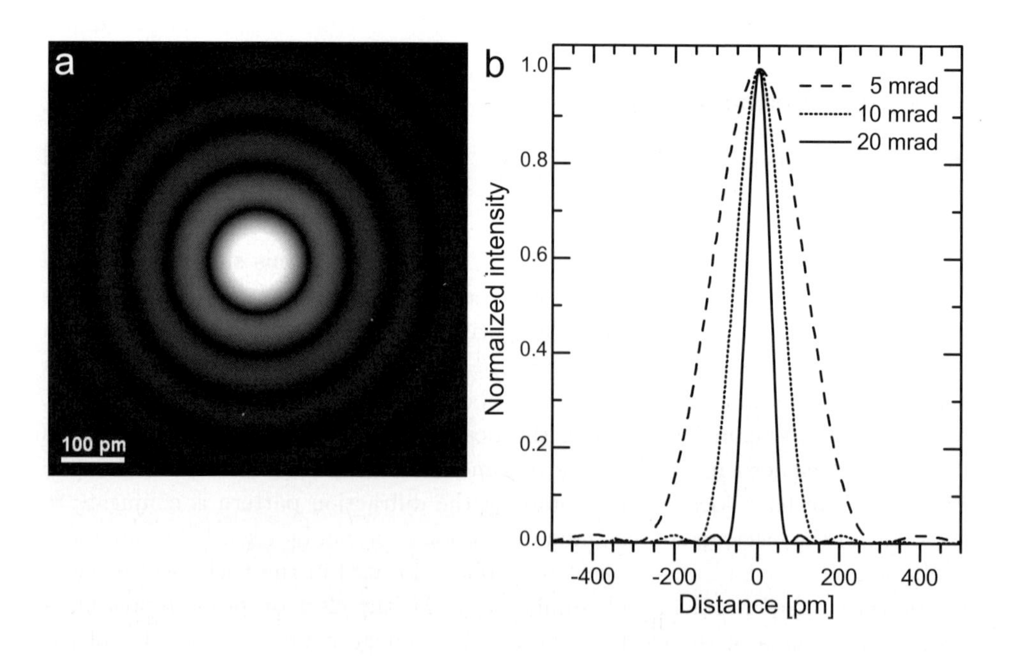


Fig. 3.2 Airy pattern. (**a**) shows an Airy pattern calculated for 200 keV electrons ($\lambda = 2.5$ pm) and an illumination semi-angle $\alpha$ of 5 mrad. In order to reveal the side lobes of the Airy pattern, it is plotted on a logarithmic scale. (**b**) shows three line profiles through Airy patterns, calculated for 200 keV electrons and illumination semi-angles of 5, 10 and 20 mrad (dashed, dotted and full lines). The first minimum of the curves defines the diffraction limit according to Eq. (3.1). For 5 mrad $\delta_D$ is 305 pm, for 10 mrad it is 153 pm and for 20 mrad it is 76 pm.

the geometry of the (coherent) illumination. The width of such an electron probe increases with increasing $\lambda$ and decreasing $\alpha$. Only for the case that $\alpha \rightarrow \infty$ or $\lambda \rightarrow 0$ is the electron probe point-like, i.e. $\delta_{\mathrm{D}} \rightarrow 0$. Figure 3.2b plots line profiles across (normalized) Airy patterns for three different illumination semi-angles. It clearly reveals that with increasing illumination semi-angle, the central maximum becomes narrower.

The limitation of the probe size due to the illumination semi-angle $\alpha$ expressed in Eq. (3.1) is called the *diffraction limit*[2]. In fact, Eq. (3.1) is the resolution criterion of an optical system which is solely limited by diffraction; it expresses the Rayleigh limit or Rayleigh criterion. The diffraction limit reveals that in order to increase the resolution in STEM imaging, one should work with a large probe illumination angle and employ electrons of high energy.

The diffraction limit in STEM imaging has an alternative, visual interpretation, which can be regarded as a complementary point of view. Let us assume we do STEM imaging with a crystalline specimen which is in some zone-axis orientation. There shall be the forward scattered beam $\mathbf{0}$ and a diffracted beam $\boldsymbol{g}$. The scattering angle of the beam $\boldsymbol{g}$ shall be $\theta$ and the illumination semi-angle of the incident electron probe is $\alpha$. Hence, instead of a sharp diffraction spot, the illumination angle of the illumination causes diffraction disks to appear in the diffraction plane; one for the forward scattered beam and one for the diffracted beam $\boldsymbol{g}$. The radius of both disks in the diffraction plane corresponds with the illumination semi-angle $\alpha$. The diffraction angle $\theta$ between $\mathbf{0}$ and $\boldsymbol{g}$, i.e. the angle in respect to the specimen plane connecting the centers of the disks in the diffraction plane, is given by Bragg's law (see, e.g. Schwartz and Cohen, 1987)

$$\lambda = 2d_{\boldsymbol{g}} \sin\left(\frac{\theta}{2}\right), \tag{3.2}$$

where $d_{\boldsymbol{g}}$ corresponds to the crystal spacing which gives rise to the diffraction disk $\boldsymbol{g}$. Now we assume that the diffraction disks are just large enough that they touch each other. Hence, the diffraction angle $\theta$ is equal to twice the illumination semi-angle $\alpha$, i.e. $\theta = 2\alpha$ (see Fig. 3.3a). Neglecting the curvature of the Ewald sphere, we can redraw the triangle ABC indicated in Fig. 3.3a and obtain the triangle shown in Fig. 3.3b. The scattering triangle ABC is an equal-sided triangle; the vector $\overrightarrow{AB}$ corresponds to the incident wave vector, $\overrightarrow{AC}$ is the scattered wave vector and $\overrightarrow{BC}$ is the scattering vector. Hence, the lengths of the two sides $AB$ and $AC$ correspond with the wave vector of the incident and elastically scattered electron, which is $\lambda^{-1}$,

[2] The diffraction limit not only affects STEM imaging but is also essential in TEM. With increasing size of the objective aperture, beams of higher spatial frequency can be transferred to the image plane where they are brought to interference. Since the diffracted beams of high spatial frequencies carry the high resolution information, the objective aperture similarly causes the HRTEM resolution to be limited by diffraction. Selecting, for instance, a very small objective aperture, which transmits only the forward scattered beam, simply implies that there is no lattice information in the micrograph. This mode, which is called bright-field zone-axis imaging, is used to map strain fields at high resolution (Matsumura *et al.*, 1990).

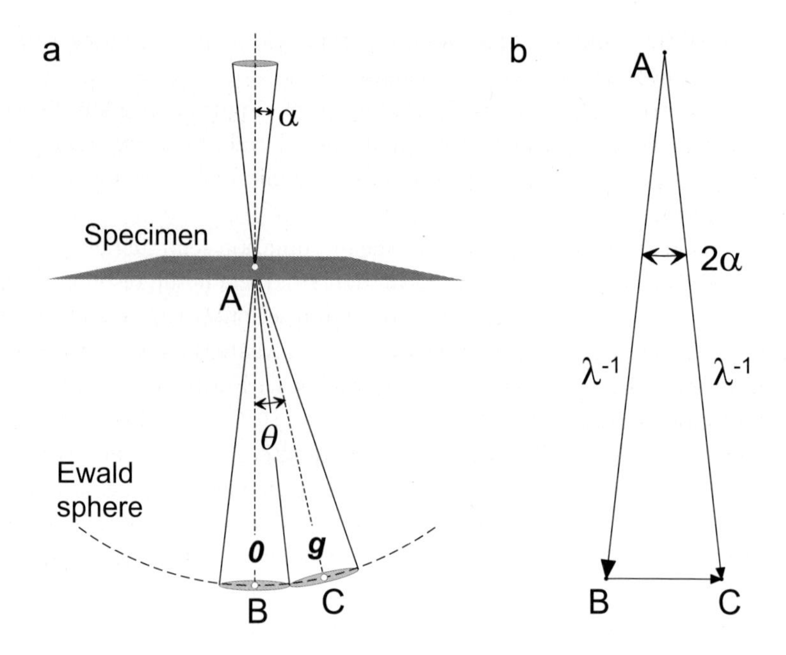


Fig. 3.3 Diffraction limit.

and the third side measures $1/d_{\boldsymbol{g}}$ (see, e.g. Schwartz and Cohen, 1987). Making the approximation for small scattering angles, Fig. 3.3 leads us to a simplified version of Bragg's law, which can be written as

$$\lambda \approx 2d_{\boldsymbol{g}}\alpha. \tag{3.3}$$

Here we used the approximation $\sin(\theta/2) \approx \theta/2 = \alpha$. Rewriting this relation yields

$$d_{\boldsymbol{g}} = \frac{1}{2}\frac{\lambda}{\alpha}. \tag{3.4}$$

Comparing this relation with the diffraction limit given in Eq. (3.1) clearly shows that $d_{\boldsymbol{g}}$ is smaller than the diffraction limit $\delta_{\mathrm{D}}$, i.e. $0.5\lambda/\alpha < 0.61\lambda/\alpha$. Hence, the geometrical setup described in Fig. 3.3 does not allow for resolving the $d_{\boldsymbol{g}}$-spacing.

From this argument we can conclude that in order to resolve a crystal spacing $d_{\boldsymbol{g}}$ in STEM mode, the corresponding diffraction disk $\boldsymbol{g}$ has to partially overlap with the diffraction disk of the forward scattered beam $\mathbf{0}$. The factor 0.61 in Eq. (3.1) essentially describes the amount of overlap that is needed. If the diffraction disks overlap by exactly the amount given by the diffraction limit, the corresponding crystal spacing is theoretically resolved such that it just fulfills the Rayleigh criterion. Provided the contribution of other effects can be ignored, the contrast of the crystal spacing in the micrograph is then 19%.

We can conclude that if diffraction disks do not overlap, the corresponding spatial distance cannot be resolved in a STEM micrograph. This is equivalent to the statement that if the illumination angle of an electron probe is too small, the (diffraction-limited) electron probe is too large to resolve a given spatial distance.

The diffraction limit thus shows that a large illumination angle is needed for a small electron probe, and that in order to resolve a certain crystal spacing, the illumination angle has to be large enough such that the diffraction disk corresponding with the resolvable crystal spacing overlaps with the diffraction disk of the forward scattered beam.

### 3.2.2 *Lens effects — spherical aberration*

In Chapter 2 we saw that the effect of the rotationally symmetric objective lens in HRTEM can be described by the constant of spherical aberration $C_3$ and by the defocus $C_1$. This is based on the assumption that other effects, like for instance astigmatism, are sufficiently small such that they do not significantly affect the imaging process. While $C_1$ is a variable parameter, $C_3$ is a fixed quantity characteristic for the lens. For TEM imaging, it is the spherical aberration $C_3$ of the post-field of the objective lens which is of importance. For the formation of the electron probe in STEM, it is the spherical aberration of the pre-field of the objective lens. Since $C_1$ can be adjusted, $C_3$ imposes the actual limit.

We can consider the pre-field of the objective lens to produce an image of a point-like electron source. The image of the source is the electron probe. An ideal lens focuses all the rays emerging from the source point in the object space in one single point in the image space. The effect of positive spherical aberration is that rays that pass the lens in a distance from the optical axis are brought to focus closer to the lens than rays that run near the optical axis. One can say that the focal distance of the lens decreases with increasing off-axial distance of the rays entering the lens field. The focal point of the rays that run in an infinitely small distance from the optical axis through the lens defines the Gaussian focal plane. The bundle of electrons in Fig. 3.4 emerges from a point source, passing an aperture, and is brought to focus by a lens suffering from spherical aberration. The aperture defines the illumination semi-angle $\alpha$. In the Gaussian focal plane, where one assumes the specimen to be located on to which the electron beam should be focused, there is a broad disk instead of a point-like image of the point-like electron source. Furthermore, in a certain distance in front of the focal plane there is an area where the envelope of all the rays forms a disk which is clearly smaller than the disk in the Gaussian focal plane. It is this disk which defines the smallest achievable electron probe limited by spherical aberration. This disk of radius

$$\delta_S = \frac{1}{4} C_3 \alpha^3 \tag{3.5}$$

is called the *disk of least confusion*. It expresses the limitation imposed by the spherical aberration $C_3$ on the achievable probe size for a given probe illumination semi-angle $\alpha$. One easily sees that the smallest disk of least confusion is obtained for vanishing $\alpha$; the smaller the illumination semi-angle, the smaller is $\delta_S$. Apparently, this trend goes in the opposite direction compared with the diffraction limit.

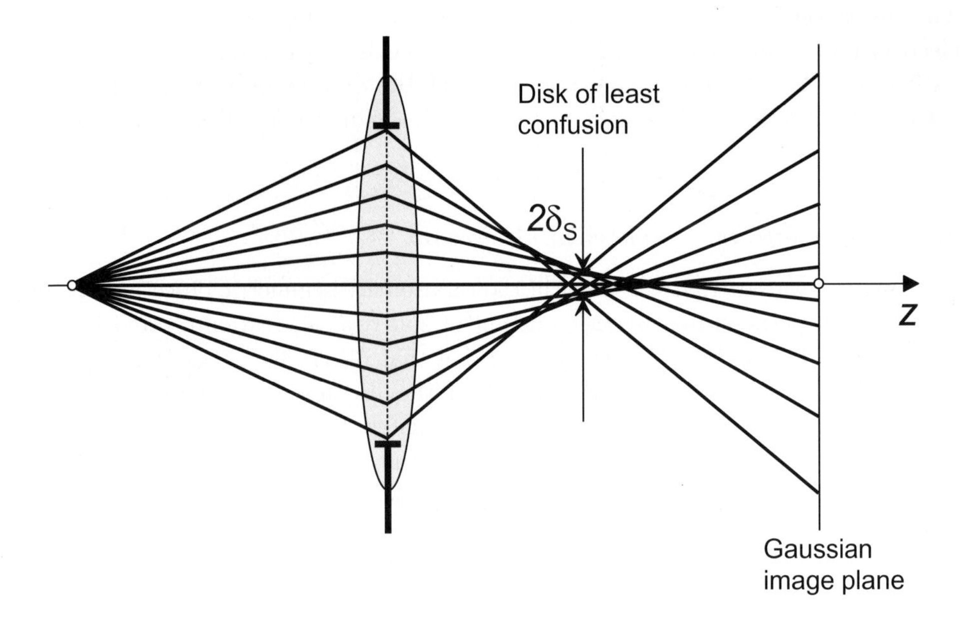


Fig. 3.4 Spherical aberration. The disk of least confusion.

Furthermore, since the defocus $C_1$ describes the deviation of the focus from the Gaussian focal plane, Fig. 3.4 clearly reveals that in order to minimize the effect of spherical aberration, one should work at a finite defocus which essentially moves the disk of least confusion onto the specimen plane.

### 3.2.3 *Partial temporal coherence — chromatic aberration*

An ideal electron source emits electrons of equal energy. However, real electron sources emit electrons of slightly varying energy and thus exhibit a characteristic energy distribution. The energy distribution of the electron beam can further be influenced by the high-tension ripple. Though not strictly valid, the energy distribution of the electrons can approximately be described by a Gauss function around a nominal electron energy $E_0 = eU$. However, the actual distribution function depends on the type of electron source and its operation condition. For field-emission electron microscopes, the width of the energy distribution, which is often quantified by the full width at half maximum, is better than 1 eV. Employing an electron monochromator, the energy width of the beam can be reduced to values below 100 meV.

The problem of using a non-monochromatic electron beam for STEM (or TEM) imaging lies in the fact that electron lenses are not achromatic. They suffer from the chromatic aberration. This means, as illustrated in Fig. 3.5, that the focal point of the lens depends on the energy of the electrons. The focal point of electrons with

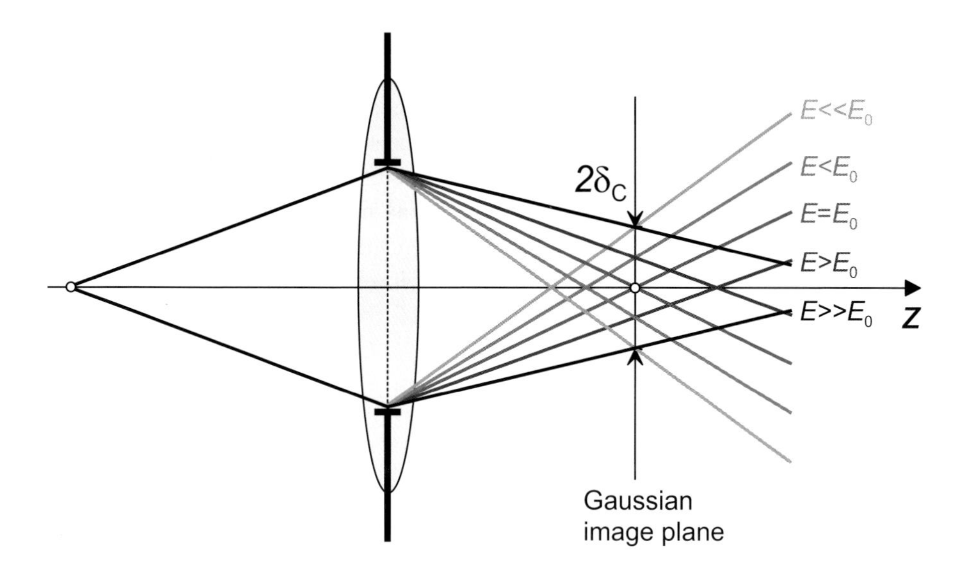


Fig. 3.5 Chromatic aberration. The disk of confusion.

the nominal electron energy $E_0$ lies in the Gaussian image plane. Electrons of energies greater than $E_0$ find a focal point behind the Gaussian focal plane, and electrons of smaller energy in front of this plane.

Hence, if there is a point source in front of a lens, and the lens is used to form a small electron probe, the electron probe is not point-like. In the Gaussian focal plane of the pre-field of the objective lens, i.e. where the specimen for the STEM investigation is located, the point source is imaged into a disk of confusion. The diameter of this disk increases with increasing energy spread. Furthermore, with decreasing opening angle of the aperture and thus increasing illumination semi-angle $\alpha$, the diameter of the disk decreases. In general, the radius of the disk of confusion $\delta_\mathrm{C}$ due to the chromatic aberration is given by

$$\delta_\mathrm{C} = C_\mathrm{C}\alpha\frac{\Delta E}{E_0}, \tag{3.6}$$

where $C_\mathrm{C}$ is the constant of chromatic aberration of the lens and $\Delta E$ is a measure for the width of the energy distribution.

Scherzer's theorem (Scherzer, 1936b) states that in rotationally symmetric and stationary electromagnetic lenses, which are free of space charges, the constant of spherical aberration $C_3$ and the constant of chromatic aberration $C_\mathrm{C}$ are finite positive and can never be nulled. Hence, in conventional electron microscopes, the disks of confusion due to spherical and chromatic aberrations are unavoidable and thus affect the achievable size of the electron probe used in STEM.

### 3.2.4 *Partial spatial coherence — the effective source size*

Real electron sources not only show a finite energy spread but are also finite in size and not point-like. Independent of how many lenses are between source and specimen, a STEM probe is always an image of the source. Because the size of the source is finite, the STEM probe can never be point-like. This is independent of the spherical and chromatic aberrations as well as independent of the diffraction limit. One can argue that sufficient, maybe even infinite demagnification can be employed to obtain a point-like electron probe. However, this leads to an electron probe of zero current. If we recall from Chapter 2 the definition of the brightness in Eq. (2.25) given as

$$B = \frac{I_s}{\Omega A_s}$$

and, by making the approximation that the solid angle $\Omega \approx \pi\alpha^2$, transform it to

$$B = \frac{I_{\text{probe}}}{\pi\alpha^2 A_s}, \tag{3.7}$$

we see that the probe current $I_{\text{probe}}$ depends linearly on the brightness $B$. We have still assumed that the source is imaged 1:1 to the specimen plane. If we demagnify the source of area $A_s$ to a circular area of radius $r_{\text{geo}}$, we see that the current $I_{\text{probe}}$ of the electron probe is

$$I_{\text{probe}} = B\frac{\pi^2\alpha^2 r_{\text{geo}}^2}{M}, \tag{3.8}$$

where we set $\pi r_{\text{geo}}^2 = MA_s$, with $M$ expressing the (de-)magnification of the source. If we apply a high demagnification ($M \ll 1$), $r_{\text{geo}} \to 0$ and we end up with a point-like electron source. However, for a given illumination semi-angle $\alpha$, Eq. (3.8) shows that if $r_{\text{geo}} \to 0$, a finite probe current $I_{\text{probe}}$ can only be maintained for the unphysical case that $B \to \infty$. Electron sources of infinite brightness do not exist. Demagnification comes at the expense of beam current. Yet, this argument explains why electron sources of high brightness are especially important for STEM. The higher the brightness of the source, the larger the probe current that can be maintained for a sufficiently small effective source radius $r_{\text{geo}}$. Sources of high brightness, like Schottky field-emission sources and cold field-emission sources, which have a greater brightness than Schottky emitters, are thus indispensible for high-resolution STEM instruments. The importance of the source brightness in STEM is underlined by the fact that atomic-resolution STEM imaging has only become feasible with field-emission sources. Indeed, it was essentially the development of the cold field-emission source that enabled the first STEM instrument to produce results comparable to the TEM mode, where brightness is also of high importance but not as crucial as it is for STEM (Crewe, 1966).

The demagnification of the electron source is usually done by employing either an (electrostatic) gun lens and/or the first condenser lens, which is the so-called spot-size lens. With increasing excitation of the lens, the demagnification is increased. A

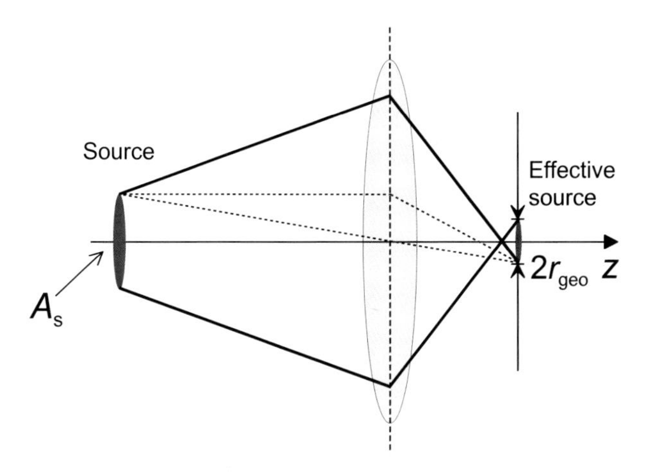


Fig. 3.6 Demagnification of the source of area $A_s$ to an effective source of radius $r_{geo}$.

higher demagnification refers to a higher spot size number. This can be seen from the schematics in Fig. 3.6. In reality, of course, the emission area $A_s$ of the source is not a top-hat function which has sharp edges. The source is usually described by a Gaussian source distribution function and $\delta_{geo}$, the full width at half maximum of the Gaussian function, is a measure for the width of the effective source size, i.e. the size of the image of the source on the specimen.

### 3.2.5 *Stability*

An additional factor which can contribute to the effective size of the source imaged onto the specimen is the overall impact of disturbances. We can distinguish between two types of disturbances; electromagnetic disturbances and mechanical vibrations. Electromagnetic fields, which can be present as stray fields in the environment of the microscope or can be caused by an instable lens or deflector, can cause the electron beam, and thus the electron probe, to jitter. On the other hand, mechanical instabilities become apparent if the specimen is instable in respect to the electron beam. This, for instance, can be caused by an instable sample holder or by thermal drift of the specimen.

If disturbances occur in periods shorter than the dwell time of the scan process, the effective electron probe that a particular scan position experiences becomes larger than the actual (instantaneous) geometrical size of the electron probe. In this case, the effective source size is enlarged by the blurring due to the disturbances. If high-frequency disturbances are present which enlarge the effective source size, $\delta_{geo}$ can be replaced by $\delta_{geo,eff}$ where $\delta^2_{geo,eff} = \delta^2_{geo} + \delta^2_{noise}$. The term $\delta_{noise}$ describes the blurring due to the disturbances. If disturbances occur in periods longer than the dwell time, they become apparent either as (periodic) noise in the image or as distortions. The latter effect becomes apparent if disturbances have periods exceeding multiples of the line time[3], or if the specimen and/or the electron beam are prone to a continuous drift. While high-frequency disturbances are difficult to detect in an image, disturbances of lower frequencies can be revealed by taking a fast Fourier transform (FFT) of an atomic-resolution STEM micrograph. Random scan noise is revealed by streaks along the slow scan direction, while periodic scan noise can lead to false crystal reflections in the FFT.

### 3.2.6 *Small electron probes*

The resolution in STEM imaging is fundamentally limited by the size of the electron probe. Two objects which are at a distance smaller than the size of the electron probe cannot be resolved. A smaller electron probe enables higher resolution. The task of optimizing the resolution of a scanning transmission electron microscope means finding an optical setting for which the overall effect of the probe-limiting factors discussed above is minimal.

For a given microscope high tension and for a given demagnification of the source, the wavelength $\lambda$ and the effective source size expressed by $\delta_{\text{geo}}$ are fixed. The remaining parameters which need to be considered in the optimization of an electron probe are the size of the illumination aperture expressed by $\alpha$, the geometrical lens parameters $C_3$ and $C_1$ and the chromatic aberration $C_\text{C}$. Each contribution has a specific dependence on the illumination semi-angle $\alpha$.

Figure 3.7 shows a log–log plot visualizing the dependencies of $\delta_{\text{geo}}$, $\delta_\text{D}$, $\delta_\text{S}$ and $\delta_\text{C}$ on the illumination semi-angle $\alpha$, assuming $C_3 = 1$ mm, $C_\text{C} = 1$ mm and $\lambda = 2.5$ pm, i.e. $E_0 = 200$ keV with $\Delta E = 1$ eV and $\delta_{\text{geo}} = 50$ pm. This set of parameters is chosen arbitrarily. Nevertheless, the parameters are quite common for conventional scanning transmission electron microscopes equipped with Schottky field-emission electron sources. Hence, though we cannot deduce general rules from a single set of parameters, we can still see which factors are of relevance for a certain range of $\alpha$. Furthermore, since the slopes of the curves in Fig. 3.7 depend on the power $n$ of $\alpha^n$ in the expressions for $\delta_\text{D}$, $\delta_\text{S}$ and $\delta_\text{C}$, one has to be aware that a change of one of the parameters only leads to a parallel shift of the corresponding line in the log–log plot.

The plot in Fig. 3.7 reveals that for the parameters selected, the diffraction limit imposes the limit at small illumination semi-angles, while for larger $\alpha$ it is the spherical aberration which becomes the probe-size limiting quantity. The impact of the chromatic aberration is not critical provided, of course, that the constant

[3]We denote the fast scan direction as the direction along the electron probe scans the first line in the frame. The slow scan direction is perpendicular, along the direction which is consecutively filled by scanned lines. The *dwell time* is the period the beam is stationary on a scan position. The *line time* is the dwell time multiplied by the amount of pixels along the fast scan direction. Multiplying the line time with the amount of scanned lines gives the approximate *frame time*. It is the approximate frame time because the actual frame time also depends on the scan synchronization. Often, frames are chosen that are squares of sides which contain $2^n$ pixels ($n = 8, 9, 10...$).

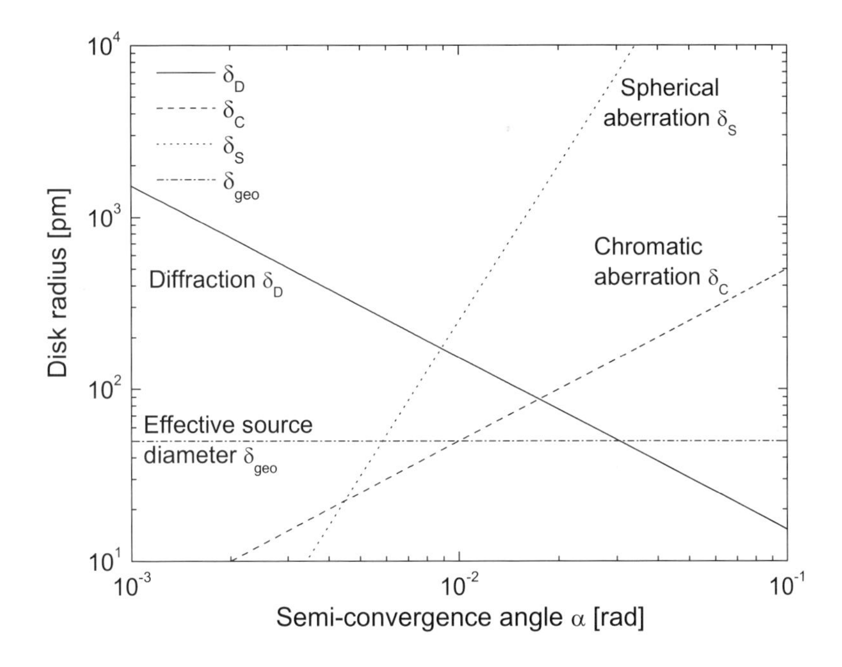


Fig. 3.7 Contributions to the STEM probe. Dependency of the diffraction limit $\delta_D$, the spherical aberration $\delta_S$, the chromatic aberration $\delta_C$ and the effective source diameter $\delta_{geo}$ on the illumination semi-angle $\alpha$, according to Eqs. (3.1), (3.5) and (3.6). The geometrical source size $\delta_{geo}$ is assumed to be 50 pm, $C_3 = C_C = 1$ mm, $\Delta E = 1$ eV, $E_0 = 200$ keV and thus $\lambda = 2.5$ pm.

of chromatic aberration $C_C$ is of the same order of magnitude as the constant of spherical aberration $C_3$ and provided that the energy spread $\Delta E$ is of the order of 1 eV. This latter condition can be considered to be fulfilled for the case of field-emission electron sources. Hence, the chromatic aberration is not the limiting factor in conventional probe-forming microscopes operated above about 100 kV. This has been investigated in detail by Shao and Crewe (1987). Furthermore, because $\delta_S \propto \alpha^3$ while $\delta_C \propto \alpha$, the impact of the spherical aberration must exceed the impact of the chromatic aberration with increasing illumination semi-angle $\alpha$. The effective source size, or the demagnification, can in principle be chosen such that the finite size of the source is not limiting the probe size. This, of course, comes at the expense of probe current.

From Fig. 3.7 we can conclude that what essentially needs to be considered in the optimization of an electron probe in a conventional scanning transmission electron microscope are the diffraction limit and the spherical aberration. These two contributions need to be balanced. Furthermore, a defocus $C_1$ needs to be chosen, similar to the Scherzer focus in Chapter 2, which translates the disk of least confusion to the specimen plane. Crewe and Salzman (1982) solved this problem and derived that for an optimum defocus $C_{1\,\mathrm{opt}}$ of

$$\boxed{C_{1\,\mathrm{opt}} = -\sqrt{\lambda C_3}} \tag{3.9}$$

and for an optimum illumination semi-angle $\alpha_{\mathrm{opt}}$ of

$$\boxed{\alpha_{\mathrm{opt}} = \left(\frac{4\lambda}{C_3}\right)^{\frac{1}{4}}}, \tag{3.10}$$

a resolution $\rho_{\mathrm{r}}$ of

$$\boxed{\rho_{\mathrm{r}} = 0.43\sqrt[4]{C_3\lambda^3}} \tag{3.11}$$

can be achieved. While experimentally the optimum defocus can be found by optimizing the contrast of an atomic-resolution STEM micrograph, the optimum illumination semi-angle given in Eq. (3.10) needs to be selected carefully in order to achieve the smallest probe size. Only the optimization of the illumination semi-angle in regard to the spherical aberration enables highest resolution in a conventional scanning transmission electron microscope.

For the set of parameters given above, we obtain for the optimum semi-illumination angle $\alpha_{\mathrm{opt}} = 10$ mrad. This value, which is quite typical for conventional STEM microscopes, roughly coincides with the $\alpha$-value of the point of intersection between the lines $\delta_{\mathrm{D}}$ and $\delta_{\mathrm{S}}$ in Fig. 3.7. Provided the defocus is set to -50 nm according to Eq. (3.9), Eq. (3.11) reveals that the electron probe formed with the optimum illumination semi-angle enables a STEM resolution of 0.15 nm.

Comparing Eqs. (3.9) and (3.11) with the equivalent expressions for HRTEM given in Eqs. (2.21) and (2.32) reveals that both expressions for the optimal focus setting and the expression for the resolution are very similar. Though the relations for an optimum STEM probe given here were derived by Crewe and Salzman (1982), these expressions, as well as the expression for the optimum illumination semi-angle, were essentially derived by Scherzer (1939, 1949). While the optimal focus setting for HRTEM given in Eq. (2.21) is known as the Scherzer focus, the conditions for an optimized STEM probe, which consists of an expression for the focus setting and a requirement for the illumination semi-angle given in Eqs. (3.9) and (3.10) respectively, are sometimes referred to as the *Scherzer incoherent conditions* (Pennycook and Jesson, 1991).

A final point about the geometrical considerations of STEM probes concerns the quantities $\delta_{\mathrm{D}}$, $\delta_{\mathrm{S}}$, $\delta_{\mathrm{C}}$ and $\delta_{\mathrm{geo}}$, which are given in Eqs. (3.1), (3.5) and (3.6). As pointed out by Crewe (1997), these quantities should be distinguished from actual resolution expressions, like the one for $\rho_{\mathrm{r}}$ given in Eq. (3.11). The values $\delta_{\mathrm{D}}$, $\delta_{\mathrm{S}}$ and $\delta_{\mathrm{C}}$ do not express the achievable STEM resolution for a given illumination semi-angle. We should regard these quantities as measures for the impact of a certain effect, rather than as resolution criteria. This can easily be seen from comparing $\delta_{\mathrm{S}}$ with $\delta_{\mathrm{D}}$. While $\delta_{\mathrm{S}}$ corresponds to the outermost radius of a caustic which in general is sharply peaked and decays fast towards the edge of the caustic, the $\delta_{\mathrm{D}}$-value provides a measure for the width of the central peak of an Airy pattern, knowing that there is intensity beyond the $\delta_D$-radius. Hence, $\delta_S$ and $\delta_D$ are not directly comparable; in the first case we measure the probe to the very end of the weak tails, while in the second case we basically ignore the tails of the probe. The important point about $\delta_D$, $\delta_S$, $\delta_C$ and $\delta_{geo}$ is that they express the basic effects of $C_3$, $C_C$ and $\alpha$ on the STEM probe. For this reason, it does not seem appropriate to define the size of the electron probe as well as the STEM resolution simply by taking the geometrical mean of all four quantities $\delta_D$, $\delta_S$, $\delta_C$ and $\delta_{geo}$ as $\sqrt{\delta_D^2 + \delta_S^2 + \delta_C^2 + \delta_{geo}^2}$, which, however, can quite frequently be found in the literature. Indeed, looking at the plot in Fig. 3.7 reveals that there is no point in the plot that would reflect such an electron probe.

Furthermore, the effects of diffraction, the spherical and chromatic aberrations and the finite size of the source can in general not be treated independently. They are highly interrelated, which, in particular, becomes apparent if the electron probe is not solely considered as a two-dimensional focused electron spot but as a three-dimensional entity, which, apart from the lateral extension, also has a longitudinal or vertical component. Hence, the above geometrical considerations about the lateral extension of an electron probe have clear limits. Nonetheless, the simplifications upon which they are based allow us to develop an understanding of the individual components that influence the spatial resolution in STEM imaging. However, in order to understand the electron probe as the result of the collective effect of all four factors mentioned above, we need to describe the electron probe on the basis of wave optics rather than on purely geometrical grounds. The wave optical description of the electron probe, which is the topic of the following section, will enable us to differentiate between the effects of the different probe contributions mentioned above. Still, it is important to note that the above considerations about the individual contributions to the STEM probe and their dependence on the illumination semi-angle qualitatively remain valid. For instance, it is still the spherical aberration and the diffraction limit that define the optimum STEM probe of a conventional scanning transmission electron microscope.

## 3.3 Wave Optical Description of an Electron Probe

In the following we derive a wave optical description of an electron probe, taking into account the four factors mentioned above. The wave optical description allows us to calculate an electron probe with any desired precision and in all three spatial dimensions, provided, of course, all relevant input parameters are known with sufficient precision. It has to be emphasized that even with very detailed knowledge about the lateral and longitudinal extension of the electron probe, the problem of relating the probe characteristics to a resolution criterion is not solved simply. The fundamental problem is that an electron probe shows in principle infinitely long tails and, as such, it is not a trivial problem to extract a measure on the basis of the calculated electron probe which directly predicts the STEM resolution we recognize in a micrograph. One could, for instance, apply the Rayleigh criterion to estimate the resolution on the basis of a calculated STEM probe. The minimum distance between two calculated STEM probes which enables a central dip between the two superimposed electron probes of 19% is known at the Rayleigh resolution. However, even this very practical but theoretical resolution criterion is not suitable, in case the electron probe has significant side lobes (Fertig and Rose, 1979) or in case the micrographs are affected by noise, which for normal experimental conditions is unavoidable (Van Aert and Van Dyck, 2006). Nevertheless, even though the relation between the geometry of the STEM probe and the actual resolution is intricate, the general rule is clear; the smaller the STEM probe, the better the optical resolution of the instrument. Furthermore, in order to be able to compare and optimize a STEM probe, detailed knowledge about the actual geometry of the STEM probe and its dependence on the set of relevant parameters is crucial.

The way we introduce the wave optical description of an electron probe is equivalent to the geometrical considerations discussed in the previous section. First, we start with the effect of the finite aperture, then we include the lens aberrations, and finally we incorporate the effects of partial spatial and partial temporal coherence. We denote the electron wave incident on the object plane by $\psi_0(\boldsymbol{r})$, and the electron wave in the aperture plane in front of the object plane by $\psi_0(\boldsymbol{q})$ (see Fig. 3.1). The vector $\boldsymbol{r}$ is a two-dimensional position vector in the object plane, where the specimen is situated, and $\boldsymbol{q}$ is a two-dimensional position vector in the aperture plane, with its unit reflecting a reciprocal distance. We recall that the wave vector $q$ is related to the scattering angle $\theta$ by $\theta = q\lambda$. Similarly, the wave vector can be related to the illumination semi-angle $\alpha$ by $\alpha = q\lambda$. These relations are valid for small angles $\theta$ and $\alpha$.

### 3.3.1 *Diffraction*

Looking at Fig. 3.1, which illustrates the basic experimental setup in STEM mode, it can be seen that the pre-field of the objective lens focuses the electron wave in the aperture plane $\psi_0(\boldsymbol{q})$ onto the specimen plane. This situation is in principle analogous to the situation in HRTEM (see Fig. 2.1): a plane wave passes through the specimen plane and the post-field of the objective lens produces a focused electron beam in the back focal plane. If a diffracting specimen is located in the specimen plane, each diffracted beam gives rise to a focused spot, i.e. a diffraction pattern is formed. However, similar to an electron probe in STEM, a focused diffraction spot is an image of the source (James and Browning, 1999; Rose, 2009a). For the case of HRTEM as well as for STEM, a lens focuses a parallel beam into a spot. For HRTEM, the focused spot lies in the back focal plane, while for STEM the focused spot lies in the specimen plane, which indeed is the back focal plane of the pre-field of the objective lens. In both modes, i.e. in HRTEM and in STEM mode, the specimen plane is described by the position coordinate $\boldsymbol{r}$. The plane of the illumination aperture and the plane of the objective aperture are aperture planes which are described by $\boldsymbol{q}$. While in HRTEM the back focal plane containing the diffraction pattern is a plane conjugate to the plane of the electron source, in STEM it is the specimen plane which is conjugate to the plane of the electron source. In order to move the plane which is conjugate to the electron source from the back focal plane to the specimen plane, i.e. in order to switch from TEM to STEM mode, the mini-condenser lens above the objective lens is changed. In STEM mode, the mini-condenser lens is optically off, while in TEM mode, it is optically on (Williams and Carter, 1996; James and Browning, 1999). From this we can conclude that similar to the case of the formation of the diffraction pattern in TEM, in STEM, the electron wave in the aperture plane and the electron wave in the specimen plane are linked to each other by a Fourier transform.

Let us thus start with the electron wave function at the aperture plane in front of the specimen (see Fig. 3.1). We assume that the phase and the amplitude of the electron wave are constant across the aperture opening. For a circular aperture, the electron wave in the aperture plane is a top-hat function described by

$$\psi_0(\boldsymbol{q}) = \begin{cases} 1 \text{ if } |\boldsymbol{q}| < |\boldsymbol{q}_\mathrm{a}| \\ 0 \text{ otherwise.} \end{cases} \tag{3.12}$$

The radius of the aperture opening is denoted by $\boldsymbol{q}_\mathrm{a}$, which is related to the illumination semi-angle $\alpha$ by $\alpha = q_\mathrm{a}\lambda$. In analogy to the previous chapter, the top-hat function can approximately be described in a closed form by employing the Fermi function. This yields

$$\psi_0(\boldsymbol{q}) = \frac{1}{1 + \exp\left\{\dfrac{|\boldsymbol{q}|^2 - |\boldsymbol{q}_\mathrm{a}|^2}{\delta_\mathrm{a}^2}\right\}}, \tag{3.13}$$

where $\delta_\mathrm{a}$ can be chosen as a small fraction of $q_\mathrm{a}$. A finite value of $\delta_\mathrm{a}$ numerically blurs the edge of the aperture, which reduces the risk of introducing artifacts in calculations (Kirkland *et al.*, 1987; Kirkland, 1998).

Now we could simply derive the Fourier transform of Eq. (3.13) to obtain the electron wave on the specimen plane, i.e. the electron probe. Doing so, we would obtain an electron probe whose geometry is determined solely by the diffraction limit; the larger $q_\mathrm{a}$, the smaller the probe (see, Fig. 3.3). Hence, the Fourier transform of Eq. (3.13) for $\delta_\mathrm{a} \to 0$ yields an Airy pattern as, for instance, illustrated in Fig. 3.2a.

### 3.3.2 *Defocus and spherical aberration*

We have already seen that owing to the fact that electron lenses are not perfect, the diffraction limit is not the only contribution to the electron probe. If we only had to consider the diffraction limit, we would simply employ the largest illumination aperture in order to obtain the smallest probe. However, the next contribution that needs our attention is the effect induced by the phase shifts caused by lens aberrations, namely by the third-order spherical aberration $C_3$, whose effect can be balanced by a proper choice of the defocus $C_1$. The effect of the lens aberrations is taken into account by considering the aberration phase-shifts in the aperture plane. This is equivalent to HRTEM, where we also included the lens aberrations in the aperture plane, which, however, for HRTEM lies behind the object plane, namely in the back focal plane of the objective lens' post-field. Furthermore, because the effect of defocus and spherical aberration is described in the aperture plane — in STEM mode as well as in HRTEM mode — such aberrations are also called aperture aberrations.

The spherical aberration $C_3$ and the defocus $C_1$ cause a modulation of the phase of the electron wave in the aperture plane. In analogy to HRTEM, the aberration function can be written as

$$\chi(\boldsymbol{q}) = \chi(q) = \frac{1}{2}q^2\lambda^2 C_1 + \frac{1}{4}q^4\lambda^4 C_3\,. \tag{3.14}$$

Since we deal with isotropic aberrations, without loss of generality we can switch from the vector notation $\boldsymbol{q}$ to the scalar notation $q$. The phase shifts $\gamma$ are given by $\gamma(q) = 2\pi/\lambda \cdot \chi(q)$.

As any wave function, the incident electron probe in the aperture plane has the general form $\psi_0 = A \cdot \exp\{\mathrm{i}B\}$, with $B$ the phase of the wave and $A$ the amplitude. So far, we set the amplitude $A = 1$ within the aperture opening, and the phase $B$ we set arbitrarily equal to zero. Incorporating the effect of the geometrical lens aberrations into the probe calculation, we have to include the phase shifts given above. With this addition, the electron wave in the aperture plane given in Eq. (3.13) becomes

$$\psi_0(q) = \exp\left\{-\frac{2\pi\mathrm{i}}{\lambda}\chi(q)\right\}\frac{1}{1+\exp\left\{\dfrac{q^2-q_\mathrm{a}^2}{\delta_\mathrm{a}^2}\right\}}\,. \tag{3.15}$$

While the first term determines the phase of the wave function in the aperture plane, the second term, i.e. the aperture function, determines its amplitude. Equation (3.15) covers all coherent contributions to the wave function. The next step is to take the Fourier transform of Eq. (3.15) to obtain the coherent electron wave on the specimen plane. This can be written as (see, e.g. Nellist and Pennycook, 1998, 2000)

$$\psi_0(\boldsymbol{r}) = \int_{-\infty}^{\infty} \frac{\exp\left\{-\dfrac{2\pi\mathrm{i}}{\lambda}\chi(q)\right\}}{1+\exp\left\{\dfrac{|\boldsymbol{q}|^2-|\boldsymbol{q}_\mathrm{a}|^2}{\delta_\mathrm{a}^2}\right\}}\exp\{-2\pi\mathrm{i}\boldsymbol{q}\cdot\boldsymbol{r}\}\,d\boldsymbol{q}. \tag{3.16}$$

The integral runs in principle from $-\infty$ to $+\infty$. However, since for $\boldsymbol{q}$ values beyond $\boldsymbol{q}_\mathrm{a}$ the integrand is zero, the integral can in principle be taken within the aperture opening. Furthermore, if the probe needs to be calculated for a specific position $\boldsymbol{r}_{\mathrm{p}}$ on the specimen plane, $\boldsymbol{r}$ can be replaced by $\boldsymbol{r} - \boldsymbol{r}_{\mathrm{p}}$ throughout Eq. (3.16). The intensity $I_0(\boldsymbol{r})$ of the electron probe on the specimen plane is then given by taking the modulus of the complex wave function $\psi_0(\boldsymbol{r})$. This is written as

$$I_0(\boldsymbol{r}) = \psi_0(\boldsymbol{r})\overline{\psi}_0(\boldsymbol{r}) = |\psi_0(\boldsymbol{r})|^2, \tag{3.17}$$

where $\overline{\psi}_0$ denotes the complex conjugate of $\psi_0$.

In the previous section, we discussed the optimization of a STEM probe by balancing the effects of defocus, spherical aberration and the diffraction limit. This is essentially expressed by the Scherzer incoherent conditions given in Eqs. (3.9) and (3.10). For a given wavelength $\lambda$ and for a given spherical aberration $C_3$, those two equations allow us to calculate optimal settings for the illumination semi-angle $\alpha$ and the defocus $C_1$, which are necessary to achieve a small probe size. The wave optical description of the electron probe according to Eq. (3.17) can be used to illustrate the optimization related to Eqs. (3.9) and (3.10). Figure 3.8 depicts three probe intensity profiles calculated according to Eq. (3.17) for 200 keV electrons ($\lambda$ = 2.5 pm), a constant of spherical aberration $C_3$ of 1 mm and an optimized illumination semi-angle $\alpha_{\mathrm{opt}}$ of 10 mrad. For the given $C_3$ and $\lambda$ we can employ Eq. (3.9) to derive the optimal setting for the defocus. This yields $C_{1\,\mathrm{opt}} = -50$ nm. The electron probes in Fig. 3.8 reveal the effect of a change of defocus on the probe intensity. One probe is calculated for $C_1 = -75$ nm, one for the optimal defocus and one for a defocus closer to the Gaussian focus, i.e. $C_1 = -25$ nm. It is clear that deviations from the optimal defocus $C_{1\,\mathrm{opt}}$ result in probe intensity profiles which have either a wider central maximum or show substantial side lobes that would significantly reduce the achievable image contrast.

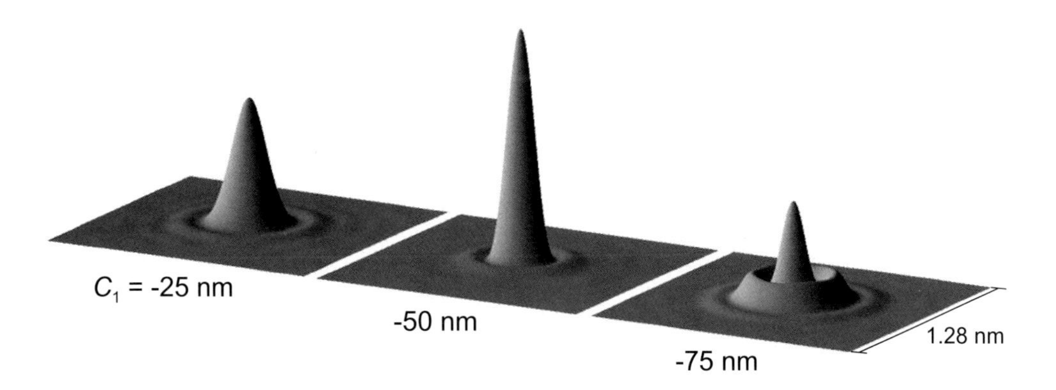


Fig. 3.8 Intensity profiles of electron probes calculated for 200 keV electrons ($\lambda$ = 2.5 pm) and $C_3$ = 1 mm. The probes were calculated according to Eq. (3.17) for a defocus of −75 nm, for an optimized defocus of −50 nm (see Eq. (3.9)), and for −25 nm. The probe illumination semi-angle is in all three cases 10 mrad, corresponding to the optimum angle (see Eq. (3.10)).

### 3.3.3 *Geometrical source size*

The coherent contributions to the wave field of the electron probe in the specimen plane are expressed in Eq. (3.16). However, we have already seen that the image formation in HRTEM is not fully coherent such that envelope functions due to partial temporal and partial spatial coherence have to be introduced, which essentially reduce the transfer function at high spatial frequencies. In STEM, the situation is indeed very similar. We also started above with a fully coherent electron probe and now we have to incorporate contributions which take into account the limited coherence of the electron beam.

Partial spatial coherence essentially means that the electron source is not point-like, but has a finite size which we described in the previous section with the source radius $r_{\mathrm{geo}}$. The value $r_{\mathrm{geo}}$ was our quantity which measures the radius of the electron source after demagnification projected onto the specimen plane. This implied that the source is homogenous across the emission area. However, real electron sources are not disk-like and they do not emit electrons homogenously over a well defined area (Swanson and Schwind, 1997). Real electron sources have emission characteristics which also include the actual shape of the emitting tip. The intensity distribution of the emission imaged onto the specimen plane is described by a source intensity distribution function (see, e.g. Dwyer *et al.*, 2008). A suitable and simple model for a source intensity distribution function $S(\boldsymbol{r})$ is a Gaussian function

$$S(\boldsymbol{r}) = S(r) = \frac{1}{\sqrt{2\pi\sigma_{\mathrm{s}}^2}} \exp\left\{-\frac{r^2}{2\sigma_{\mathrm{s}}^2}\right\}. \tag{3.18}$$

This function describes the distribution of the emitting points of the source which contribute to the electron probe. The magnitude of $S$ models their relative contribution to the total intensity. Though $S(\boldsymbol{r})$ is called the source intensity distribution function, it is indeed the source intensity distribution projected and demagnified onto the object plane. The standard deviation $\sigma_{\mathrm{s}}$ quantifies the width of the demagnified source on the specimen plane. However, often one is interested in the full width at half maximum (FWHM) of the distribution function, which we identify with $\delta_{\mathrm{geo}}$. The standard deviation and the FWHM of the Gaussian source intensity distribution function are related to each other by $\delta_{\mathrm{geo}}^2 = 8\ln 2\,\sigma_{\mathrm{s}}^2 \approx 2.355^2\sigma_{\mathrm{s}}^2$. We call $\delta_{\mathrm{geo}}$ the geometrical (or effective) source size.

The finite size of the source intensity distribution means that the electron beam does not emerge in a single point, as assumed in Eq. (3.16), but it is emitted from a multitude of points described by $S(\boldsymbol{r})$. Each element of the source gives rise to a coherent electron probe as described in Eq. (3.17). The electron probes due to each source element incoherently combine to yield the complete electron probe. Hence, in order to incorporate the source intensity distribution in the probe intensity calculation, we have to convolute the intensity of the coherent probe wave field given in Eq. (3.17) with the source intensity distribution given in Eq. (3.18). We deal solely with the intensity of the electron probe and not with its phase, because the finite size of the source is an incoherent contribution to the electron probe. Hence, we assume that there is no interference between electrons which are emitted from different points of the source. Or, in other words, the source distribution diminishes the magnitude of the interference between different $\boldsymbol{q}$ vectors within the illumination aperture.

With the addition of the source distribution function, the intensity of the electron probe can now be written as

$$I_0(\boldsymbol{r}) = |\psi_0(\boldsymbol{r})|^2 \otimes S(\boldsymbol{r}), \tag{3.19}$$

where $\otimes$ denotes the convolution. Equation (3.19) shows clearly that the effect of the finite source size results in an incoherent blurring of the electron probe.

In analogy to the previous section, the impact of high frequency noise can be incorporated by replacing $\delta_{\text{geo}}$ with an effective geometrical source size $\delta_{\text{geo,eff}}$, which includes the blurring of the electron probe by $\delta^2_{\text{geo,eff}} = \delta^2_{\text{geo}} + \delta^2_{\text{noise}}$.

### 3.3.4 *Energy spread of the electron beam*

The last step that needs to be done is to include the effect of partial temporal coherence. As we have already seen in Chapter 2, the finite energy spread of the electron beam in combination with the chromatic aberration of the lens essentially leads to a spread of focus (see Eqs. (2.22) and (2.23)). For STEM, this is essentially the same.

Because of the variation $\delta E$ of electron energies around the nominal electron energy $E_0$, the chromatic aberration $C_\text{C}$ causes a variation $\delta C_1$ of the defocus $C_1$. An energy off-set $\delta E = E - E_0$, where $E_0$ is the nominal electron energy and $E$ the actual energy of a particular electron, results in a change of focus of $\delta C_1$ which is proportional to the constant of chromatic aberration $C_\text{C}$; $\delta C_1 = C_\text{C}\delta E/E_0$. Hence, it is the spread of focus induced by the finite energy spread of the beam and the chromatic aberration which needs to be incorporated in the probe calculation.

The focus is blurred. This means that for a given nominal defocus $C_1$, the intensity of the electron probe is the incoherent superposition of electron probes integrated over a certain defocus range, which is determined by $C_\text{C}$ and $\Delta E$. The focus blur is incorporated via the aberration function $\chi$ given in Eq. (3.14), which is a function of the defocus $C_1$.

Let us assume that the energy distribution $T(E)$ of the electron beam is a Gaussian function. This of course is a simplified model which does not take into account the emission characteristics of a given electron source. For our purpose, however, it is sufficiently accurate. We thus can model the energy distribution by

$$T(E) = \frac{1}{\sqrt{2\pi\sigma_\text{t}^2}} \exp\left\{-\frac{(E - E_0)^2}{2\sigma_\text{t}^2}\right\}, \tag{3.20}$$

with the standard deviation $\sigma_\text{t}$ related to the full width at half maximum of the Gaussian energy distribution $\Delta E$ by $\Delta E^2 = 8 \ln 2\, \sigma_\text{t}^2 \approx 2.355^2 \sigma_\text{t}^2$ (see Fig. 3.9a).

The electron probe can now be calculated by (see, e.g. Haider *et al.*, 2000)

$$I_0(\boldsymbol{r}) = \int_{-\infty}^{\infty} \left[|\psi_0(\boldsymbol{r}, E)|^2 \otimes S(\boldsymbol{r})\right] T(E)dE, \tag{3.21}$$

which describes the incoherent superposition of electron probes, weighted by $T(E)$, spread over a certain focus range which is determined by the energy spread of the beam. This is illustrated in Fig. 3.9.

It is important to emphasize the dependence of $\psi_0(\boldsymbol{r}, E)$ on the energy $E$ in Eq. (3.21). For each electron energy $E$ within the energy distribution $T(E)$, an electron probe $\psi_0(\boldsymbol{r}, E)$ needs to be calculated where the effective defocus $C_1$ in Eq. (3.14) includes an offset $\delta C_1$ with respect to the nominal focus of the electron probe. This focus offset $\delta C_1$ is given by $\delta C_1 = C_C(E - E_0)/E_0 = C_C\delta E/E_0$.

### 3.3.5 *Concluding remarks*

As already mentioned above, an electron probe should be considered as a three-dimensional intensity distribution (see, e.g. Erni *et al.*, 2009). The lateral extension of the electron probe essentially determines the lateral STEM resolution, while the

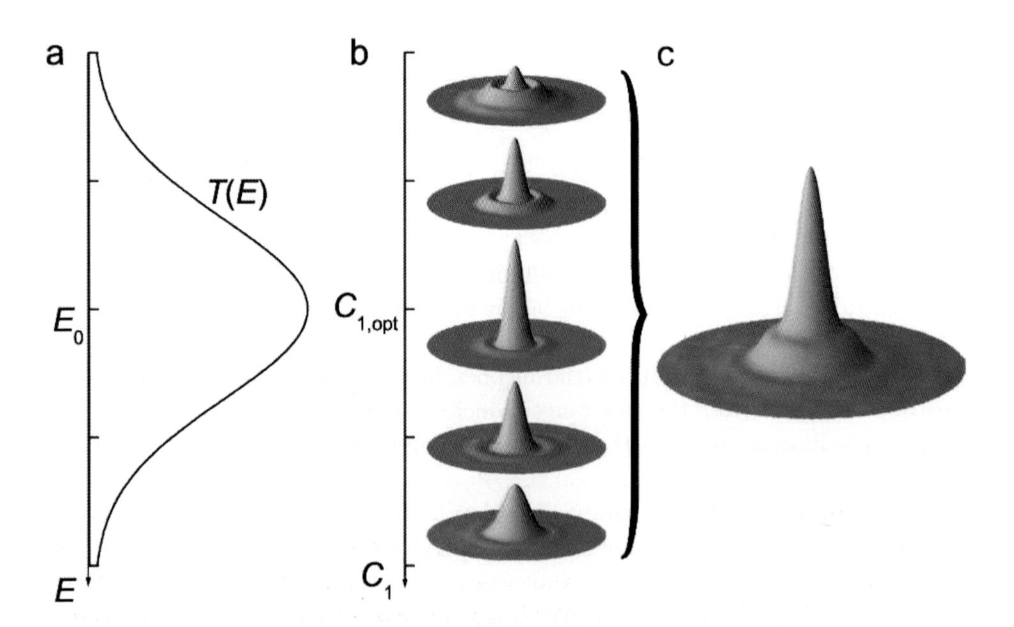


Fig. 3.9 Effect of the finite energy spread of the electron beam and the chromatic aberration $C_C$ on the probe intensity. (**a**) The energy distribution of the electron beam is described by a Gaussian function $T(E)$ with FWHM $\Delta E$ centered around the nominal electron energy $E_0$. (**b**) and (**c**) The energy spread leads to a focus spread of the electron probe which can be considered as an incoherent superposition of electron probes, weighted by $T(E)$ or $T(C_1)$ respectively, over the defocus range with FWHM of $\Delta C_{1\,\mathrm{FWHM}} = C_C\Delta E/E_0$.

longitudinal extension of the electron probe, i.e. the extension along the optical axis of the microscope, defines the depth resolution or the depth of field. The chromatic aberration and thus partial temporal coherence affects the electron probe in the lateral and in the longitudinal direction.

Firstly, due to the incoherent superposition of electron probes over a finite defocus range $\Delta C_1$ (see Fig. 3.9), the lateral extension of the electron probe is enlarged. However, for conventional scanning transmission electron microscopes, this effect is not the probe size-limiting factor (Shao and Crewe, 1987). Furthermore, it is not so much the lateral extension of the central maximum of an electron probe that is affected by the chromatic aberration, but the side lobes of the electron probe (see Fig. 3.9). This leads to an increased background intensity and thus to a reduced image contrast (see, e.g. Fertig and Rose, 1979).

The longitudinal extension of the electron probe is essentially determined by the geometrical restriction imposed by the finite probe illumination angle. For electron probes of conventional field-emission scanning transmission electron microscopes, the depth of field is typically larger than the thickness of the specimen. However, as already pointed out by Rose (1975), with increasing probe illumination angle and thus improved resolution, the depth of field becomes smaller. This opens the way to access not just projected information about the specimen, but three-dimensional information. Three-dimensional information can be obtained from the specimen by recording focal series on specimens whose thickness exceeds the depth of field of the electron probe. The depth of field $\Delta_{C_1}$ can be expressed as (Born and Wolf, 2001)

$$\Delta_{C_1} \approx \frac{\lambda}{\alpha^2}. \tag{3.22}$$

For our virtual electron microscope with $C_3 = 1$ mm operated at 200 keV ($\lambda = 2.5$ pm), we obtain for an optimized illumination semi-angle $\alpha = 10$ mrad a depth of field $\Delta_{C_1}$ of 25 nm. This value is in the order of magnitude of the thickness of a typical high-resolution STEM specimen. Hence, we do not expect to see a distinct depth dependence on a conventional STEM instrument. Moreover, the chromatic aberration further contributes to the blurring of the defocus. However, if we consider a spherical aberration-corrected instrument, which can be used to form an electron probe with an illumination semi-angle of 25 mrad, the depth of field estimated by Eq. (3.22) is 4 nm for 200 keV electrons. This can be significantly smaller than the thickness of a specimen. Indeed with the advent of aberration-corrected scanning transmission electron microscopes, depth sectioning by means of recording through focal series in STEM mode has become feasible and has been brought to application (van Benthem *et al.*, 2005, 2006).

This topic will be discussed in more detail in the last part of this book, where we deal with applications of small electron probes formed with aberration-corrected scanning transmission electron microscopes. For now, we shall be concerned solely with the effect of partial temporal coherence on the spatial extension of the electron probe. The spread of focus, which is caused by the energy spread of the beam and the finite value of $C_C$, increases the depth of field and thus decreases the achievable depth resolution in STEM imaging. However, the focus spread due to partial temporal coherence is clearly smaller than the effect given by the geometry of the electron probe (see, e.g. Eq. (3.22)). Hence, we do not expect that on conventional scanning transmission electron microscopes the focus blur due to partial temporal coherence will be significant.

We conclude that for the case of conventional probe-forming instruments operated between about 100 kV and 300 kV employing field-emission sources, the impact of the chromatic aberration is not critical. This is in agreement with the purely geometrical considerations summarized in Fig. 3.7. The electron probe in conventional field-emission scanning transmission electron microscopes is limited by the spherical aberration. The defocus and the aperture opening are set according to the Scherzer incoherent conditions given in Eqs. (3.9) and (3.10), in order to minimize the impact of the spherical aberration $C_3$ and thus to obtain a small electron probe.

Figure 3.10 summarizes the individual contributions to the electron probe discussed above. We employ the microscope parameters from above, i.e. the acceleration voltage is 200 kV and the constant of spherical aberration of third order is $C_3 = 1$ mm. Equation (3.10) allows us to calculate the optimum illumination

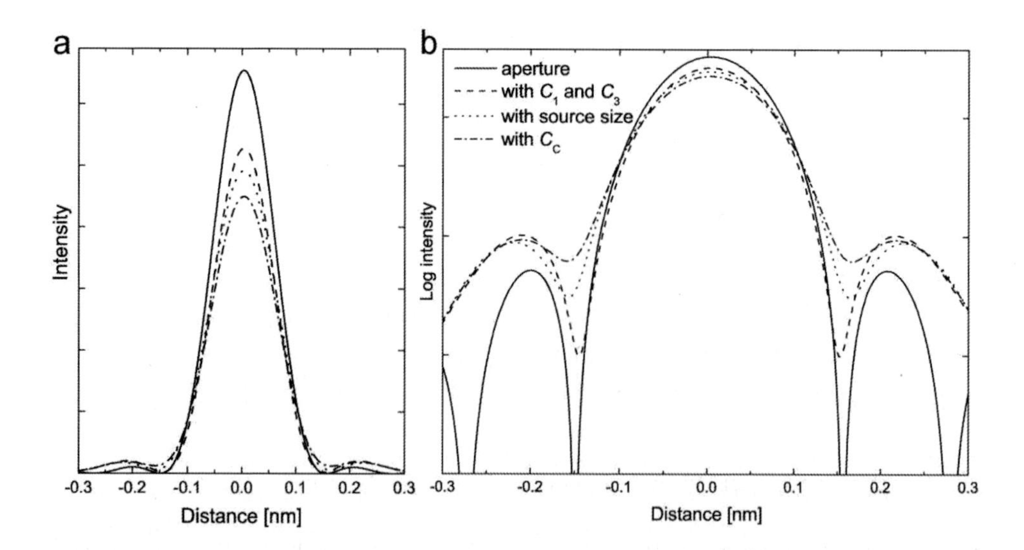


Fig. 3.10 Electron probe intensity profiles; (**a**) linear scale, (**b**) logarithmic scale. The full line considers only the finite size of the aperture, the dashed line includes aperture aberrations $C_1$ and $C_3$ according to Eq. (3.17), the dotted line considers in addition a finite source size $\delta_{\rm geo}$ of 0.08 nm, and the dashed-dotted line is calculated according to Eq. (3.21) considering $C_C$ of 2 mm and $\Delta E = 5$ eV in order to amplify the effect of partial temporal coherence. The other parameters are $C_3 = 1$ mm, $E_0 = 200$ keV ($\lambda = 2.5$ pm) and $\alpha = \alpha_{\rm opt} = 10$ mrad.

semi-angle, which is $\alpha_{\mathrm{opt}} = 10$ mrad. The full line in Fig. 3.10 (**a** and **b**) shows a probe-intensity profile calculated according to Eq. (3.13), where the only contribution to the probe is the effect of the finite aperture size. This is essentially a line profile through an Airy pattern, as already illustrated in Fig. 3.2. The dashed line in Fig. 3.10 corresponds to a probe intensity profile incorporating, in addition to the finite aperture angle, the aperture abberations $C_1$ and $C_3$ according to Eq. (3.16). The defocus $C_1$ was set to $-50$ nm in agreement with the optimum defocus which can be derived from Eq. (3.9). The finite source size of 0.08 nm is taken into account as an additional factor in the dotted curve in Fig. 3.10. Since the effect of partial temporal coherence is small, in order to illustrate its impact on the electron probe, we considered an energy spread of $\Delta E$ of 5 eV and a constant of chromatic aberration $C_{\mathrm{C}}$ of 2 mm. This yields the dashed-dotted curves in Fig. 3.10.

Figure 3.7a shows clearly that incorporating stepwise the individual contributions to the electron probe, the intensity of the central maximum of the electron probe decreases. However, none of the effects substantially increases the width of the central maximum. Even the unrealistically large energy spread of 5 eV for a field-emission source combined with a $C_{\mathrm{C}}$ of 2 mm does not substantially degrade the width of the central maximum. The diffraction limit $\delta_{\mathrm{D}}$ due to $\alpha = 10$ mrad would in principle allow for a resolution of 0.15 nm (see Eq. (3.1)). The FWHM of the full curve measures 0.13 nm. Though the FWHM is smaller than the $\delta_{\mathrm{D}}$, we use the FWHM to compare the probe profiles in Fig. 3.10. Incorporation of $C_3$ including $C_{1\,\mathrm{opt}}$ leads to a FWHM of 0.133 nm; including $\delta_{\mathrm{geo}}$ of 0.08 nm leads to a FWHM of 0.14 nm, and including partial temporal coherence, i.e. the chromatic aberration, leads to a FWHM of 0.14 nm. Hence, provided the defocus $C_1$ and the probe illumination semi-angle $\alpha$ are set according to the optimum settings given in Eqs. (3.9) and (3.10), neither the spherical aberration, the chromatic aberration nor the finite source size of 0.08 nm substantially reduces the probe size and thus the achievable STEM resolution. The overall effect due to $C_3$, $\delta_{\mathrm{geo}}$ and $C_{\mathrm{C}}$ is less than 10 pm. This, of course, depends strongly on the assumption of the effective source size, but 0.08 nm is not an unrealistically small value (see, e.g. LeBeau *et al.*, 2009; Dwyer *et al.*, 2008). Hence, the Scherzer incoherent conditions make it possible that for the optimum illumination semi-angle the effect of the spherical aberration is minimized such that the resolution is nearly determined by the diffraction limit imposed by $\alpha_{\mathrm{opt}}$.

Although the intensity of the central maximum decreases when incorporating the different contributions to the electron probe (see Fig. 3.10a), the total probe intensity needs to be a constant. Hence, the intensity which is lost in the central maximum is transferred to side lobes. This can be seen in the logarithmic plot in Fig. 3.10b. The main impact of the spherical aberration under optimized probe conditions and the influence of partial coherence is to increase the intensity of the tails of the electron probe. The tails of the probe do not primarily impact the achievable STEM resolution, but the image contrast. Because the central maximum is hardly affected by the individual probe contributions, the probe remains sensitive to nearly the same object spacings. However, with growing tails of the probe, the background intensity of a micrograph increases and, as a result, the contrast decreases. Fertig and Rose (1979) show that for an electron probe which provides zero contrast due to its side lobes, the Rayleigh criterion can still be fulfilled. This clearly shows that relating the probe characteristics to an achievable resolution is not a trivial issue. Fertig and Rose (1979) suggest relating the diameter $d_{59}$ which contains 59% of the total probe intensity with the STEM resolution. This is essentially the diameter corresponding to the first minimum of the electron probe and is thus equivalent to the diffraction limit, which also relates the first minimum of the Airy pattern with the achievable resolution.

## 3.4 Summary

This chapter focused on one particular aspect of STEM imaging; the formation of the electron probe. Contributions which affect the characteristics of the electron probe were discussed from a purely geometrical as well as from a wave optical point of view. Keywords addressed in this chapter are: STEM imaging modes; electron probe; diffraction limit; isotropic aberration function; chromatic aberration; effective source size; resolution and depth of field.

## Chapter 4

# Limits of Conventional Atomic-Resolution Electron Microscopy

In the previous two chapters we discussed the imaging process in HRTEM and the formation of the electron probe in STEM mode for conventional transmission electron microscopes. We saw that the spherical aberration $C_3$ limits the point resolution in both modes. The present chapter discusses the consequences of the presence of a finite spherical aberration for high-resolution imaging. As will be shown, the spherical aberration does not prevent us from observing object information that lies beyond the point resolution of the microscope. However, such information is not directly accessible. Indeed, the presence of spatial frequency components beyond the resolution limit severely complicates the image interpretation.

## 4.1 The Case of Transmission Electron Microscopy

For conventional HRTEM imaging, it is the spherical aberration of the objective lens $C_3$, which limits the point resolution (see Eq. (2.32)), while the chromatic aberration $C_C$ essentially imposes a limit on the information transfer (see Eq. (2.33)). In order to optimize the phase contrast imaging condition, the defocus needs to be adjusted in such a way that it partially balances the effect of the positive spherical aberration. This condition is referred to as Scherzer focus (Scherzer, 1949). Characteristic of this optimization is that over a large spatial frequency range the phase contrast transfer function is nearly constant. This enables a direct image interpretation for the spatial frequencies that fall into this passband. The first zero-crossing of the phase contrast transfer function at Scherzer focus defines the point resolution. This is expressed in Eq. (2.32). The resolution depends on the spherical aberration $C_3$ and on the electron wavelength $\lambda$.

For transmission electron microscopes which are equipped with thermionic electron sources, like a $LaB_6$ or a tungsten cathode, the coherence of the beam, determined by its energy spread and the source size, does not enable substantial information transfer beyond the point resolution. The damping of the phase contrast transfer function caused by the envelope functions due to temporal and spatial partial coherence imposes a limit on the information transfer. One can say that the point resolution nearly coincides with the real limit of the observable information content of a given micrograph. However, for field-emission electron microscopes this is different. The increased coherence of the beam, and thus the reduced impact of the damping envelope functions, makes it possible to have significant information transfer beyond the point resolution. There is a pronounced gap between point resolution and information limit, which is due to the enhanced brightness of the electron source. Figure 2.10 shows that the image frequencies beyond the point resolution are transferred by a highly oscillating transfer function. These oscillations are a direct consequence of the presence of $C_3$, which induces a phase shift that increases with the fourth power of the spatial frequency. The exponential function in Eq. (2.13) translates this phase shift into an oscillating phase contrast transfer function. While at lower spatial frequencies these oscillations can largely be prevented by a suitable choice of the defocus, for high spatial frequencies this is not the case. The reason for this is simple; the phase shifts induced by the spherical aberration increase with the fourth power of the spatial frequency, whereas the phase shifts due to defocus increase with the square of the spatial frequency (see Eq. (2.15)). Hence, the parabolic behavior of the defocus can only partially compensate the effect of $C_3$. The frequency range wherein $C_1$ partially compensates for $C_3$ can also be shifted to higher spatial frequencies as, for example, shown by O'Keefe *et al.* (2001a). However, the higher this passband appears, the smaller its width. The largest passband can be obtained if the defocus is set to Scherzer focus. In any case, for spatial frequencies that are transferred by a highly oscillating transfer function, one expects that within narrow frequency intervals the phase contrast switches from positive to negative. This, of course, complicates the image interpretation, even for a purely kinematic specimen that does not show any dynamic scattering effects.

Besides the complicated interpretation of such high-resolution micrographs, the spatial frequencies, which are transferred by a highly oscillating phase contrast transfer function, are image-delocalized. *Image delocalization* means that image details are displaced from their true location in the specimen. This effect can be prominent in HRTEM micrographs and makes direct image interpretation nearly impossible (Otten and Coene, 1993). For a perfectly homogeneous defect-free crystalline specimen of uniform thickness, translational symmetry is fulfilled and therefore image delocalization is not necessarily apparent. On the other hand, image delocalization is readily observable at interfaces, specimen edges or at the surface of nanoparticles (Coene and Jansen, 1992). Figure 4.1a shows a high-resolution micrograph of a gold foil imaged in a $\langle 110 \rangle$ zone-axis orientation, using a 200 kV electron microscope equipped with an objective lens of $C_3 = 0.5$ mm. The defocus was adjusted in order to visually minimize the delocalization. Still, at the edge of the specimen there are faint modulations visible which reach nearly 1 nm into the vacuum. Hence, without employing an objective aperture which limits the maximum spatial frequency contributing to the image, image delocalization cannot be prevented in a conventional field-emission electron microscope.

As can be deduced from the above argument, the importance of image delocalization in HRTEM is closely related to the coherence of the electron source (Coene and Jansen, 1992). For an electron beam of high coherence, the damping effect of the envelope functions due to partial coherence is reduced and even high spatial frequencies can be transferred. Yet, such high spatial frequencies are transferred by an oscillating transfer function. Image delocalization increases with the gradient of the transfer function, i.e. the delocalization is proportional to the oscillation speed of the transfer function in a particular frequency range (Coene and Jansen, 1992). The effect of image delocalization is therefore most severe for spatial frequencies beyond the point resolution. This is the frequency regime where a field emission microscope still transfers information. Furthermore, since the gradient of the transfer function can be different for each spatial frequency, there is not a fixed delocalization distance that would describe the delocalization for all spatial frequencies under certain imaging conditions.

Coene and Jansen (1992) derive an expression for the spatial information shift $\boldsymbol{\Delta r}$, i.e. the image delocalization expected for a spatial frequency $\boldsymbol{q}$ related back to the scale of the specimen

$$\boldsymbol{\Delta r} = \frac{1}{\lambda}\nabla\chi(\boldsymbol{q}) = \lambda\boldsymbol{q}\left(C_1 + C_3\lambda^2\boldsymbol{q}^2\right), \tag{4.1}$$

which for the isotropic case $\lambda q = \theta$ can be expressed in terms of the scattering angle $\theta$. Alternatively, we can employ the complex scattering angle $\omega$ as introduced in Chapter 2 such that $|\omega| = \theta$. As already mentioned in Chapter 2, the complex scattering angle $\omega$ reflects the complex coordinate in the aperture plane. With this substitution and Eq. (2.16), Eq. (4.1) can be rewritten as

$$\Delta r = \nabla\chi(\theta) = \theta C_1 + \theta^3 C_3 = |\omega| C_1 + |\omega|^3 C_3. \tag{4.2}$$

Equation (4.1) shows that the effect of image delocalization can be reduced by decreasing the wavelength $\lambda$, i.e. by increasing the high tension of the microscope or by using an objective lens of lower spherical aberration $C_3$. For an aberration-corrected microscope, where $C_3 \to 0$, image delocalization can essentially be annulled provided that the defocus is $C_1 \to 0$ and that other aberrations are also minimized. This is demonstrated in Fig. 4.1b.

However, for a conventional microscope of finite spherical aberration $C_3$, image delocalization can never be fully annulled. As an example, for the virtual 200 kV transmission electron microscope with $C_3 = 1$ mm, which we introduced in Chapter 2, a spatial frequency corresponding to a distance of 0.20 nm would be imaged in a distance of 1.2 nm away from its actual position in the object. This applies for the case in which the microscope is operated at Scherzer focus. Though the delocalization is substantial, the spatial distance of 0.20 nm is only a little smaller than the actual point resolution of this microscope, which is 0.23 nm. Hence, image delocalization can severely influence the image formation in conventional high-resolution field-emission electron microscopes. Each spatial frequency suffers from a specific delocalization distance. Therefore, image delocalization complicates, if not even prohibits, a direct image interpretation (see Fig. 4.1c). The deleterious effect of image delocalization is particulary critical for micrographs where analysis of the edges of wedge-shaped samples, grain boundaries or the surfaces of nanoparticles is attempted.

Equation (4.2) reveals that by choosing a negative defocus $C_1$, the image delocalization caused by the spherical aberration $C_3$ can be reduced. In fact, for *one* particular scattering angle $|\omega|$, i.e. for *one* particular spatial frequency $q$, there is always one (negative) $C_1$ setting such that the image delocalization induced by $C_1$ fully compensates the delocalization of the positive spherical aberration $C_3$. This is illustrated in Fig. 4.1d.

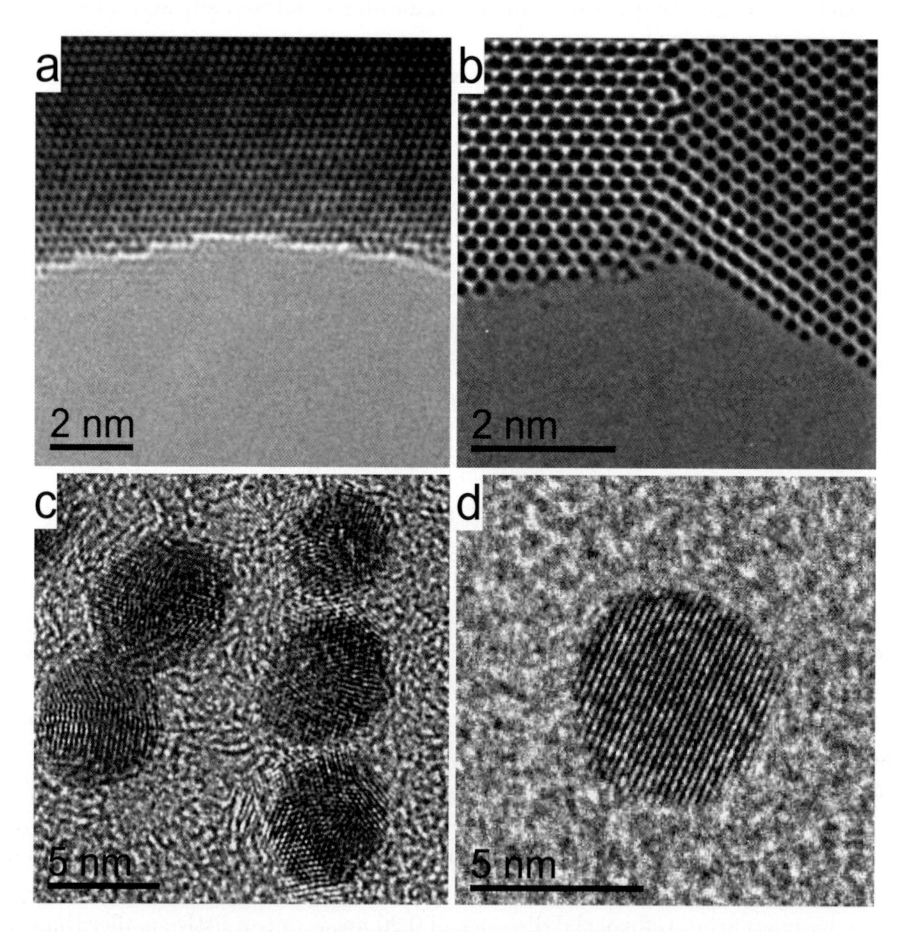


Fig. 4.1 Image delocalization. (**a**) Conventional HRTEM micrograph of Au⟨110⟩; (**b**) aberration-corrected HRTEM micrograph of Au⟨110⟩; (**c**) conventional HRTEM micrograph of Au particles on a carbon support film; and (**d**) conventional HRTEM micrograph of an Au particle with only one set of lattice planes visible.

Figure 4.1d shows a high-resolution micrograph of a nanoparticle recorded with the same microscope as used for Figs. 4.1a and 4.1c. Due to the orientation of the nanoparticle on the carbon support film, only one set of lattice planes, i.e. only one spatial frequency $q$, contributes to the structural information of the nanoparticle. Since for one particular $q$ there is always one particular $C_1$ setting which balances the image delocalization caused by the spherical aberration $C_3$, the defocus $C_1$ can be adjusted such that no delocalization is observable, i.e. $\Delta r$ is annulled for the spatial frequency which describes the set of lattice planes of the nanoparticle.

As will be shown in Chapter 7, Eq. (4.2) simply expresses the superposition of the *image aberrations* of $C_1$ and $C_3$ referred back to the object plane. Indeed, any residual coherent aberrations, which as a matter of fact are also present in the case of aberration-corrected microscopes, lead to characteristic (anisotropic) modulations of the phase contrast transfer function. Since the image delocalization is essentially defined by the gradient of the (generalized) aberration function (see Chapter 7), any modulation of the phase contrast transfer function leads to image delocalization. Hence, even in the case of aberration-corrected microscopes (where $C_3 \approx 0$), image delocalization is feasible, though in general on a much smaller scale. Moreover, for a finite defocus, image delocalization still occurs even if $C_3 = 0$, i.e. a finite defocus leads to an image aberration which is observed as an image delocalization similar to the effect of $C_3$.

Although image delocalization can be annulled for a given spatial frequency $q$, image delocalization cannot be prevented for a finite $q$-range (provided $C_3 \neq 0$). However, the impact of image delocalization can be minimized for a given frequency range. Hannes Lichte (1991) derived an expression for an optimal defocus $C_{1\,\mathrm{Lichte}}$, which minimizes delocalization in HRTEM micrographs over the entire spatial frequency range that contributes to the image. This so-called *Lichte focus* is given by

$$C_{1\,\mathrm{Lichte}} = -\frac{3}{4}\,C_3\left(q_{\mathrm{max}}\lambda\right)^2 = -\frac{3}{4}\,C_3|\omega_{\mathrm{max}}|^2, \tag{4.3}$$

with $q_{\mathrm{max}}$ the highest spatial frequency which contributes to the image. The frequency $q_{\mathrm{max}}$ is determined either by the opening of the objective aperture ($q_{\mathrm{max}} = q_{\mathrm{a}}$) or, in case no aperture is used, by the information limit of the microscope, i.e. $q_{\mathrm{max}} = 1/\rho_{\mathrm{t}}$ (see Eq. (2.33)).

Comparing the Scherzer focus given in Eq. (2.21) and the Lichte focus in Eq. (4.3) reveals that there are two (different) values for an optimal defocus. Each one of these defocus values optimizes the imaging conditions according to a specific measure; while at Scherzer focus a flat transfer passband up to the point resolution of the microscope is established, which essentially allows for a direct image interpretation up to the point resolution, at Lichte focus image delocalization is minimized. In general, these two defocus values do not coincide. Moreover, the larger the gap between point resolution and information limit, the larger the gap between Scherzer and Lichte focus. Hence, it is not possible to simultaneously optimize the imaging conditions in respect of image delocalization *and* resolution.

The Lichte focus $C_{1\,\text{Lichte}}$ shows a $\lambda^2$ dependence, whereas the Scherzer focus $C_{1\,\text{Scherzer}}$ is proportional to $\sqrt{\lambda}$. Therefore, reducing the electron wavelength $\lambda$ brings the two focus values closer together. Working at higher acceleration voltages thus makes it easier to optimize the imaging conditions, especially when using a field-emission transmission electron microscope. Furthermore, if $C_3 \rightarrow 0$, as it is feasible on aberration-corrected microscopes, the Lichte focus $C_{1\,\text{Lichte}} \rightarrow C_{1\,\text{Scherzer}}$. Hence, only in the case of aberration-corrected microscopes, the imaging conditions can be optimized simultaneously in respect to the image delocalization and the point resolution. For aberration-corrected instruments, the impact of additional (anisotropic) aberrations also need to be considered in the analysis. However, the image delocalization in conventional field-emission transmission electron microscopes, which is unavoidable, imposes a fundamental limit in using such instruments in direct ultra high-resolution imaging.

As explained above, image delocalization stems from the fact that spatial frequencies are allowed to contribute to the image which fall into a frequency regime where the transfer function is oscillating. Using an objective aperture to block high spatial frequencies can substantially reduce the amount of image delocalization. However, in doing so one would lose contributions of high spatial frequencies transferred to the image, which indeed can be of some advantage. Yet, the high-resolution object information, which falls into the spatial frequency range between point resolution and information limit, is not directly accessible or interpretable. It is for this reason that during the decade that passed between the implementation of field-emission electron sources and the first successful realization of an aberration corrector for HRTEM imaging, many research activities were initiated to solve the problem of how to retrieve the delocalized high-resolution information in HRTEM micrographs. While a field-emission source in HRTEM imaging enables a higher information limit, the point resolution, defined by the wavelength and the spherical aberration, remained unaffected by this change.

The efforts to retrieve the delocalized image information led, for instance, to the realization of algorithms which allow for iteratively retrieving the complex electron wave at the exit plane of the specimen from a series of micrographs recorded under varying imaging conditions (see, e.g. Fig. 2.2). The idea of reconstructing the exit-plane wave goes back to Peter Schiske (1968)[1], who first proposed this technique. Since then, the method has been developed over more than four decades (see, e.g. Saxton, 1978; Kirkland, 1982, 1984; Kirkland *et al.*, 1985; Saxton and Smith, 1985; Coene and Jansen, 1992), incorporating various types of modifications and improvements[2]. It has only become a widely applicable technique within the

[1]Schiske's original text written in German has been translated into English by E. Zeitler. It is published in Schiske (2002).

[2]Saxton (1994) provides a review about the different approaches and the evolution of this technique.

last ten to fifteen years, when field-emission electron microscopes have become more popular (see, e.g. Coene *et al.*, 1996; Op de Beeck *et al.*, 1996; Thust *et al.*, 1996).

Though the reconstruction of the electron wave at the exit plane of the specimen is feasible by recording a series of images under varying tilt illumination angle (Kirkland *et al.*, 1995), the most common reconstruction technique is the so-called focal series reconstruction, which is also called the through focal series reconstruction technique or the focus variation method. This approach originally proposed by Schiske (1968) allows for reconstructing the exit-plane wave from a series of micrographs recorded under varying focus condition.

However, independent of how the exit-plane wave is reconstructed, the important point about these reconstruction algorithms is that they allow retrieval of the electron wave at the exit plane of the specimen. Since the exit-plane wave is the product of the (dynamic) diffraction of the electrons on transmitting the specimen, the exit-plane wave is unaffected from the coherent transfer function of the microscope, and as such does not suffer from image delocalization. Indeed, with this technique it is possible to retrieve the object information which is transferred between point resolution and information limit, i.e. the object information that suffers from image delocalization. Still, the exit-plane wave is in general the result of dynamic scattering within the specimen and as such does not necessarily provide a direct image of the projected crystal structure.

Since the reconstructed exit-plane wave is a complex function, which is described by a phase and an amplitude, focal series reconstruction (or tilt series reconstruction) can be considered as a holographic technique which essentially solves the phase problem, i.e. the problem that the phase information is lost on forming an (intensity) image of the complex image wave function. Particularly because of this reason, focal-series and tilt-series reconstructions are also applied on data recorded in aberration-corrected microscopes (see, e.g. Rossell *et al.*, 2009; Haigh *et al.*, 2009b). Hence, the reconstruction of the exit-plane wave allows decoupling of the contribution of the microscope to the image, and makes it possible to solve the phase problem, which, except for off-axis electron holography (see, e.g. Tonomura *et al.*, 1995), is inherent for any kind of electron microscopy imaging. The only parameter that is crucial for the resolution is the information limit of the microscope. Independent of the spherical aberration and the electron wavelength, the reconstruction limit is determined by the information limit of the instrument. However, it is worth mentioning that the acquisition conditions have to be optimized and that the experimental parameters, like for instance the defocus range or the applied focal step, need to be known with high precision — with a precision that is usually only determinable by iteratively optimizing the nominal reconstruction parameters (Tang *et al.*, 1996; Erni *et al.*, 2010).

Because the reconstruction allows for retrieving information which in a single micrograph is lost due to the oscillating transfer function, and thus for overcoming the limit imposed by the spherical aberration, reconstruction of the exit-plane wave has also been considered as a type of *numerical* or *indirect* aberration correction (Hetherington *et al.*, 2008). With the focal-series reconstruction technique it was, for instance, possible to access sub-Ångström specimen information on a microscope which has a $C_3$-limited point resolution of about 1.7 Å (see, e.g. O'Keefe *et al.*, 2001a, 2001b; Kisielowski *et al.*, 2001; O'Keefe, 2008). Hence, with the focal series reconstruction technique it is in principle possible to numerically extend the optical point resolution of a microscope. However, severe data analysis is involved.

In summary, the problem of image delocalization is essentially caused by the presence of the spherical aberration which results in a highly oscillating transfer function. Employing an electron source of high brightness, information beyond the point resolution can be transferred to the image. Since the spatial frequencies between point resolution and information limit fall into a region where the transfer function is highly oscillating, it is this information which is most vulnerable to image delocalization. In a single micrograph, this high frequency object information can be considered to be lost. There are reconstruction algorithms available which can be applied to restore the object information between point resolution and information limit. However, these techniques, which can be regarded as a kind of numerical aberration correction, involve extensive data analysis, which, as a matter of fact, makes them vulnerable to artifacts.

## 4.2 The Case of Scanning Transmission Electron Microscopy

For STEM, the impact of the spherical aberration is similar to the situation of HRTEM. In order to reveal its impact on conventional probe forming instruments, we have to distinguish the different effects which can potentially limit the size of the electron probe and thus the resolution achievable in STEM mode. The four effects we discussed in Chapter 3 are the diffraction limit, the spherical aberration, the chromatic aberration and the effective source size.

The impact of the effective source size is not inherent for conventional electron microscopes. Indeed, if the demagnification of the electron source is insufficient or if instabilities are present which cause an incoherent blurring of the demagnified electron source, the size of the electron probe becomes larger and as a result the STEM resolution is degraded. This can happen in conventional electron microscopes as well as in aberration-corrected instruments. However, here we shall assume that the microscope is sufficiently stable and that the demagnification of the electron source is sufficiently high such that the effective source size is not the resolution-limiting factor. Indeed, this is a reasonable approximation for conventional electron microscopes. For aberration-corrected microscopes, where the achievable resolution is in general higher than in conventional microscopes, the requirements for the stability and the demagnification increase such that the size of the effective source becomes a crucial parameter in the setup of the electron probe.

Although the finite energy spread of the electron beam in combination with the chromatic aberration of the objective lens can impair the size of the electron probe, in general, the blurring due to the chromatic aberration can be neglected in conventional electron microscopes. This was explained in Chapter 3 (see, e.g. Fig. 3.7). Provided that the microscope is equipped with a field-emission electron source with an energy spread smaller than about 1 eV, the chromatic aberration is not the resolution-limiting factor in conventional high-resolution scanning transmission electron microscopes. However, similar to the impact of the effective source size, it has to be pointed out that in case of aberration-corrected instruments, the chromatic aberration can severely impact the STEM performance. The main effect of the chromatic aberration is that the contrast of the micrographs is reduced (see, e.g. Fertig and Rose, 1979). This will be discussed in more detail in Chapter 9.

The factors that are of crucial importance for the STEM resolution in conventional scanning transmission electron microscopes are the diffraction limit and the spherical aberration. The diffraction limit is not the actual limiting factor; one could always choose an aperture size such that the diffraction limit is not impairing the resolution. However, in order to reduce the deleterious effect of the spherical aberration on the shape of the electron probe, one needs to strongly confine the illumination angle of the electron probe such that with a suitable defocus, the remaining effect of the spherical aberration can be properly balanced within the limited beam illumination angle. This is expressed by the Scherzer incoherent conditions, summarized in Eqs. (3.9) and (3.10), which can be rewritten as (Scherzer, 1949; Crewe and Salzman, 1982; Pennycook and Jesson, 1991)

$$C_{1\,\mathrm{opt}} = -\sqrt{\lambda C_3} \qquad \text{and} \qquad \alpha_{\mathrm{opt}} = \sqrt[4]{\frac{4\lambda}{C_3}}\,.$$

The two conditions expressed in the above equations describe the impact of the spherical aberration on the optimization of the probe size in a conventional scanning transmission electron microscope. Because of the spherical aberration, the defocus needs to be adjusted slightly negative in order to compensate for the effect of the spherical aberration[3]. This is equivalent to the case of HRTEM imaging where a negative defocus, i.e. Scherzer focus, optimizes the imaging conditions (Scherzer, 1949). However, in case of STEM, the restriction imposed by the spherical aberration demands another adjustment; the illumination semi-angle $\alpha$ must match the optimum illumination semi-angle $\alpha_{\mathrm{opt}}$.

The limitation of the probe illumination semi-angle has a twofold consequence: the size of the STEM probe is essentially determined by the diffraction limit imposed by the limited illumination semi-angle $\alpha_{\mathrm{opt}}$; and secondly, the limited beam illumination angle significantly reduces the current of the electron probe (see Eq. (3.8)). Assuming that for a sufficiently high demagnification of the electron source the

[3]Figure 3.4 shows that in order to move the disk of least confusion towards the Gaussian image plane where the specimen is located, one needs to reduce the strength of the lens. This is meant by a negative focus setting $C_1 < 0$.

intensity in the aperture plane in front of the specimen is homogeneous, a smaller aperture simply means that more electrons are blocked out and thus cannot contribute to the current of the STEM probe. Typically, one ends up with probe currents $I_{\mathrm{probe}}$ of atomic-resolution electron probes of the order of 10 pA or even less. This complicates analytical measurements. In addition, a small probe current leads to noisy micrographs.

Considering the presence of noise, it is possible that the experimentally measurable resolution is limited by noise rather than by the actual optical setup of the microscope. Experimental data are always susceptible to noise. In many cases, one measures a certain signal or records a certain micrograph which shows a certain, but uncritical, noise level. As long as the features of interest can be identified, the presence of noise seems to be harmless. However, in cases where the noise level is high, it can happen that the statistical significance of a certain image feature is no longer warranted, i.e. the strength of the signal drops below the noise level. While experimental data are always prone to a certain level of noise, classical resolution criteria, like, for example, the Rayleigh criterion (Born and Wolf, 2001), do not consider the noise in experimental data. Analyzing the information content of experimental data, such as a STEM micrograph, the noise level always needs to be evaluated and compared with the expected data, and only if the statistical significance of the structural data exceeds the noise level can a resolution claim can be made (see, e.g. Erni *et al.*, 2009). Hence, the information content of a micrograph and its resolution need to be considered in relation to the experimental noise (Van Aert and Van Dyck, 2006). It is the signal-to-noise ratio which can be decisive.

Since under similar acquisition conditions the noise level increases with decreasing current of the electron probe, the current of the electron probe can be crucial for the resolution in STEM imaging. For a fixed demagnification of the electron source, the limited illumination semi-angle $\alpha_{\mathrm{opt}}$ imposes a severe barrier on the achievable current of the electron probe, and thus on the achievable resolution.

As mentioned above, the size of an electron probe, set up according to the Scherzer incoherent conditions, is essentially limited by the diffraction limit imposed by the restricted illumination semi-angle which has to be chosen to minimize the deleterious effect of the spherical aberration. Let us consider the STEM resolution of a 300 kV scanning transmission electron microscope ($\lambda = 1.969$ pm) which is equipped with an objective lens of a spherical aberration $C_3$ of 1.2 mm. According to Eq. (3.10), the optimum illumination semi-angle $\alpha_{\mathrm{opt}}$ is then 9.0 mrad, and according to Eq. (3.11), the achievable STEM resolution is 0.133 nm. This is the point resolution for STEM imaging of this 300 kV electron microscope. The diffraction limit of a 300 keV electron probe of 9 mrad is 0.133 nm as well (see Eq. (3.1)). Since the impact of the spherical aberration is compensated by the defocus within the limited illumination angle, the Scherzer incoherent conditions thus enable the formation of an electron probe which is equivalent to a diffraction limited electron probe with $\alpha = \alpha_{\mathrm{opt}}$.

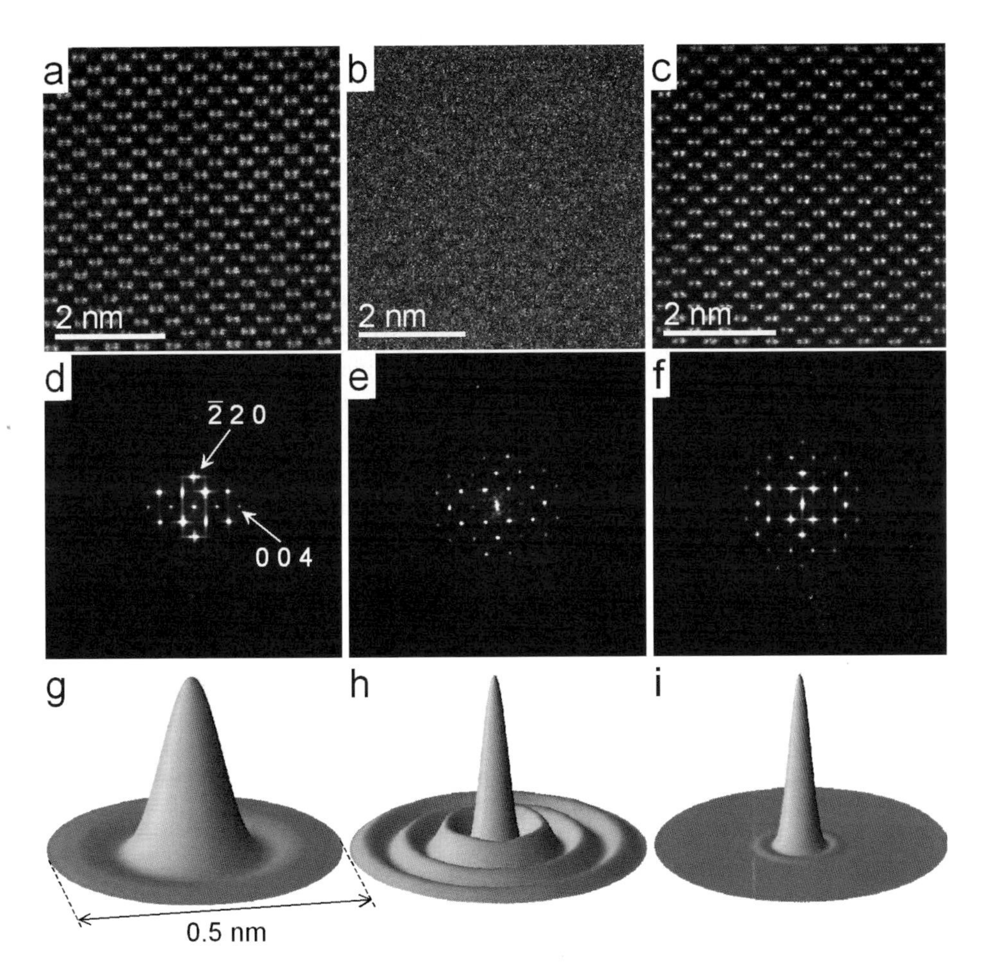


Fig. 4.2 HAADF STEM of Si⟨110⟩. (**a**) Conventional 300 kV microscope with $C_3 = 1.2$ mm employing Scherzer incoherent conditions, (**b**) Super resolution mode with $\alpha = 15$ mrad, and (**c**) aberration-corrected 300 kV microscope. (**d**), (**e**) and (**f**) show the corresponding power spectra, and (**g**), (**h**) and (**i**) the corresponding (calculated) electron probes.

Figure 4.2a shows a HAADF STEM micrograph of a silicon specimen in ⟨110⟩ zone-axis orientation. The distance between the dumbbell atoms for Si⟨110⟩ is 0.136 nm. The 300 keV electron probe with $C_3 = 1.2$ mm enables, under Scherzer incoherent conditions, a resolution of 0.133 nm and thus allows for resolving the dumbbell atoms. The corresponding power spectrum of the micrograph in Fig. 4.2d clearly shows the presence of the 004 reflection, which corresponds to the 0.136 nm dumbbell interatomic distance. Figure 4.2g shows the calculated Scherzer electron probe, which consists of one dominant intensity maximum.

In the previous section we discussed the impact of image frequencies that are beyond the point resolution contributing to the HRTEM image intensity. Provided that the information limit exceeds the point resolution, we saw that the image frequencies between the point resolution and the information limit give rise to image delocalization, i.e. the image is no longer directly interpretable. We can consider the equivalent case in STEM imaging. What happens if we choose a probe illumination semi-angle larger than the optimum angle $\alpha_{\text{opt}}$ such that the applied defocus can no longer fully compensate the impact of the spherical aberration?

Figure 4.2b shows a HAADF STEM micrograph recorded using a 300 keV STEM probe of $\alpha = 15$ mrad and $C_3$ of 1.2 mm. The probe illumination semi-angle clearly exceeds $\alpha_{\text{opt}} = 9$ mrad. Hence, the defocus $C_1$, which was chosen to be about $-200$ nm, can no longer homogeneously balance the effect of $C_3$. Still, in order to take into account the enhanced impact of the spherical aberration within the larger illumination cone, its value was chosen more negatively than the Scherzer incoherent focus. On first sight, the micrograph of Si$\langle 110 \rangle$ shown in Fig. 4.2b looks like random noise. However, deriving its power spectrum shows that there is structural information of the Si$\langle 110 \rangle$ specimen contained in the micrograph (see Fig. 4.2e). Furthermore, the reflections in the power spectrum reach higher spatial frequencies than in the case of the electron probe set up according to the Scherzer incoherent conditions. The highest spatial frequency corresponds to a spacing of 0.1 nm. The calculated electron probe shown in Fig. 4.2h gives an explanation for this unexpected observation.

As described in Chapter 3, HAADF STEM imaging can be considered as an incoherent imaging technique. The incoherent imaging model states that the image intensity $I(\boldsymbol{r})$ is given by the convolution of the intensity of the electron probe $|\psi_0(\boldsymbol{r})|^2$ and an object function $O(\boldsymbol{r})$, which, for a crystal in zone-axis orientation, is peaked at the sites of the atomic columns, i.e.

$$I(\boldsymbol{r}) = |\psi_0(\boldsymbol{r})|^2 \otimes O(\boldsymbol{r}) \,. \tag{4.4}$$

The incoherent image is the object function blurred by the intensity profile of the electron probe (see, e.g. Nellist and Pennycook, 2000). With this imaging model in mind, the impact of the shape of the probe on the achievable image contrast can be explained straightforwardly. The width of the central maximum of the STEM probe is essentially determined by the diffraction limit, which, for a 15 mrad STEM probe at 300 keV, is 0.08 nm. Hence, in principle, object information down to 0.08 nm can be sampled with the electron probe shown in Fig. 4.2h. However, because of the spherical aberration (and the applied defocus), the STEM probe contains significant side lobes. On the basis of the incoherent imaging model, the HAADF STEM micrograph can be considered to represent the convolution of the intensity profile of the STEM probe and the projected crystal potential. The projected crystal potential is strongly peaked at the sites of the atomic columns. Hence, for a sufficiently small STEM probe mainly consisting of one central maximum, like, for example, the one shown in Fig. 4.2g, each peak of the projected crystal potential gives rise to an intensity peak in the HAADF STEM micrograph. We can associate each peak with an atomic column. However, for a STEM probe which contains significant side lobes like the one in Fig. 4.2h, image intensity can occur at positions where there are no atomic columns, i.e. the side lobes essentially create image intensity between the atomic columns. If this side-lobe intensity is high, the contrast of the atomic structure information can degrade to a level where it is no longer recognizable. However, since there is a small distinct peak in the electron probe (see Fig. 4.2h), high frequency object information can still be present. Such high-resolution information can be uncovered by calculating the power spectrum of the micrograph. Hence, a too large an illumination angle of a STEM probe in a conventional scanning transmission electron microscope leads to high frequency image information. Owing to the unavoidable side lobes of such an electron probe, the contrast of the structural information is strongly reduced. Because of the side lobes of the electron probe, the corresponding micrographs might no longer be directly interpretable (see Fig. 4.2b).

The fact that an electron probe formed under too large an illumination semi-angle can lead to artificial spots in HAADF STEM imaging was analyzed by, for example, Yamazaki *et al.* (2001). Based on detailed simulations, Yamazaki *et al.* (2001) concluded that artificial spots in HAADF STEM imaging are indeed due to side lobes of the electron probe. Under Scherzer incoherent conditions, the STEM probe has a simple shape which enables a direct image interpretation in terms of an incoherent imaging model of a self-luminous atomic structure. If, however, deviations from such a simple electron probe profile occur, the image intensity does not necessarily represent a simple projection of the atomic columns. This makes it necessary to carry out simulations which can be compared with the experimental data and which then allow for interpreting the micrographs (see, e.g. Kirkland *et al.*, 1987; Kirkland, 1998; Watanabe *et al.*, 2001, 2003). Hence, a wrongly chosen electron probe for STEM imaging leads to a loss of interpretability. This is indeed very similar to the loss of interpretability in HRTEM imaging due to image delocalization.

Nevertheless, another aspect of a too large illumination angle is that it makes it possible to observe object details which might not be resolvable with an electron probe optimized according to the Scherzer incoherent conditions. Of course, if the illumination angle chosen is much too large, the loss in contrast makes it impossible even to properly recognize object information. This is the case in Fig. 4.2h. Yet, if the illumination semi-angle is chosen only a bit larger (1–2 mrad) than $\alpha_{\text{opt}}$, then it is possible that the object information can still be recognized and that even object information that is not accessible by the Scherzer electron probe can be observed. Working with a slightly too large illumination angle in order to enhance the diffraction limit, combined with a slightly too large negative defocus ($C_1 < C_{1\,\text{opt}}$), is known as *super resolution* mode (Nellist and Pennycook, 1998). Employing this super resolution mode, Nellist and Pennycook (1998) show an information transfer in a STEM micrograph down to 0.078 nm using a dedicated scanning transmission electron microscope whose Scherzer point resolution is 0.136 nm. However, the 0.078 nm object information is not recognizable in the corresponding micrograph.

Watanabe *et al.* (2003) employ the super resolution mode in order to resolve the 0.136 nm Si⟨110⟩ dumbbells on a 200 kV microscope with a spherical aberration $C_3$ of 1.0 mm. According to the Scherzer incoherent conditions, this microscope would allow for a STEM resolution of 0.152 nm. Hence, in this second case, the increased resolution makes it possible to resolve an object feature which is not accessible by a Scherzer electron probe. While for the 200 kV microscope used by Watanabe *et al.* (2003), the optimum illumination semi-angle $\alpha_{\text{opt}}$ is 10 mrad, an enlarged illumination semi-angle of 12 mrad enabled the transfer of the 0.136 nm object feature. However, since under such imaging conditions side lobes in the electron probe cannot be prevented, the increased resolution comes at the expense of reduced image contrast.

The super resolution mode can be beneficial if an object detail needs to be resolved which is just a bit smaller than the STEM point resolution (Watanabe *et al.*, 2003). However, one has to be careful in applying this super resolution mode, particularly on specimens whose atomic structure is unknown; the side lobes of the probe, as pointed out above, can give rise to artificial spots at locations where there are no atoms present (Yamazaki *et al.*, 2001).

The side lobes of the electron probe which complicate the image interpretation and reduce the image contrast in super resolution mode are caused by the spherical aberration. If a too large illumination semi-angle is chosen, i.e. $\alpha > \alpha_{\text{opt}}$, the effect of the spherical aberration cannot be balanced by the defocus. For scanning transmission electron microscopes equipped with a corrector for the spherical aberration, on the other hand, it is possible to increase the illumination semi-angle without introducing side lobes in the electron probe. Though the Scherzer incoherent conditions as described in Eq. (3.9) and (3.10) are not valid any longer once $C_3 \rightarrow 0$, the increased beam illumination angle essentially reduces the impact of the diffraction limit. An aberration-corrected microscope thus enables a smaller electron probe. Furthermore, for a given effective source size, the increased beam illumination angle provides a higher probe current. Figure 4.2c shows a HAADF STEM micrograph of the Si⟨110⟩ specimen, recorded employing a 300 kV aberration-corrected STEM microscope. Comparing Fig. 4.2a with Fig. 4.2c clearly shows the enhanced resolution and the enhanced image contrast which are feasible on an aberration-corrected scanning transmission electron microscope. The corresponding power spectrum in Fig. 4.2f shows reflections corresponding to object details below 0.1 nm. The aberration-corrected electron probe is shown in Fig. 4.2i. Though the probe illumination semi-angle employed is 25 mrad, no side lobes are introduced which would complicate the image interpretability and reduce the image contrast. One could expect that once the spherical aberration is corrected, it is solely the diffraction limit which defines the size of the electron probe. For the 300 keV STEM probe in Fig. 4.2i with $\alpha = 25$ mrad, the diffraction limit is 0.048 nm. Yet, such high-resolution information is not present in the power spectrum of the aberration-corrected micrograph. Hence, other factors, like the effective source size, instabilities and the chromatic aberration, can become crucial once $C_3 \to 0$. This will be discussed in more detail in Chapter 9.

## 4.3 Summary

The spherical aberration limits the resolution of conventional electron microscopes. Furthermore, in HRTEM as well as in STEM mode, the spherical aberration can lead to a loss of the interpretability of the respective micrographs. If spatial frequencies are allowed to contribute to the image intensity in HRTEM, which are beyond the point resolution, image delocalization is unavoidable (see e.g. Fig. 4.1). Hence, it is a too-large scattering angle below the specimen which impairs the interpretability of the micrograph. If an objective aperture is used which blocks electrons scattered to high angles, image delocalization can in principle be prevented. For STEM, it is a too-large beam illumination angle which can give rise to artificial spots in HAADF STEM micrographs. In both cases, it is the spherical aberration that complicates the imaging process. In HRTEM, the spherical aberration leads to oscillations in the transfer function, and in STEM the spherical aberration leads to side lobes in the STEM probe. Both the oscillations and the side lobes can be reduced if a negative defocus is applied. Full compensation of the impact of the spherical aberration is, however, only possible within a given scattering angle range in HRTEM mode and within a given beam illumination angle in STEM mode. These limited angular ranges ultimately limit the resolution in conventional electron microscopes. Transfer of object information beyond the point resolution is possible in HRTEM and in STEM mode. However, in general, such information is not directly interpretable.

instrument and the chromatic aberration can be [illegible]. This [illegible] will be discussed in more detail in Chapter 9.

## 4.3 Summary

[illegible]

# PART II
# Electron Optics

## Chapter 5

# Basic Principles of Electron Optics

The duality between the particle and the wave characteristics of electrons is a phenomenon that manifests itself in daily work with electron microscopes. While the formation of an electron hologram or the information transfer in high-resolution phase contrast imaging are aspects that most naturally are explained by assuming the electron is a wave, there are phenomena which are conceptually better understandable by emphasizing the particle properties of the electron. An example of the latter is the creation of point defects by radiation damage, where the electron is more suitably described by a particle that elastically transfers part of its momentum to an atom in the solid. The elastic interaction and exchange of momentum between balls on a billiard table is a very simple visualization of this situation. One could thus overstate that electron microscopists have accepted the particle-wave duality of electrons, and adopt either of these two concepts in order to keep explanations simple. In cases where interference effects are dominant, it is mainly the wave characteristic that is of importance, and in most other cases it is sufficient and adequate to deal with electrons that are point-like charged particles.

An important branch of electron optics, which focuses in particular on the technological implementation of electron lenses and their imaging properties, deals with particle electrons that are exposed to stationary electromagnetic fields. The fundamental problem to solve lies in the derivation of the trajectories along which electrons can travel in the electromagnetic fields, which are determined by the optical elements of the microscope. It is the geometrical characteristics of the electron propagation that are of central importance, and not so much the temporal evolution. Knowing the trajectories from source to the detector allows one to predict where the electrons impinge the detector. The approach used to describe electron propagation on the basis of purely geometrical particle trajectories is called *geometrical electron optics*.

Although the great potential of novel electron microscopes in materials and life sciences is being widely noted, the modern literature about electron optics is certainly not as voluminous as the literature that deals with its practical applications in science and technology. There are, however, very valuable textbooks available that cover electron optics in depth, i.e. the physics, which, of course, is fundamental to any kind of electron microscopy. Early texts, which nonetheless instructively and illustratively elucidate the fundamentals of electron optics, include, for instance, *Optique électronique et corpusculaire* by De Broglie (1950) and *Grundlagen der Elektronenoptik* by Walter Glaser (1952). Though Glaser's comprehensive text about the optics and the instrumentation had been the standard literature in electron optics for nearly thirty years, it was extended and to some extent also replaced by other books. *Electron Optics* by Pierre Grivet (1972) is a text which addresses many aspects of electron microscopy. It can be considered as a forerunner of the most comprehensive work, entitled *Principles of Electron Optics*, by Hawkes and Kasper (1989a, 1989b, 1994). This modern three-volume collection covers in detail the concepts of geometrical optics, wave optics and instrumentation. A more recent textbook by Harald Rose (2009a), entitled *Geometrical Charged-Particle Optics*, considers novel electron optical devices and treats in detail the electron optical concepts that are essential for current and future aberration-corrected electron optical instruments. There are, of course, many more textbooks about electron or charged-particle optics; for a detailed overview see for example, Hawkes and Kasper (1989a).

Part II of this book does not intend to cover comprehensively the fundamentals of electron optics. For this, we refer to the texts mentioned above. The present part of this book aims at providing a brief overview of some of the very fundamental electron optical concepts without addressing aberration-corrected electron optics in detail. The goal of the present chapter is to give an overview of selected topics and as such to provide an introduction to the basic electron optics, which we think is of importance for the understanding and practical application of aberration-corrected imaging in (scanning) transmission electron microscopy.

## 5.1 The Electron Microscope

Although the previous chapters discussed two different high-resolution operation modes in electron microscopy, we did not introduce the very basic setup of a (scanning) transmission electron microscope. A transmission electron microscope is an instrument that consists of an electron source, an accelerator and a series of condenser lenses in front of the specimen, which are used to form the illumination of the specimen. In most present-day microscopes, the specimen is located within an objective lens which consists of two pole pieces; one in front and one behind the specimen. The specimen is immersed in the magnetic field formed by the two pole pieces. The projector lenses behind the objective lens are used to magnify and transfer the electron beam that interacts with the specimen to the detector or recording device. This is the fundamental setup of a transmission electron microscope, as already depicted by Knoll and Ruska (1932).

While the microscope column and the lenses are of macroscopic size, the electron beam is propagated within a narrow tube referred to as the liner tube. The

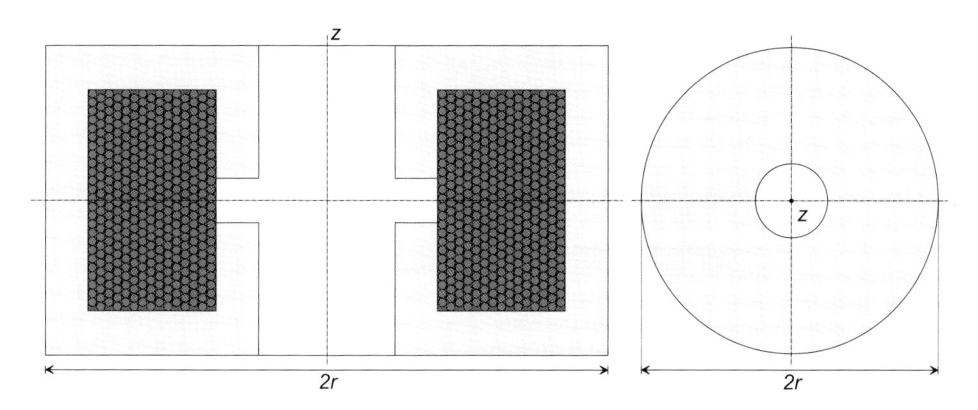


Fig. 5.1 A magnetic lens. Left: side view cross-section. Right: top view in a scale 1:2 compared to the side view. The coil carrying an electric current is surrounded by a ferromagnetic yoke, which, due to the material's permeability, enhances the field strength of the lens (Tsuno, 2009).

action of the lenses on the electron beam is based on electromagnetic fields. Nowadays, transmission electron microscopes mostly employ magnetic electron lenses (see e.g. Fig. 5.1). Electrostatic lenses are normally used in the gun area only. This is reasoned by the fact that the focusing power of magnetic lenses in transmission electron microscopes, typically operated between 100 to 300 kV, is superior compared to the focusing effect of electrostatic lenses (Rose, 2009a). Hence, once the electrons are accelerated to the target energy of the microscope, the magnetic fields of the lenses are used to guide the electron beam in a well defined manner through the microscope column. Besides the lenses, stigmators and deflectors are used to correct for mechanical misalignments of the lenses, as well as for inhomogeneities of the materials the lenses are made of.

Knowing the geometry of a lens, the current that runs through its coil and the ferromagnetic material the yoke and the pole-pieces are made of, it is possible to (approximately) calculate the magnetic field associated with the lens (for details see, e.g. Hawkes and Kasper, 1989a). In the present case, we assume that the fields are known and that they are stationary, i.e. the field strength does not change with time. What remains to be solved in order to characterize the microscope's optical properties is the derivation of the electron trajectories in the microscope column.

## 5.2 Newton's Second Law of Motion and Electron Optics

Let us consider a setup which clearly is an oversimplification of an electron microscope, but which nonetheless provides valuable insight into the functionality of an electron microscope. This setup, depicted in Fig. 5.2, shall consist of an idealized point source which emits electrons of equal energy. The emitted electrons, whose initial momenta we shall know, are then accelerated by an electric field. The electric

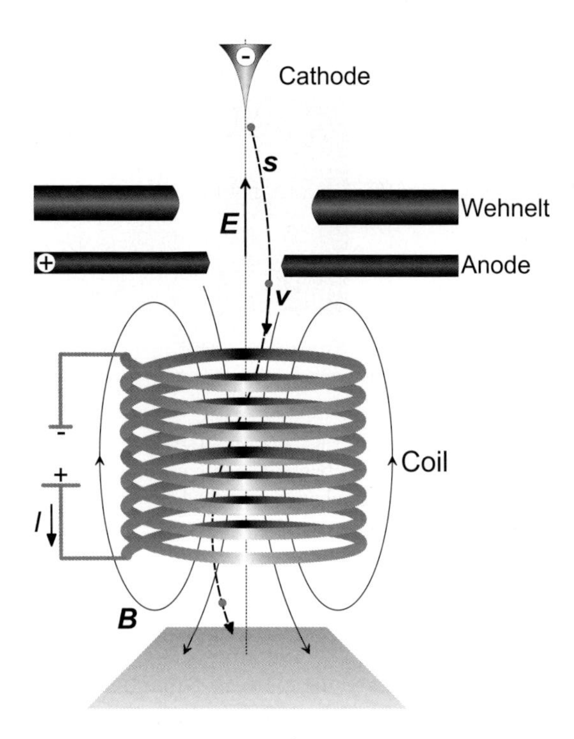


Fig. 5.2 A simple electron-transfer instrument. The fields acting on the electrons emitted by an ideal electron source are the electric field $\boldsymbol{E}$ determined by the voltage between cathode and anode, and the magnetic field described by the magnetic induction $\boldsymbol{B}$. The Lorentz force given in Eq. (5.1) determines the path of an electron through this system.

field shall be defined by the voltage between the source acting as the cathode and the anode. The opening in the anode limits the angular distribution of the electrons, and the Wehnelt on zero potential makes sure that not too many electrons impinge the anode. Beneath the anode, the electrons enter a magnetic field which transfers them to an electron detector. This simple "electron-transfer instrument" can thus be characterized by the electric field $\boldsymbol{E}(\boldsymbol{r})$, which acts as the accelerator, and by the magnetic field described by the magnetic induction $\boldsymbol{B}(\boldsymbol{r})$.

For now, we shall not be concerned with interference effects. Under the assumption that we deal with charged particles rather than with waves, we would like to find out what forces determine the impact positions of the electrons on the detector. For this, we would basically need to know the trajectories along which the electrons propagate towards the detector. We consider the electrons to be point-like particles which are characterized by the rest mass $m_0$ and the charge $q = -e$, where $e$ is the elementary charge. Newton's second law of motion tells us that the trajectory of a particle is determined by the forces acting on the particle. For an electron travelling at a velocity $\boldsymbol{v}$ at a position $\boldsymbol{r} = (x, y, z)$ in space[1], which is exposed to an electric field $\boldsymbol{E}$ and a magnetic induction $\boldsymbol{B}$, it is the Lorentz force $F_{\mathrm{L}}$

$$\boldsymbol{F}_{\mathrm{L}}(\boldsymbol{r}) = q\left[\boldsymbol{E}(\boldsymbol{r}) + \boldsymbol{v}(\boldsymbol{r}) \times \boldsymbol{B}(\boldsymbol{r})\right] \tag{5.1}$$

which is of relevance (Jackson, 1998). It determines the change of the *kinetic momentum* $\boldsymbol{p} = m_0\boldsymbol{v}$ of an electron of mass $m_0$ at a point of its trajectory that it passes at the time $t$

$$\frac{d\boldsymbol{p}}{dt} = \boldsymbol{F}_{\mathrm{L}}. \tag{5.2}$$

For fast electrons where the velocity of the electron becomes a significant fraction of the speed of light $c$, we need to replace the rest mass $m_0$ in Eq. (5.1) by the relativistically corrected mass $m$ given by

$$m = \frac{m_0}{\sqrt{1-\beta^2}} = \gamma m_0, \tag{5.3}$$

with

$$\beta := \frac{|\boldsymbol{v}|}{c} = \frac{v}{c} \quad \text{and} \quad \gamma = \frac{1}{\sqrt{1-\beta^2}}. \tag{5.4}$$

The simple electron optical instrument depicted in Fig. 5.2 illustrates that classical mechanics based on the Lorentz force can be employed to describe the behavior of electrons exposed to electromagnetic fields. The approach to describe the imaging characteristics of a system on the basis of individual trajectories is particularly useful if the characteristics of a few distinguished trajectories are of interest, and especially if the chronological evolution of the electron propagation is of importance, i.e. when the time $t$ an electron passes a point $\boldsymbol{r}$ along a given trajectory is needed. Often, however, the time dependence of the electron propagation is not explicitly needed. Hence, instead of deriving (time-dependent) equation of motions to describe the trajectories of the electrons, it is then more useful to derive path equations which do not depend on time. A solution of a path equation is a geometrical curve which reflects a possible electron trajectory.

## 5.3 The Hamiltonian Analogy

There is a conceptually alternative way to grasp the effect of the electromagnetic fields on the electrons. This alternative approach is based on the analogy between the trajectory of charged particles exposed to electromagnetic fields and the path of light rays in an optical medium. According to its discoverer, William Rowan Hamilton (1831), the principle is known as the *Hamiltonian analogy*. The Hamiltonian

[1] While in the previous chapters $\boldsymbol{r}$ was a two-dimensional vector, here it is a three-dimensional vector.

analogy reasons why the problem of electrons moving in electromagnetic fields can be reduced to a purely optical problem that finds its counterpart in light optics. In fact, it is only for this analogy that we can talk about electron optics.

The Hamiltonian analogy, which is the fundament of the Hamiltonian optics, does not imply that the laws of light optics can directly be applied in order to describe the motion of electrons exposed to electromagnetic fields. The analogy with light optics means that the attention is drawn to the global description of a system and to its geometrical characteristics. Hence, applying light-optical concepts to the investigation of electron motion requires a phenomenological translation. The discussion of the refractive index of electrons, which finds its counterpart in the light-optical refractive index, will exemplarily illustrate this translation.

## 5.4 Geometrical Electron Optics

A large variety of light optical phenomena which are not governed by interference effects can be explained reasonably well by neglecting the wave characteristics of the radiation. The optical problem is then simply reduced to a purely geometrical problem. The branch of optics that deals with such phenomena is called geometrical optics.

In geometrical optics, the wavelength of the radiation is neglected and the energy, which in principle is carried by waves or wave packets within an optical system, is considered to be transported along certain curves that are referred to as rays. We describe a ray by a vector $\boldsymbol{s}(\boldsymbol{r})$, which is a unit vector in direction of the tangent of the curve representing the ray at point $\boldsymbol{r}$ (see Fig. 5.3). For a ray that follows a straight line, the direction of propagation $\boldsymbol{s}(\boldsymbol{r})$ = constant. For a general curve as

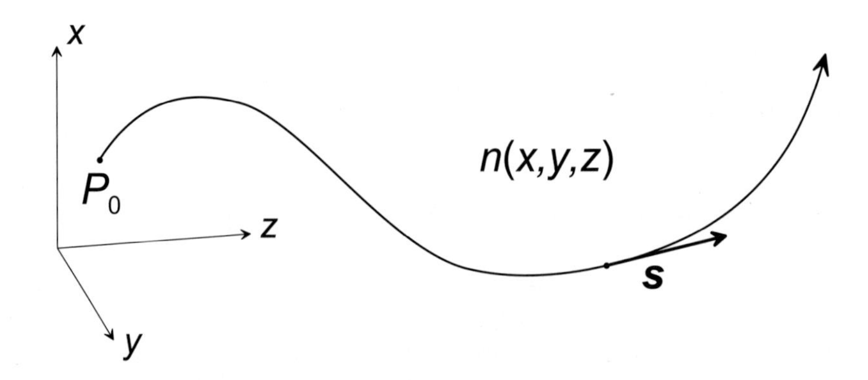


Fig. 5.3 A ray $\boldsymbol{s}(\boldsymbol{r})$ emerging from point $P_0$ in a medium described by its optical refractive index $n(\boldsymbol{r})$. In light optics, a ray describes the path of a photon. In electron optics, a ray describes the geometrical curve along which an electron travels.

in Fig. 5.3, however, $\boldsymbol{s}$ becomes a function of the spatial coordinate $\boldsymbol{r} = (x, y, z)$. The ray depicted in Fig. 5.3 emerges from point $P_0$ and propagates in an optical medium that is characterized by its optical refractive index $n$. For an isotropic inhomogeneous optical medium, the refractive index $n$ is a scalar function of the spatial coordinate $n = n(\boldsymbol{r})$.

On the basis of the Hamiltonian analogy it can be expected that similar to light optics, a variety of electron optical phenomena can be explained by neglecting the wave characteristic of the electron. Such an approach, of course, has its limitations, particularly when it comes to explaining interference effects that arise through the interplay of individual waves or wave packets. But similar to light optics, for a variety of phenomena in electron optics it turns out that neglecting the wave characteristics of the radiation is a suitable and sufficiently accurate approach to describe the global and, in particular, the geometrical characteristics of an optical system. Furthermore, geometrical optics turns out to be convenient for transmission electron microscopy where the wavelength of the electrons is at least one order of magnitude smaller than the spatial extension of the objects that are investigated. The perception of a particle electron travelling along a certain path or trajectory, referred to as a ray, is an acceptable point of view. A point of view which, of course, neglects the wave characteristic of the electron.

Considering a trajectory along which an electron travels implies that there is a time dependence that describes the time $t$ at which the electron passes a point $\boldsymbol{r}$. Hence, the trajectory could be described by an ensemble of points $\boldsymbol{r}(t)$. However, rays, i.e. the fundamental curves of geometrical optics, do not depend on time. A ray described by $\boldsymbol{s}(\boldsymbol{r})$ is a stationary curve that can be imagined as being the trace of an electron.

## 5.5 Electrons in the Electrostatic Field

We start with the treatment of electrons which are exposed solely to an electrostatic field $\boldsymbol{E}(\boldsymbol{r})$. This implies that the Lorentz force in Eq. (5.1) with $\boldsymbol{B} = \boldsymbol{0}$ reduces to the electrostatic force given by

$$\boldsymbol{F}(\boldsymbol{r}) = q\boldsymbol{E}(\boldsymbol{r}) = -e\boldsymbol{E}(\boldsymbol{r}). \tag{5.5}$$

This is the force of an electrostatic field $\boldsymbol{E}$ acting on a charge $q = -e$.

The reason why we differentiate between an electromagnetic and an electrostatic field lies in the fact that an electro*magnetic* field (with $\boldsymbol{B} \neq \boldsymbol{0}$) represents a substantially different electron optical medium compared to an electrostatic field. The differentiation between different types of optical media is not specific to electron optics and finds its analogy in light optics. While for an *inhomogeneous* light optical medium the refractive index $n(\boldsymbol{r})$ is a real scalar function of $\boldsymbol{r}$, a *homogeneous* optical medium is an optical medium whose refractive index does not change with the spatial coordinate, i.e. $n(\boldsymbol{r}) = n =$ constant. Glass of constant density, for instance, is a homogeneous light optical medium. An optical lens of varying density, such as for instance the lens of the human eye, is an inhomogeneous optical medium.

The distinction between isotropic and anisotropic reflects another characteristic of an optical medium. An *isotropic* optical medium is characterized by a refractive index which is independent of the direction a light ray propagates within this medium. Again, glass as an amorphous material is in general an isotropic medium. On the other hand, there are *anisotropic* optical media such as, for example, a calcite crystal whose refractive index depends on the direction $\boldsymbol{s}$ along a light ray propagates within the crystal. In such cases, the optical index is described by a matrix, called indicatrix, which takes account of the anisotropy of the refractive index (Kleber *et al.*, 1990). Hence, for an anisotropic optical medium, the refractive index depends on the direction of the ray, i.e. $n = n(\boldsymbol{s})$, where $\boldsymbol{s}$ is the unit vector describing the tangent and thus the direction of the ray $\boldsymbol{s}$ in a point $\boldsymbol{r}$ (see Fig. 5.3).

Going back to electron optics; while an electron exposed to an electrostatic field behaves like a light ray interacting with an inhomogeneous isotropic optical medium, an electron exposed to an electromagnetic field with $\boldsymbol{B} \neq \boldsymbol{0}$ resembles the situation of a light ray interacting with an inhomogeneous anisotropic optical medium.

### 5.5.1 *Snell's law of refraction*

The parameter that determines the path of a light ray in an optical system is the optical refractive index $n(\boldsymbol{r})$. For now, we assume $n$ to be isotropic. The fact that the refractive index determines the path of a light ray is expressed in Snell's law of refraction, which can be discussed on the basis of Fig. 5.4. A light ray emerging from point $A$ in a homogeneous optical medium, which is described by the refractive index $n_1$, enters a second homogeneous optical medium $n_2$ at point $B$. Snell's law tells us that the angle of incidence $\alpha_1$ and the angle of refraction $\alpha_2$ are related to each other by

$$\boxed{\frac{\sin\alpha_1}{\sin\alpha_2} = \frac{n_2}{n_1}}\,. \tag{5.6}$$

Hence, knowing the boundary plane $z = 0$ and the points $A$ and $B$, i.e. the angle of incidence, as well as the refractive indices $n_1$ and $n_2$, the angle of refraction, which defines the direction $\overrightarrow{BC}$ along the light ray propagates in the second medium, is determined by Eq. (5.6). Describing the motion of electrons in optical terms thus raises the question about the quantity in electron optics which corresponds to the refractive index in light optics.

### 5.5.2 *Snell's law of refraction for electrons*

In order to find the corresponding quantity for an electron exposed to an electrostatic field $\boldsymbol{E}$, we try to picture a situation for an electron that is analogous to the

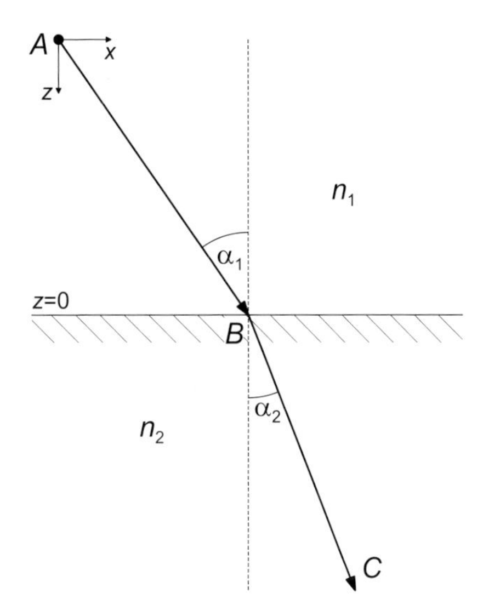


Fig. 5.4 Snell's law of refraction. A light ray emerging from point $A$ in a homogeneous optical medium of refractive index $n_1$ enters a second homogeneous optical medium $n_2$ at point $B$. Given the angle of incidence $\alpha_1$ and the refractive indices $n_1$ and $n_2$, the angle of refraction $\alpha_2$ is given by Snell's law of refraction in Eq. (5.6).

situation of a light ray illustrated in Fig. 5.4. Following the train of thought described by Glaser (1952), we assume that there are two metallic grids at a distance $d$ to each other. The grids are on different electrostatic potentials $\phi_1$ and $\phi_2$ with $\Delta\phi = \phi_2 - \phi_1$. The grids thus generate an electric field $\boldsymbol{E} = \Delta\phi/d$ which shall point in negative $z$-direction (see Fig. 5.5).

An electron travelling in a field-free space approaches the first grid at point $B'$ under an angle of incidence $\alpha_1$. The electron passes from the first grid to point $B''$ of the second grid. While between the grids, the electron experiences the force $-e\boldsymbol{E}$ according to Eq. (5.5). The electrostatic force accelerates the electron in positive $z$-direction. The $x$-component of its velocity remains unaffected by the electrostatic field. However, because the electron gains velocity in $z$-direction between the grids, the electron leaves the second grid at point $B''$ under an angle $\alpha_2$, which is smaller than $\alpha_1$.

For the case that the distance $d \to 0$, Fig. 5.5 approaches the situation depicted in Fig. 5.6, which in its very basic ray characteristics is equivalent to Snell's law of refraction illustrated in Fig. 5.4. Hence, in order to uncover the electron optical quantity that corresponds to the light optical refractive index, the only thing we need to find out is how the two angles $\alpha_1$ and $\alpha_2$ in Fig. 5.6 are related to each other.

Let us start with a more quantitative discussion of Fig. 5.5. An electron of rest mass $m_0$ travels with a velocity $\boldsymbol{v}_1$ from point $A$ along the direction $\overrightarrow{AB'}$ towards

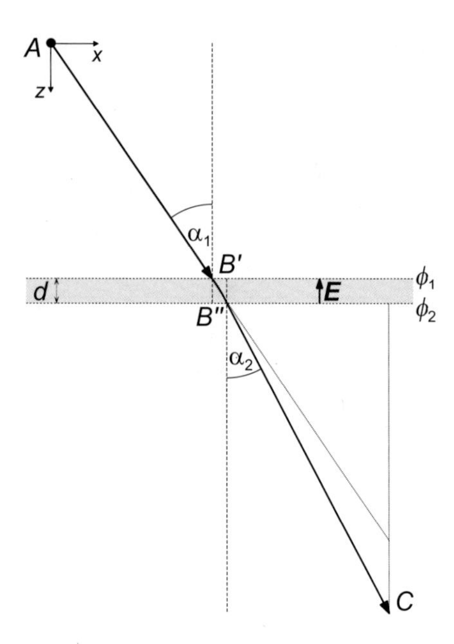


Fig. 5.5 An electron travelling in a field-free space approaches two parallel metallic grids that are on different electrostatic potentials. The voltage between the grids creates an electric field $\boldsymbol{E} = \Delta\phi/d$, which accelerates the electron according to Eq. (5.5). After leaving the second grid and entering in the field-free space again, the direction of the electron's momentum is changed. The angle of incidence on the first grid is $\alpha_1$; the exit angle on the second grid is $\alpha_2$.

point $B'$ on the first grid. The kinetic momentum of the electron is $\boldsymbol{p}_1 = m_0\boldsymbol{v}_1$, which shall correspond to the distance between $A$ and $B'$, i.e. $\boldsymbol{p}_1 = \overrightarrow{AB'}$. Since there is no force acting on the electron, according to Eq. (5.2) $d\boldsymbol{p}_1/dt = 0$, i.e. the momentum of the electron does not change along the trajectory between point $A$ and point $B'$. Once the electron reaches the first grid at point $B'$ under an angle of incidence $\alpha_1$, it is exposed to the electrostatic field $\boldsymbol{E}$. Hence, while travelling to the second grid, the electron experiences the electrostatic force given in Eq. (5.5), which changes the electron's momentum by

$$\frac{d\boldsymbol{p}}{dt} = -e\boldsymbol{E}. \tag{5.7}$$

The acceleration the electron experiences is thus given by $\boldsymbol{a}_z = -e\boldsymbol{E}/m_0$. Since $\boldsymbol{E}$ is directed in negative $z$-direction, the electron is accelerated in positive $z$-direction. Furthermore, the force acts along the $z$-direction only, therefore the acceleration affects only the $z$-component of the electron's momentum. The change of momentum is given by

$$\Delta\boldsymbol{p}_z = -e\boldsymbol{E}\Delta t, \tag{5.8}$$

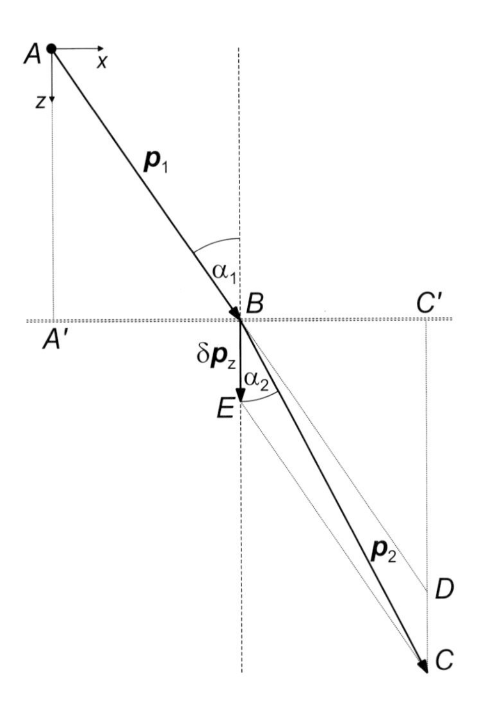


Fig. 5.6 An electron travelling in a field-free space with momentum $\boldsymbol{p}_1$ experiences a sudden change of momentum $\delta\boldsymbol{p}_z$ at point $B$. The momentum after experiencing the momentum change is $\boldsymbol{p}_2 = \boldsymbol{p}_1 + \delta\boldsymbol{p}_z$. This illustrates the same situation as depicted in Fig. 5.5 for the case that $d \to 0$. The angle of incidence on the plane $A'BC'$ is $\alpha_1$ and the exit angle, or the angle of refraction, is $\alpha_2$.

with $\Delta t$ being the period of time the electron needs to travel from point $B'$ to $B''$[2]. The electron's momentum in point $B''$ is then $\boldsymbol{p}_2 = \boldsymbol{p}_1 + \Delta\boldsymbol{p}_z$.

Let us now assume that in Fig. 5.5 $d \to 0$, such that $B' \to B$ and $B'' \to B$, as depicted in Fig. 5.6. Hence, instead of having a continuous change of the electron's momentum between the first and the second grid, the electron experiences an instantaneous change of momentum at point $B$. Similar to the case illustrated in Fig. 5.4, where a light ray changes its direction abruptly at the interface between two optical media $n_1$ and $n_2$, the direction of the electron's trajectory is abruptly changed in point $B$.

[2]The period of time is not of crucial importance for our discussion. Just for completeness, we shall give its value. It can be deduced by analyzing the evolution of the velocity of the electron exposed to the electric field, making the assumption that the initial velocity in $z$-direction as well as the field strength $E$ is known. One obtains that the period of time is given by

$$\Delta t = \frac{1}{a_z}\left(-v_{1,z} + \sqrt{v_{1,z}^2 + 2da_z}\right), \tag{5.9}$$

which depends on the initial momentum $\boldsymbol{p}_1$, i.e. on the $z$-component of the electron's velocity $v_{1,z}$ at point $B'$ and on the acceleration $a_z$.

The change of momentum directed in positive $z$-direction at point $B$ is now $\delta \boldsymbol{p}_z = -e\boldsymbol{E}\delta t$, i.e. $\delta \boldsymbol{p}_z = \overrightarrow{BE}$. The momentum after leaving point $B$ is $\boldsymbol{p}_2 = \overrightarrow{BC}$, which is given by adding the momentum vectors $\boldsymbol{p}_2 = \boldsymbol{p}_1 + \delta \boldsymbol{p}_z$, i.e. $\overrightarrow{BC} = \overrightarrow{AB} + \overrightarrow{BE}$. Since only the $z$-component of the electron's momentum changes in point $B$, the distances $A'B$ and $BC'$ have to be equal. The distance $A'B$ is $|\boldsymbol{p}_1| \sin \alpha_1$, and $BC'$ is $|\boldsymbol{p}_2| \sin \alpha_2$. With $p_1 = |\boldsymbol{p}_1|$ and $p_2 = |\boldsymbol{p}_2|$, we obtain that $p_1 \sin \alpha_1 = p_2 \sin \alpha_2$. This yields the following relation

$$\frac{\sin \alpha_1}{\sin \alpha_2} = \frac{p_2}{p_1} \,. \tag{5.10}$$

Equation (5.10) is equivalent to Snell's law in Eq. (5.6). The analogy between Eq. (5.10) and Eq. (5.6) allows us to write

$$\frac{p_1}{p_2} = \frac{n_{\boldsymbol{E},1}}{n_{\boldsymbol{E},2}} \,, \tag{5.11}$$

from which we obtain that the refractive index of electrons exposed to an electrostatic field $\boldsymbol{E}$ is proportional ($\propto$) to its kinetic momentum $p$

$$\boxed{n_{\boldsymbol{E}} \propto p.} \tag{5.12}$$

The subscript $\boldsymbol{E}$ shall indicate that this relation holds for electrons in the electrostatic fields and that there is no magnetic field present.

For fast electrons we have to consider the relativistic kinetic momentum of the electron given by

$$p = \gamma m_0 v \tag{5.13}$$

with $\gamma$ from Eq. (5.4). This yields again

$$n_{\boldsymbol{E}} \propto p,$$

and since the rest mass $m_0$ is a constant, it can be omitted such that

$$n_{\boldsymbol{E}} \propto \gamma v. \tag{5.14}$$

Equation (5.12) describes the analogy between the light optical refractive index and the corresponding quantity in electron optics. The refractive index of electrons exposed to an electrostatic field is proportional to its kinetic momentum. The constant of proportionality between $n_{\boldsymbol{E}}$ and $p$ is not of importance, since according to Eq. (5.10), only the ratio of the refractive indices is needed to describe the trajectory of an electron. Hence, in the foregoing discussion of electrons exposed to the electrostatic field, we omit the constant of proportionality such that, without loss of generality, $n_{\boldsymbol{E}} = p$. However, it is common to choose a suitable value $p_0$, like for instance the kinetic momentum of the electron in the field-free space of the image plane, in order to have $n_{\boldsymbol{E}}$ dimensionless, i.e. as is the case for the light optical refractive index.

### 5.5.3 *Fermat's principle: The shortest light optical path*

In Secs. 5.5.1 and 5.5.2 we employed Snell's law of refraction to find an analogy between the light optical refractive index and the corresponding quantity in electron optics. What we did not do is actually derive the law of refraction as given in Eq. (5.6). In the following subsection, we show that Snell's law of refraction can be derived from a more general optical principle called Fermat's principle, which finds a counterpart in electron optics. Fermat's principle provides a means to geometrically describe a ray in a general, i.e. not necessarily in a homogeneous, optical medium. Finding then a corresponding relation for electrons exposed to an electrostatic field makes it possible to base the analogy between particle optics and light optics on a deeper level.

Let us start once again with two homogeneous optical media described by their refractive indices $n_1$ and $n_2$, as depicted in Fig. 5.7. The two media have a common boundary plane, which we describe by $z = 0$. A light ray emerges from point $P_0$ in $n_1$ and travels to point $P_1$ at the boundary. From $P_1$ it proceeds to point $P_2$ in $n_2$. For now, we treat the location of point $P_1$ on the plane $z = 0$ as a variable parameter which we are meant to determine. The segment between $P_0$ and $P_1$ in the medium $n_1$ is $s_1$ and the segment between $P_1$ and $P_2$ in $n_2$ is $s_2$ (see Fig. 5.7). Hence, $s_1$ and $s_2$ are the geometrical distances the ray travels in medium $n_1$ and $n_2$, respectively.

Figure 5.7 allows us to introduce a quantity which is called the *optical path length* $S$; the optical path length of a ray in a homogeneous optical medium $i$ is the geometrical distance $s_i$, multiplied by the corresponding refractive index $n_i$. According to Fig. 5.7, the optical path length in the first medium is $n_1 s_1$ and the

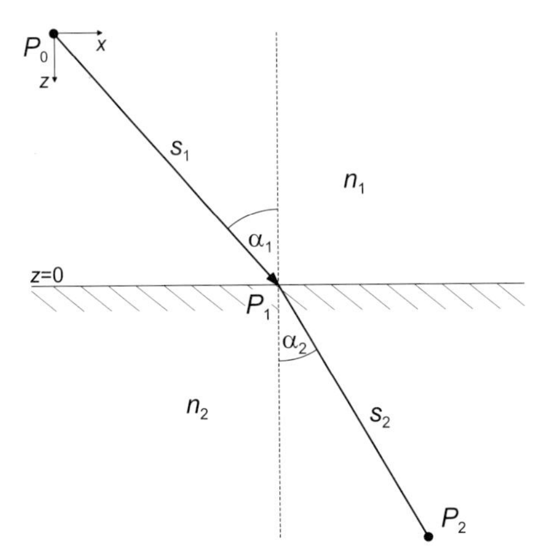


Fig. 5.7 Definition of the optical path length $S$. While the distance between point $P_0$ and point $P_1$ in the optical medium $n_1$ is $s_1$, the optical path length is defined as $n_1 s_1$. The total optical path between point $P_0$ and point $P_2$ is $S = n_1 s_1 + n_2 s_2$.

optical path length in the second medium is $n_2 s_2$. The total optical path length of the ray connecting point $P_0$ and $P_2$, i.e. $P_0P_1P_2$ is thus

$$S = n_1 s_1 + n_2 s_2. \tag{5.15}$$

With the definition of the optical path length $S$, *Fermat's principle* can be formulated as follows:

**Fermat's principle**: a light ray follows the geometrical curve that corresponds to the smallest optical path length $S$:

$$S \rightarrow \text{minimum}. \tag{5.16}$$

Because the optical path length is the geometrical distance weighted by the index of refraction, the minimum of the optical path length is in general different from the shortest distance connecting two points in different optical media[3]. Fermat's principle thus imposes a strong condition on the optical path of a light ray, i.e. the geometrical curve of a light ray is in fact defined by Eq. (5.16). What does Fermat's principle imply for the ray depicted in Fig. 5.7? Under the assumption that both points $P_0$ and $P_2$ are fixed and the optical indices $n_1$ and $n_2$ are known, we employ Fermat's principle to derive the position of $P_1$ on the plane $z = 0$. Figure 5.7 illustrates that solving the position of point $P_1$ on $z = 0$ is equivalent to finding the angles $\alpha_1$ and $\alpha_2$.

The optical path length between point $P_0$ and $P_2$ is given in Eq. (5.15). It can be seen that $S$ is a function of $s_1$ and $s_2$, i.e. $S = S(s_1, s_2)$. The segment $s_1$ is given by the distance between $P_0P_1$ and $s_2$ is given by $P_1P_2$. Fermat's principle tells us that that a light ray follows the curve that corresponds to an extremum of the optical path $S$ such that $S = S(s_1, s_2) \rightarrow$ minimum. Since $P_0$ and $P_2$ are considered to be fixed, we can vary the position of $P_1$ on $z = 0$. Let us compare the optical path length related to a point $P_1$ on $z = 0$ with the optical path length corresponding to a situation where $P_1$ is replaced by $P_1'$, which, according to Fig. 5.8, shall be located at a distance $\delta x$ from $P_1$.

Because $S$ is a function of $s_1$ and $s_2$, we can find an extremum of $S$ by setting the total differential of $S(s_1, s_2)$ equal zero

$$dS = \frac{\partial S}{\partial s_1} ds_1 + \frac{\partial S}{\partial s_2} ds_2 \equiv 0. \tag{5.17}$$

Employing the partial derivatives according to Eq. (5.15) yields

$$dS = n_1 ds_1 + n_2 ds_2 = 0. \tag{5.18}$$

[3]Fermat's principle can be formulated in an alternative way: a light ray always follows the path that minimizes the time. Hence, since the speed of light in an optical medium $i$ is $c/n_i$, where $c$ is the speed of light in vacuum, the time needed to travel a certain geometrical distance depends on $n_i$.

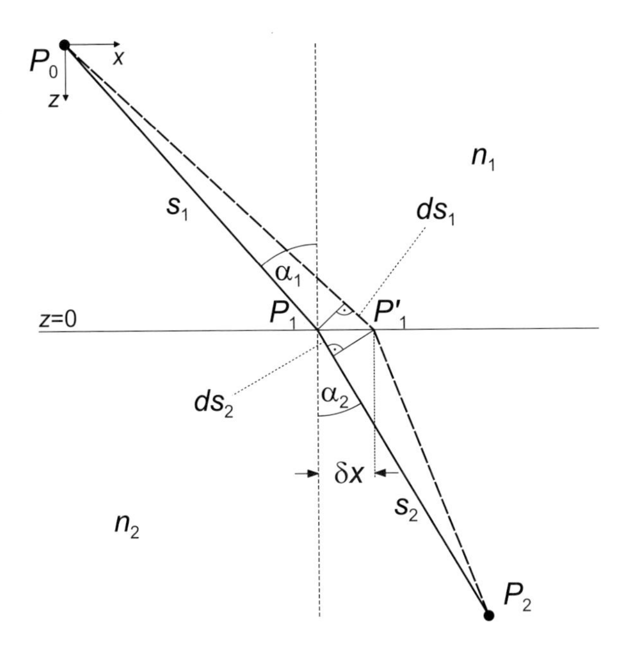


Fig. 5.8 Application of Fermat's principle to determine the optical path between point $P_0$ and point $P_2$. The shortest optical path goes through point $P_1$. Variation of the position of $P_1$ to $P_1'$ increases the optical path length by $dS$. The optical path length is minimized if the relation between the angle of incidence $\alpha_1$ and the exit angle $\alpha_2$ is in agreement with Snell's law, given in Eq. (5.6).

On the other hand, Fig. 5.8 illustrates that by moving point $P_1$ by $\delta x$ to $P_1'$, the change of the geometrical distances $s_1$ and $s_2$ are

$$\begin{aligned} ds_1 &= \delta x \sin \alpha_1 \\ ds_2 &= -\delta x \sin \alpha_2. \end{aligned} \tag{5.19}$$

The negative sign in $ds_2$ reflects the fact that the geometrical distance $s_2$ becomes shorter when $P_1$ is moved to $P_1'$, whereas the segment $s_1$ increases and thus $ds_1 > 0$. Substituting the relations in Eq. (5.19) in Eq. (5.18) yields

$$dS = \delta x (n_1 \sin \alpha_1 - n_2 \sin \alpha_2) = 0. \tag{5.20}$$

Setting the term in brackets equal zero allows us to derive the condition for the extremum of $S$:

$$\frac{\sin \alpha_1}{\sin \alpha_2} = \frac{n_2}{n_1}.$$

This result is identical to Snell's law of refraction given in Eq. (5.6). Hence, Fermat's principle of the shortest optical path length $S$ reveals Snell's law of refraction[4].

[4]Using the alternative formulation of Fermat's principle, i.e. that the ray is given by the path that minimizes the time, yields the same result. For the case of Fig. 5.7, the time needed for the light ray to travel from $P_0$ to $P_2$ is $t = s_1 \frac{n_1}{c} + s_2 \frac{n_2}{c}$ (see footnote 3). Minimizing the time leads to the relation given in Eq. (5.18) and eventually to Snell's law in Eq. (5.6).

Admittedly, the situations depicted in Figs. 5.6 and 5.7 are of a rather exemplary style. What else apart from Snell's law can we learn from Fermat's principle?

Often, optical media are not homogeneous. This is particularly true for electron optics. Electromagnetic fields — the optical media of electrons — do not have sharp borders and the forces they exert on the electrons strongly depend on the position of the electron in the field. Hence, if we intend to expand Fermat's principle to electron optics, where an electromagnetic field rather than a material of well defined surface corresponds to the optical medium, it seems to be necessary to derive Fermat's principle for an inhomogeneous optical medium. How can we apply Fermat's principle of the shortest optical path to a light ray in an inhomogeneous optical medium?

An inhomogeneous optical medium with $n = n(\boldsymbol{r})$ can be approached by a medium which has a locally constant refractive index that discretely changes at virtual interfaces between areas wherein $n$ is constant. This is depicted in Fig. 5.9. The light ray, which connects point $P_0$ with $P$, runs trough an optical medium whose refractive index is locally constant. The light ray is thus determined by the series of locally homogeneous optical areas $n_i$ and by the geometrical length $\Delta s_i$ of each of the segments. In analogy to Eq. (5.15), the optical path length or simply the

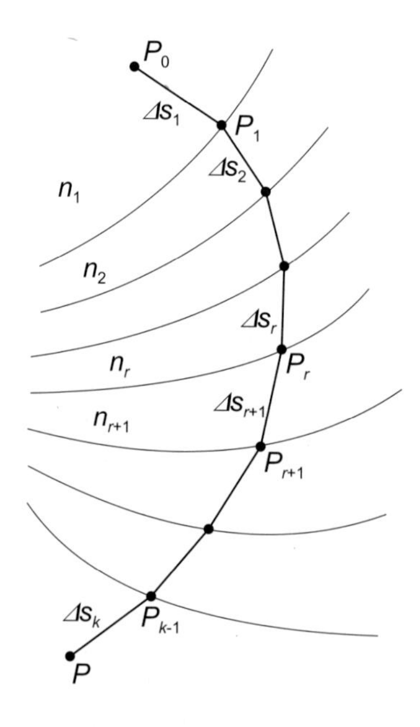


Fig. 5.9 Derivation of the optical path length of an inhomogeneous optical medium. An inhomogeneous optical medium is modelled by a locally homogeneous optical medium whose refractive index changes stepwise. The shortest optical path through this optical medium is determined by locally applying Snell's law of refraction at the boundaries where the refractive index changes.

optical length $S$ of the ray depicted in Fig. 5.9 can thus be written as

$$S = n_1 \Delta s_1 + n_2 \Delta s_2 + ... + n_k \Delta s_k = \sum_{i=1}^{k} n_i \Delta s_i, \tag{5.21}$$

where each segment $\Delta s_i$ is weighted by the corresponding index of refraction $n_i$. According to Fermat's principle, the ray connecting point $P_0$ with $P$ corresponds to the geometrical curve of shortest optical path length. This curve can be derived by locally applying Snell's law of refraction at each of the virtual interfaces. Starting with the first segment $\Delta s_1$ connecting point $P_0$ with $P_1$, the angle of incidence in respect to the boundary between the areas $n_1$ and $n_2$ shall be given as a boundary condition. Applying Snell's law determines the first angle of refraction, which also defines the angle of incidence in respect to the second boundary. This reasoning can be repeated for all the boundaries $n_r$ $n_{r+1}$ until point $P$ is reached.

Making the transition to a real inhomogeneous optical medium, i.e. an optical medium of steadily varying refractive index $n = n(\boldsymbol{r}) = n(x, y, z)$, the optical path length along a curve $\boldsymbol{s}(\boldsymbol{r})$ between a point $P_0$ and an arbitrary point $P$ can be written analogously to Eq. (5.21), replacing, however, the sum with an integral

$$S(P_0, P) = \int_{P_0}^{P} n(\boldsymbol{r}) ds. \tag{5.22}$$

Equation (5.22) is a general description of the optical path length in an inhomogeneous medium connecting point $P_0$ with $P$. With the expansion of the optical path for a inhomogeneous optical medium, Fermat's principle can be formulated as

$$\boxed{S(P_0, P) = \int_{P_0}^{P} n(\boldsymbol{r}) ds \rightarrow \text{minimum.}} \tag{5.23}$$

This summarizes Fermat's principle: from all the possible curves connecting point $P_0$ with $P$, a light ray follows the path of shortest optical path length. In the following we will see that Fermat's principle of the shortest optical path, which uniquely defines the geometrical curve of a light ray, has a counterpart in charged-particle and electron optics. This is a direct consequence of the Hamiltonian analogy.

### 5.5.4 *The reduced principle of least action*

Following the discussion of Fermat's principle in light optics, we can take the Hamiltonian analogy a step further. The electron optical quantity that corresponds to the optical path length in light optics is given by a quantity called *action* or *action integral*, which we also denote by $S$. The light-optical refractive index $n(\boldsymbol{r})$ in Eq. (5.22) is replaced by its electron optical counterpart, which essentially is the kinetic momentum of the electron $p(\boldsymbol{r}) = p(x, y, z)$ given in Eq. (5.13). This simple replacement yields the action integral[5]

$$\boxed{S(P_0, P) = \int_{P_0}^{P} p(\boldsymbol{r})ds.} \tag{5.24}$$

As the kinetic momentum $\boldsymbol{p}$ of the electron is parallel to the ray $\boldsymbol{s}$, we can omit the vectorial notation for $\boldsymbol{p}$ and $d\boldsymbol{s}$ in the above equation. While Fermat's principle, given in Eq. (5.23), defines the path of the light ray as the minimum of the optical path length, the corresponding principle in electron optics, which defines the path of an electron, is called the *reduced principle of least action*. In analogy to Eq. (5.23), it can be written as

$$\boxed{S(P_0, P) = \int_{P_0}^{P} p(\boldsymbol{r})ds \rightarrow \text{minimum.}} \tag{5.25}$$

Equation (5.25) expresses that from all the curves connecting point $P_0$ with $P$ an electron follows the path of least action.

The (reduced) principle of least action is not specific to electrons. Indeed, it was discovered by Leonhard P. Euler in the 18th century. Euler deduced that the trajectory a particle follows is distinguished from all other curves by the fact that the integration of the velocity $\boldsymbol{v}(\boldsymbol{r})$ along this curve is minimal, i.e.

$$S(P_0, P) = \int_{P_0}^{P} v(\boldsymbol{r})ds \rightarrow \text{minimum.} \tag{5.26}$$

Provided that the speed of the particle is small compared to the speed of light, i.e. $v \ll c$, this relation is equivalent to Eq. (5.25).

The action integral $S$ in Eq. (5.24) corresponds to the integral along any possible curve connecting $P_0$ with $P$. It is a function of the source point $P_0 = (x_0, y_0, z_0)$ and the end point $P = (x, y, z)$. For the case that $S \rightarrow$ minimum, the corresponding curve reflects the electron trajectory which connects point $P_0$ with $P$. Generally, a trajectory describes the location of a particle as a function of time. Hence, the actual action integral is thus an integral over the period of time the particle needs to travel from point $P_0$ to $P$. However, since we deal with static fields, the integral along the trajectory in Eq. (5.24) is a reduced, time-independent form of the action (Hawkes and Kasper, 1989a), and the corresponding condition as written in Eq. (5.25) is therefore called the *reduced* principle of least action. Hence, if $S \rightarrow$ minimum, the corresponding curve $\boldsymbol{s}$ along which the integral is taken simply corresponds to the geometrical curve along the particle moves. As above, we call this time-independent geometrical curve associated with a particle trajectory a ray $\boldsymbol{s}$.

[5] We emphasize that this "definition" of the refractive index is limited to the case of electrons in the electrostatic field, and that we arbitrarily set the constant of proportionality between $n_E \propto p$ equal one. Similarly well can we define the constant of proportionality as $1/p_0$, as explained in Sec. 5.5.2.

### 5.5.5 *The point eikonal*

The (reduced) action integral in Eq. (5.24) characterizes any curve which connects point $P_0$ with a point $P$, with a quantity $S$. In optical terms, this time-independent function $S(P_0, P)$ can be called the *point characteristic function* or the *point eikonal*. However, it needs to be pointed out that in order to translate Eq. (5.24) into optical terms, one would need to normalize the kinetic momentum $p(\boldsymbol{r})$ by the kinetic momentum $p_0$ of the electron in the field-free space, like for instance in the image plane (Hawkes and Kasper, 1989a). This was pointed out in Sec. 5.5.2. Only this normalization warrants that the refractive index $n_{\boldsymbol{E}} \propto p$ is dimensionless. Yet, the normalization does not influence the basic relation between $S$ and $p(\boldsymbol{r})$.

Figure 5.10 illustrates a point $P_0$ from which a set of rays $\boldsymbol{s}_1, \boldsymbol{s}_2, ...\boldsymbol{s}_k, ...\boldsymbol{s}_n$ emerges. Each of the rays $\boldsymbol{s}_k$ connects the source point $P_0$ with an endpoint $P$, which is specific for $\boldsymbol{s}_k$, such that the corresponding curve minimizes the action integral between $P_0$ and $P$. This, in fact, is the condition that the curves $\boldsymbol{s}_k$ are rays. Such a set of rays $\boldsymbol{s}_k$ as depicted in Fig. 5.10 shall be referred to as a *pencil*. A pencil shall be continuous in such a way that there is no space between individual rays.

According to Eq. (5.24), a point $P$ on a ray $\boldsymbol{s}_k(\boldsymbol{r})$ is characterized in respect to the source point $P_0$ by a certain value of the point characteristic function $S$. The value of the point characteristic function $S$ relating point $P_0$ with point $P$ on the ray $\boldsymbol{s}_k$ shall have a specific value denoted by $S_1$. Starting again in $P_0$ but following now another ray $\boldsymbol{s}_{k+1}$, the same value of $S$, i.e. $S_1$, is obtainable for a point $P'$ on the adjacent ray $\boldsymbol{s}_{k+1}$. Indeed, each ray contains a distinct point $P$ for which the point characteristic function in Eq. (5.24) yields $S = S_1$. The set of points $P$ for which $S = S_1$ describes a surface, which, according to Eq. (5.24), is fully determined

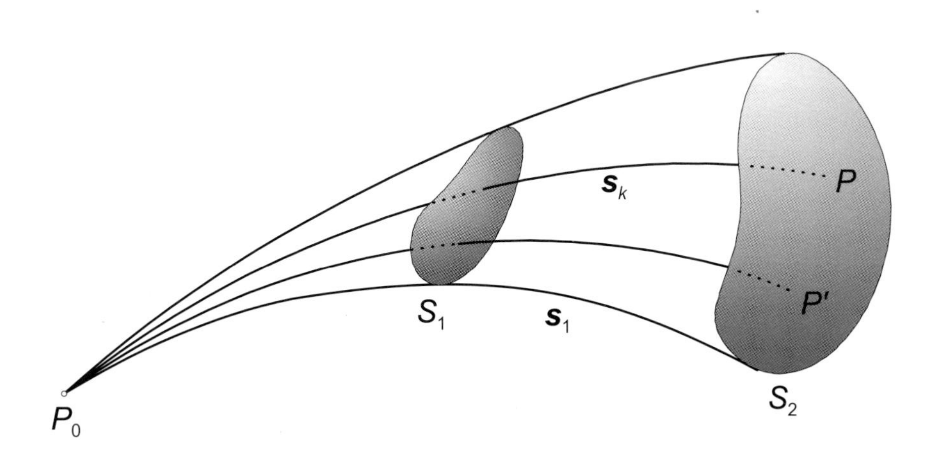


Fig. 5.10 A pencil of rays $\boldsymbol{s}_k(\boldsymbol{r})$ emerging from point $P_0$. $S_1$ and $S_2$ are surfaces of constant action corresponding to geometrical wave surfaces.

by the requirement that the integral $S$ in Eq. (5.24) from point $P_0$ along any ray $\boldsymbol{s}_k$ to this surface is constant. Such a surface, which is defined by

$$\boxed{S(P_0, P) = \text{constant},} \tag{5.27}$$

is called a surface of constant (reduced) action. In optical terms, relating $S$ with the point eikonal and the point characteristic function, the surfaces $S$ = constant are referred to as geometrical wave surfaces. Each value of $S$, i.e. $S_1, S_2, ...$ defines a specific *geometrical wave surface*.

In the following we would like to find a connection between the geometrical wave surfaces and rays that form a pencil. For this, let us assume there is a surface $S_1$ which contains a point $P_1'$ and an adjacent surface referred to as $S_2$, with $S_2 = S_1 + \Delta S$. The surface $S_2$ shall contain a point $P_2$ (see Fig. 5.11). We consider the point $P_2$ to be a fixed point given by the intersection of ray $\boldsymbol{s}'$ with $S_2$. The electron's momentum on $S_1$ is $\boldsymbol{p}_1$ and on $S_2$ it is $\boldsymbol{p}_2$, with $\boldsymbol{p}_2 = \boldsymbol{p}_1 + \Delta\boldsymbol{p}$. We would like to determine the location of the point $P_1'$ on the surface $S_1$, i.e. the location where the ray $\boldsymbol{s}'$ intersects $S_1$. What determines point $P_1'$ on $S_1$ if point $P_2$ on $S_2$ is known?

The geometrical curves that correspond to actual rays are given by the condition in Eq. (5.25). This implies that the integral in Eq. (5.24) along $\boldsymbol{s}$ between two points on a ray $\boldsymbol{s}$ is a minimum. Hence, if we ask for a ray that connects $S_1$ with $S_2$, the increment of $\Delta S$ must be a minimum. According to Eqs. (5.21) and (5.24), the increment $\Delta S$ of the characteristic function $S$ for the transition from $S_1$ to $S_2$ is given by (see Fig. 5.11)

$$\Delta S = \boldsymbol{p}_2 \Delta \boldsymbol{s}' = (\boldsymbol{p}_1 + \Delta \boldsymbol{p})\Delta \boldsymbol{s}'. \tag{5.28}$$

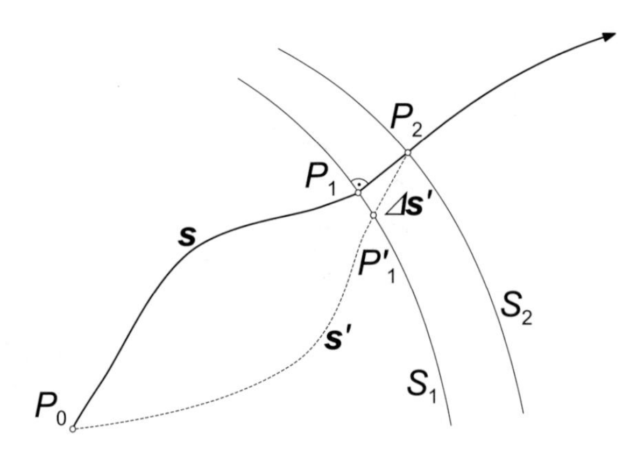


Fig. 5.11 Two geometrical wave surfaces. The optical path between the surfaces is given by minimizing the segment $\Delta \boldsymbol{s}'$. This is fulfilled if the optical path corresponds to the normal of $S_1$ in $P_1$.

Neglecting terms of higher order, i.e. $\Delta \boldsymbol{p} \Delta \boldsymbol{s}' \approx 0$, Eq. (5.28) can be rewritten as

$$\Delta S = \boldsymbol{p}_1 \Delta \boldsymbol{s}'. \tag{5.29}$$

Since the vector $\boldsymbol{s}$ in a point at position $\boldsymbol{r}$ corresponds to the normalized tangent of the corresponding particle trajectory in $\boldsymbol{r}$, $\boldsymbol{s}$ is parallel to the kinetic momentum $\boldsymbol{p}$. Therefore, we can omit the vectors and write

$$\Delta S = p_1 \Delta s', \tag{5.30}$$

with $p_1 = |\boldsymbol{p}_1|$ and $\Delta s' = |\Delta \boldsymbol{s}'|$. The (reduced) principle of least action tells us that the increment $\Delta S$ must be a minimum if the curve connecting $S_1$ with $S_2$ shall be a ray, i.e. $\Delta S \to$ minimum. The minimum of $\Delta S$ is obtained for the smallest value of the segment $\Delta s'$. This is the case if the point $P_1'$ is the base point of the normal of $S_1$ through $P_2$. The answer to the question about the location of point $P_1'$ is depicted in Fig. 5.11. The ray that goes through point $P_2$ on $S_2$ goes through point $P_1'$ if $P_1' \to P_1$ on $S_1$, and $\boldsymbol{s}' \to \boldsymbol{s}$. Hence, the solution is ray $\boldsymbol{s}$ in Fig. 5.11. It needs to be emphasized that the ray $\boldsymbol{s}$ between the surfaces $S_1$ and $S_2$ is determined by the surface normal of $S_1$ in $P_1$. Since we derived this result without any restrictions concerning its validity, it applies for any geometrical wave surface $S$ and any point $P$. In other words, in an electrostatic field the rays are orthogonal to the geometrical wave surfaces. Furthermore, since in an electrostatic field the direction of the kinetic momentum coincides with the direction of $\boldsymbol{s}$, the kinetic momentum of the electron $\boldsymbol{p}$ and the surfaces $S$ are related by

$$\boxed{\boldsymbol{p} = \nabla S,} \tag{5.31}$$

where $\nabla$ denotes the gradient of $S$ in respect to $\boldsymbol{r} = (x, y, z)$ (Glaser, 1952). Eq. (5.31) relates the kinetic momentum of the electrons in an electrostatic field to the geometrical wave surfaces. The unit vectors $\boldsymbol{s}$ describing a ray are thus given by

$$\boldsymbol{s} = \frac{\nabla S}{|\nabla S|}. \tag{5.32}$$

Moreover, Eq. (5.31) provides us with a mean writing the cartesian components of the kinetic momentum of the electron in point $P$ in a position $(x, y, z)$ as follows

$$p_x = \frac{\partial S}{\partial x}, \quad p_y = \frac{\partial S}{\partial y} \quad \text{and} \quad p_z = \frac{\partial S}{\partial z}. \tag{5.33}$$

In order to derive the components of the kinetic momentum of the electron in point $P_0$ (see Fig. 5.11), we can perform an analysis as shown above, however, in the opposite direction, i.e. from $S_2$ to $S_1$. The cartesian components of the kinetic momentum $\boldsymbol{p}_0$ of the electron in point $P_0$ can then be written as

$$p_{0,x} = -\frac{\partial S}{\partial x}, \quad p_{0,y} = -\frac{\partial S}{\partial y} \quad \text{and} \quad p_{0,z} = -\frac{\partial S}{\partial z}. \tag{5.34}$$

The fact that the variation $\Delta S$ is performed in the opposite direction of $\boldsymbol{s}$ justifies the minus signs in the expression of Eq. (5.34). Equations (5.33) and (5.34) express the kinetic momentum of the electron as a function of the point eikonal, i.e. as a function of an optical quantity. Furthermore, the total derivative of $S(P_0, P) = S(x_0, y_0, z_0, x, y, z)$ is given by

$$dS = \frac{\partial S}{\partial x}dx + \frac{\partial S}{\partial y}dy + \frac{\partial S}{\partial z}dz + \frac{\partial S}{\partial x_0}dx_0 + \frac{\partial S}{\partial y_0}dy_0 + \frac{\partial S}{\partial z_0}dz_0\,, \tag{5.35}$$

which yields by substituting the partial derivatives by the $p_i$-terms from Eqs. (5.33) and (5.34)

$$dS = p_x dx + p_y dy + p_z dz - (p_{0,x} dx_0 + p_{0,y} dy_0 + p_{0,z} dz_0)\,. \tag{5.36}$$

This relation can be rewritten in vectorial notation, yielding

$$dS = \boldsymbol{p} d\boldsymbol{r} - \boldsymbol{p}_0 d\boldsymbol{r}_0\,. \tag{5.37}$$

This is the reduced, i.e. time-independent, form of Hamilton's central equation. The above argumentation shows that it is closely related to Eq. (5.31). Indeed, because it makes a connection between the refractive index and the geometrical wave surfaces, Eq. (5.31) is of fundamental importance in geometrical (electron) optics. Replacing the kinetic momentum $p$ with the refractive index $n$ and setting the constant of proportionality in $p \propto n_{\boldsymbol{E}}$ equal one, we directly obtain from Eq. (5.31)

$$\boxed{(\nabla S)^2 = n_{\boldsymbol{E}}^2(\boldsymbol{r})\,.} \tag{5.38}$$

This relation is known as the *eikonal equation* (Born and Wolf, 2001). Equation (5.38) shows that the refractive index of an (electron) optical medium determines the geometrical wave surfaces.

An important aspect about Eqs. (5.31)–(5.38) is that they relate quantities of mechanics, i.e. the kinetic momentum (5.31) or the geometrical curve associated with an electron trajectory (5.32), with an optical quantity, namely the geometrical wave surface which for its part describes the geometrical behavior of an optical system. As shown above, in order to describe the rays emerging from point $P_0$ in Fig. 5.10, it is feasible to determine the curve of each individual ray $\boldsymbol{s}$ by applying classical mechanics. The family of geometrical wave surfaces $S =$ constant are then given by the set of planes that intersect the rays perpendicularly. Yet, the reciprocal argument is equally legitimate; the eikonal or point characteristic function $S(P_0, P)$ defines the pencil in such a way that the individual rays are normal to the surfaces $S(P_0, P) =$ constant. This is the optical description of the pencil. The function $S$ thus fully characterizes the pencil and it is for this reason that it is called a characteristic function.

Instead of dealing with individual electron trajectories, the geometrical wave surfaces allow us to describe the geometrical evolution of an entire pencil. An illustration of this concept is given in Figs. 5.12 and 5.13. An idealized monochromatic point source in point $P_0$ emits electrons isotropically in radial direction. Considering the individual rays denoted by $\boldsymbol{s}_k$, Fig. 5.12 reveals that these rays have the point $P_0$ in common. The rays, being normal to the geometrical wave surface, define a spherical wave surface (or wave front) $S$. For the case that electrons are only

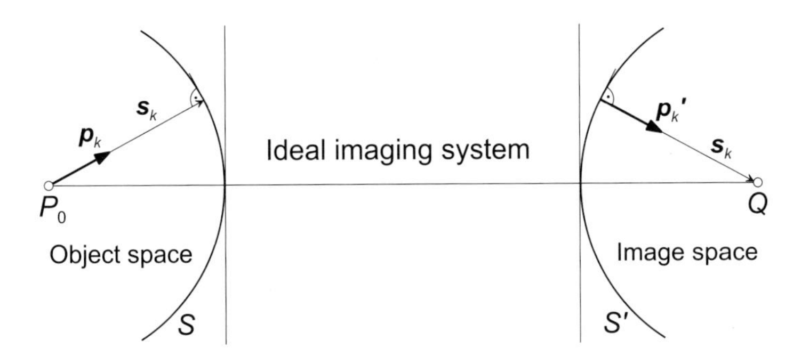


Fig. 5.12 An ideal imaging system images point $P_0$ into a point $Q$. The wave surfaces $S$ and $S'$ in the object and in the image space are spherical.

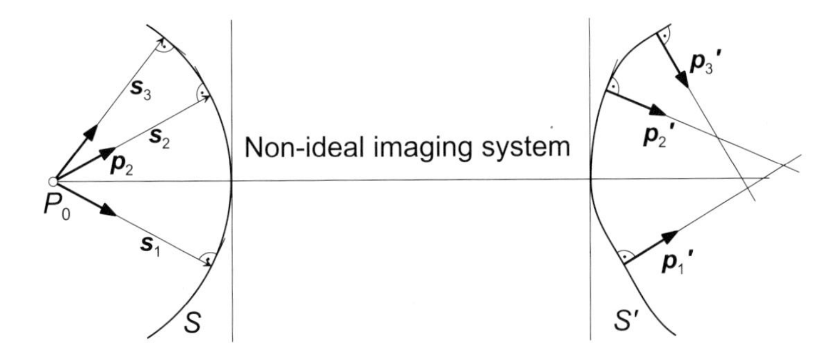


Fig. 5.13 A non-ideal imaging system does not image point $P$ into a corresponding stigmatic image point $Q$. The wave surface $S'$ in the image space is not spherical.

emitted into a limited solid angle, as depicted in Fig. 5.12, the wave surface is a segment of a sphere.

Let us assume there is a perfect imaging system which translates all the rays emitted from point $P_0$ into an image space where they intersect in one single point. This common point of intersection $Q$ is the image of $P_0$. Under the condition that all the rays in the image space go through point $Q$, point $Q$ is a stigmatic image of the source point $P_0$. The condition that warrants that there is a stigmatic image point $Q$ is equivalent to the requirement that the wave surface $S'$ in the image space is spherical. If this condition is violated, the rays do not intersect in one point. In this case, there is no distinct point in the image space that is a stigmatic image of the source point $P_0$. This is illustrated in Fig. 5.13. Since the wave surface in the image space is not spherical, the individual rays intersect in a set of points rather than in a single one. It is clear that this is not an ideal imaging system.

## 5.6 Electrons in the Stationary Electromagnetic Field

So far, we have dealt with electrons in an electrostatic field $\boldsymbol{E}(\boldsymbol{r})$. We saw that the electrostatic field can be considered as an isotropic optical medium where the kinetic momentum vectors of the electrons are orthogonal to the geometrical wave surfaces. Since most lenses in present-day transmission electron microscopes are magnetic lenses, we would like to expand this brief introduction about geometrical electron optics to a general, but still stationary, electromagnetic field defined by the electrostatic field vector $\boldsymbol{E} \neq \boldsymbol{0}$ and the magnetic induction $\boldsymbol{B} \neq \boldsymbol{0}$. We will see that the conceptual expansion from electrostatic to electromagnetic fields is quite straightforward. However, the physical interpretation changes in some very fundamental aspects.

First, we need to introduce two additional quantities; the electrostatic potential $\phi$ and the (stationary) vector potential $\boldsymbol{A}$. The magnetic induction $\boldsymbol{B}$ is related to the vector potential $\boldsymbol{A}$ by

$$\boldsymbol{B}(\boldsymbol{r}) = \mathrm{curl}\boldsymbol{A}(\boldsymbol{r}) \tag{5.39}$$

and the electrostatic potential is related to the electrostatic field by

$$\boldsymbol{E}(\boldsymbol{r}) = -\nabla\phi(\boldsymbol{r}). \tag{5.40}$$

The operation curl describes the operator $\vec{\nabla}\times$, which in cartesian coordinates can be expressed as $\vec{\nabla} = (\partial/\partial x, \partial/\partial y, \partial/\partial z)$. The symbol $\times$ denotes the cross product. Since the fields $\boldsymbol{E}$ and $\boldsymbol{B}$ are related to $\phi$ and $\boldsymbol{A}$ by different forms of differentiations, Eqs. (5.39) and (5.40) do not uniquely define $\phi$ and $\boldsymbol{A}$. Adding a constant to $\phi$ and/or $\boldsymbol{A}$ does not change $\boldsymbol{E}$ or $\boldsymbol{B}$. Hence, whereas $\boldsymbol{E}$ and $\boldsymbol{B}$ have a clear physical meaning, which for instance is expressed in the Lorentz force Eq. (5.1), the potentials $\phi$ and $\boldsymbol{A}$ do not have a physical significance, i.e. they are not gauge-invariant[6].

### 5.6.1 *Principle of Maupertius*

Fermat's principle of geometrical light optics given in Eq. (5.23) and the reduced principle of least action in Eq. (5.25) are two different forms of variational principles. While Fermat's principle provides an intuitively comprehensible way of describing the path of a light ray, the reduced principle of least action, although based on the same formalism, is conceptually more abstract. This, in particular, becomes apparent when we relate the index of refraction to the kinetic momentum of the electron. Here, we would like to go a step further. In order to include the magnetic field in our introduction to geometrical electron optics, it is necessary to expand the reduced principle of least action.

[6]In the dynamic case when the fields depend on the time, Eq. (5.39) remains valid but Eq. (5.40) would need to be rewritten as follows:

$$\boldsymbol{E}(\boldsymbol{r},t) = -\nabla\phi(\boldsymbol{r},t) + \frac{\partial}{\partial t}\boldsymbol{A}(\boldsymbol{r},t).$$

In its very basic form, similar to Eqs. (5.23) and (5.25), the time-dependent version of the principle of least action can be written as

$$S_t := \int_{t_0}^{t} L(\boldsymbol{r}, \dot{\boldsymbol{r}}, t')dt' \to \text{extremum}, \tag{5.41}$$

where the variational function $L$ depends on the position of the particle $\boldsymbol{r}$, its velocity $\dot{\boldsymbol{r}} = d\boldsymbol{r}/dt = \boldsymbol{v}$ in point $\boldsymbol{r}$ and on the time $t$ it passes the position $\boldsymbol{r}$. We employ $S_t$ to describe the time-dependent action, and thus differentiate it from the (time-independent) reduced action integral $S$. Equation (5.41) is based on a particle trajectory described by $\boldsymbol{r}(t)$, which describes the position $\boldsymbol{r}$ of the charged particle as a function of time $t$. The function $L$ is known as the *Lagrangian* or the Lagrange function, and can be written as

$$L = m_0c^2\left(1 - \sqrt{1 - \left(\frac{v}{c}\right)^2}\right) + q(\boldsymbol{v} \cdot \boldsymbol{A} - \phi), \tag{5.42}$$

where the charge $q$ has to be set $q = -e$ for an electron. The Lagrangian consists of two terms; one that depends solely on the velocity of the particle, and thus on its kinetic energy, and one that depends on the electromagnetic potentials $\phi$ and $\boldsymbol{A}$. The general form of the variational principle as given in Eq. (5.41) demands the action $S_t$ of a particle trajectory to be an extremum. Since we refer to the principle as the principle of *least* action, and not the principle of *stationary* action, in the foregoing text we shall identify the extremum with a minimum. This limits our treatment to trajectories that do not cross a caustic[7] (Rose, 2009a).

We now introduce the Hamilton function $H(\boldsymbol{r}, \boldsymbol{p}, t)$ which can be written as

$$H(\boldsymbol{r}, \widetilde{\boldsymbol{p}}, t) := \widetilde{\boldsymbol{p}} \cdot \boldsymbol{v} - L(\boldsymbol{r}, \widetilde{\boldsymbol{p}}, t), \tag{5.43}$$

with $\widetilde{\boldsymbol{p}}$ representing the *canonical momentum* of the particle. The canonical momentum can be expressed as

$$\widetilde{\boldsymbol{p}} = \boldsymbol{p} + q\boldsymbol{A}, \tag{5.44}$$

which for an electron with $q = -e$ becomes

$$\widetilde{\boldsymbol{p}} = \boldsymbol{p} - e\boldsymbol{A}. \tag{5.45}$$

This relation is illustrated in Fig. 5.14. In the presence of a magnetic field ($\boldsymbol{A} \neq \boldsymbol{0}$), the canonical momentum $\widetilde{\boldsymbol{p}}$ is different in amplitude and direction from the kinetic momentum $\boldsymbol{p}$. The canonical momentum does not have a direct or intuitive physical

[7] A non-ideal imaging system does not form a stigmatic image point of a source point in the object space. Instead, neighboring rays intersect in the image space in a continuous set of points which form a surface. This surface is called the *caustic surface* or the focal surface. Any ray of a pencil is tangent to the caustic surface in the point it intersects another ray. Caustic surfaces are the surfaces of maximal electron density; they represent singularities of the point eikonal. For an ideal imaging system, the caustic surface degenerates to a stigmatic image point. The shape and intensity distribution of the caustic reflect the aberrations of the instrument (see, e.g. Born and Wolf, 2001; Glaser, 1952).

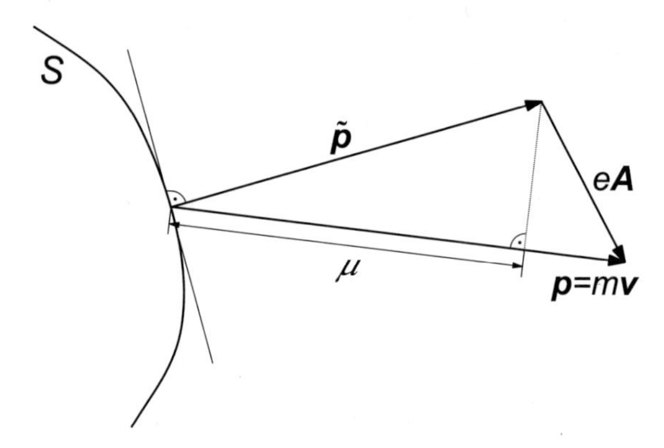


Fig. 5.14 Relation between the wave surface $S$, the canonical momentum $\widetilde{\boldsymbol{p}}$ and the kinetic momentum $\boldsymbol{p}$. The projection of the canonical momentum $\widetilde{\boldsymbol{p}}$ onto the ray vector $\boldsymbol{s}$ is $\mu$.

interpretation. However, as will be shown in the foregoing of the text, $\widetilde{\boldsymbol{p}}$ is the quantity that enables us to adopt the previously introduced electron optical description to electromagnetic fields without having to adjust the formalism developed for the case of electrostatic fields.

Having introduced the Hamilton function $H$ in Eq. (5.43) and the canonical momentum, we can rewrite the principle of least action given in Eq. (5.41) as follows:

$$S_t = \int_{t_0}^{t} (\widetilde{\boldsymbol{p}} \cdot \boldsymbol{v} - H)\, dt' \rightarrow \text{minimum}. \tag{5.46}$$

For stationary fields, we can deduce the reduced principle of least action for a charged particle exposed to an electromagnetic field. It can be shown that in a stationary system, the contribution of the Hamilton function to the action integral in Eq. (5.46) is a constant, i.e.

$$\int_{t_0}^{t} H dt' = \text{constant}. \tag{5.47}$$

The above relation is independent of the path of the particle. Therefore, the $H-$term in Eq. (5.46) does not affect the (stationary) variational principle. With this, Eq. (5.46) can be simplified to

$$S = \int_{t_0}^{t} \widetilde{\boldsymbol{p}} \cdot \boldsymbol{v} dt' \rightarrow \text{minimum}, \tag{5.48}$$

where we again employ the reduced action $S$ instead of the time-dependent form $S_t$. In the time-independent description of the trajectory, we used $\boldsymbol{s}(\boldsymbol{r})$ to describe a ray, i.e. the geometrical curve associated with a trajectory, rather than $\boldsymbol{r}(t)$. Performing a substitution of variables by employing $\boldsymbol{v} = d\boldsymbol{r}/dt$, which yields $\boldsymbol{v} = d\boldsymbol{s}/dt$ if $\boldsymbol{s}$ is used to describe the ray, leads us to

$$S = \int_{t_0}^{t} \widetilde{\boldsymbol{p}} \cdot \boldsymbol{v} dt' = \int_{t_0}^{t} \widetilde{\boldsymbol{p}} \cdot \frac{d\boldsymbol{s}}{dt} dt' = \int_{P_0}^{P} \widetilde{\boldsymbol{p}} \cdot d\boldsymbol{s}. \tag{5.49}$$

Comparing this expression with Eq. (5.41), we obtain Fermat's principle for electrons, or the so-called principle of Maupertius (Rose, 2009a)

$$\boxed{S = \int_{P_0}^{P} \widetilde{\boldsymbol{p}} \cdot d\boldsymbol{s} = \int_{P_0}^{P} (m\boldsymbol{v} - e\boldsymbol{A}) \cdot d\boldsymbol{s} \rightarrow \text{minimum.}} \tag{5.50}$$

Furthermore, since $S \rightarrow$ minimum demands a stationary value of $S$, we can write that the variation of $S$, i.e. $\delta S$ is

$$\delta S = \delta \int_{P_0}^{P} \widetilde{\boldsymbol{p}} \cdot d\boldsymbol{s} = \delta \int_{P_0}^{P} (m\boldsymbol{v} - e\boldsymbol{A}) \cdot d\boldsymbol{s} = 0. \tag{5.51}$$

Similar to the case of electrons exposed to an electrostatic field (see Eq. (5.27)), the geometrical wave surfaces of electrons exposed to a stationary electromagnetic field are determined by

$$S = \text{constant},$$

and in analogy to Eq. (5.31),

$$\widetilde{\boldsymbol{p}} = \nabla S. \tag{5.52}$$

This, on first sight, looks similar to the relation between the geometrical wave surface $S$ and the trajectory of an electron in the electrostatic field (Eq. (5.31)). However, in an electromagnetic field, an electron follows a trajectory that is not simply described by its kinetic momentum $\boldsymbol{p} = m\boldsymbol{v}$. In contrast, Eq. (5.52) reveals that the surfaces $S$ are no longer orthogonal to the kinetic momentum vector $\boldsymbol{p}$ but to the canonical momentum vector $\widetilde{\boldsymbol{p}} = \boldsymbol{p} - e\boldsymbol{A}$. Since $\boldsymbol{A}$ is not gauge-invariant, this point reflects a major difference between an electron exposed to an electrostatic field and an electron exposed to an electromagnetic field. Of course, for the case $\boldsymbol{A} \rightarrow \boldsymbol{0}$, Eq. (5.52) has the same geometrical interpretation as Eq. (5.31).

### 5.6.2 *The refractive index of electrons*

In electrostatic fields, the refractive index of electrons $n_{\boldsymbol{E}}$ is proportional to the kinetic momentum $p$, i.e. $n_{\boldsymbol{E}} \propto p$ (see Eq. (5.14)). Here, we would like to derive the equivalent quantity for electrons exposed to electro*magnetic* fields.

We start with the principle of Maupertius. Equation (5.50) contains the scalar product $\widetilde{\boldsymbol{p}} \cdot d\boldsymbol{s}$ of the canonical momentum $\widetilde{\boldsymbol{p}}$ and $d\boldsymbol{s}$. The vectorial increment $d\boldsymbol{s}$ in direction of the electron trajectory is parallel to the local tangent of the trajectory $\boldsymbol{s}(\boldsymbol{r})$. Since $\boldsymbol{s}$ is a unit vector, we can rewrite $d\boldsymbol{s}$ by $d\boldsymbol{s} = \boldsymbol{s}ds$, where $\boldsymbol{s}$ gives the direction and $ds$ the length of the vectorial increment $d\boldsymbol{s}$. Replacing $d\boldsymbol{s}$ with $\boldsymbol{s}ds$ in Eq. (5.50) yields an integrand of the form $\widetilde{\boldsymbol{p}} \cdot \boldsymbol{s}$. This scalar product reflects the projection of the canonical momentum $\widetilde{\boldsymbol{p}}$ onto the electron trajectory $\boldsymbol{s}$ (see Fig. 5.14). Denoting the scalar product $\widetilde{\boldsymbol{p}} \cdot \boldsymbol{s}$ by $\mu$, we can rewrite the principle of Maupertius as

$$\boxed{S = \int_{P_0}^{P} \mu ds \rightarrow \text{minimum.}} \tag{5.53}$$

Comparing this with Fermat's principle in Eq. (5.22), we can identify $\mu$ as a quantity which is proportional to the refractive index of electrons in a stationary electromagnetic field. The refractive index $n$ is thus given by[8]

$$n \propto \mu = \widetilde{\boldsymbol{p}} \cdot \boldsymbol{s} = p - e\boldsymbol{s} \cdot \boldsymbol{A}. \tag{5.54}$$

In order to indicate that this is no longer limited to the case of electrostatic fields, we omit the subscript $\boldsymbol{E}$ in the index of refraction. Of course, for the limiting case that $\boldsymbol{A} \rightarrow \boldsymbol{0}$, $n = n_{\boldsymbol{E}}$. However, the distinction between $n_{\boldsymbol{E}}$ and $n$ is important. Equation (5.54) shows that in the presence of a magnetic field, i.e. if $\boldsymbol{A} \neq \boldsymbol{0}$, $n$ depends on the direction $\boldsymbol{s}$ of the electron trajectory. This is in contrast to the situation where the electron is exposed solely to an electrostatic field (see Eq. (5.14)). Whereas $n_{\boldsymbol{E}}$ is a function of the position $n_{\boldsymbol{E}}(\boldsymbol{r})$, $n$ depends on both the position $\boldsymbol{r}$ and the direction $\boldsymbol{s}$ of the electron, i.e. $n = n(\boldsymbol{r}, \boldsymbol{s})$. In optical terms, an electromagnetic field acts as an inhomogeneous anisotropic optical medium. The inhomogeneity is simply given by the magnetic field distribution. The anisotropy results from the magnetic contribution to the refractive index. For instance, in case the direction of the electron trajectory $\boldsymbol{s}(\boldsymbol{r})$ in point $\boldsymbol{r}$ is perpendicular to the stationary vector potential $\boldsymbol{A}(\boldsymbol{r})$, the electron is not affected by the presence of $\boldsymbol{A}$. This is equivalent to the situation where the magnetic contribution of the Lorentz force (see Eq. (5.1)) vanishes because the electron's velocity $\boldsymbol{v}$ is parallel to the magnetic induction $\boldsymbol{B}$. Hence, it is the Lorentz force which is responsible for the anisotropy of the refractive index of electrons.

It is common to express $n$ in a normalized form. We first have to eliminate the gauge dependence of the electrostatic potential $\phi$. For this, we arbitrarily define that the electrostatic potential at the cathode, i.e. at the electron source, is zero; $\phi(\boldsymbol{r}_{\text{Cathode}} = \boldsymbol{0}) = 0$. Under the assumption of a point source, we can set the origin of the coordinate system to the cathode such that $\boldsymbol{r}_{\text{Cathode}} = \boldsymbol{0}$. An electron emitted from the cathode located at an arbitrary point $\boldsymbol{r}$ in the microscope column thus possesses the potential energy $E_{\text{p}} = -e(\phi(\boldsymbol{r}) - \phi(\boldsymbol{0})) = -e\phi(\boldsymbol{r})$. For an electron that reaches the field-free space in the image plane, we write $E_{\text{p}} = -e(\phi(\boldsymbol{r}_{\text{i}}) - \phi(\boldsymbol{0})) = -e\phi(\boldsymbol{r}) \equiv -eU$, and $U$ shall be the acceleration voltage of the microscope.

We further assume that the kinetic energy $E_{\text{k}}$ of the electrons at the surface of the cathode is zero. Since the potential energy $E_{\text{p}}$ of the electrons is also (arbitrarily) set to be zero at the cathode, the total energy $E$ is zero. From the (non-relativistic) conservation of energy

$$E = E_{\text{p}} + E_{\text{k}} \tag{5.55}$$

[8] It also applies here that in order to have the refractive index dimensionless, it can be normalized by dividing $\widetilde{\boldsymbol{p}}$ by the electron's kinetic momentum $p_0$ in the field-free space.

we obtain

$$e\phi(\boldsymbol{r}) = \frac{m_0 v^2(\boldsymbol{r})}{2}\,. \tag{5.56}$$

This yields for the (non-relativistic) kinetic momentum $p = m_0 v$ of the electron at position $\boldsymbol{r}$

$$p(\boldsymbol{r}) = \sqrt{2m_0 e\phi(\boldsymbol{r})}\,, \tag{5.57}$$

and at the field-free image plane

$$p_0 = \sqrt{2m_0 eU}\,. \tag{5.58}$$

We can now express the index of refraction $n$ as the quantity $\mu$ normalized to the kinetic momentum of the electron in the image plane, i.e.

$$n = \frac{\mu}{p_0} = \frac{p}{p_0} - \frac{e}{p_0}\boldsymbol{s}\cdot\boldsymbol{A}. \tag{5.59}$$

This finally yields for the index of refraction $n$ of an electron at position $\boldsymbol{r}$ in an electromagnetic field described by the electrostatic potential $\phi$ and the stationary vector potential $\boldsymbol{A}$

$$n = \sqrt{\frac{\phi(\boldsymbol{r})}{U}} - \frac{e}{p_0}\boldsymbol{s}\cdot\boldsymbol{A}. \tag{5.60}$$

For the relativistic case, the electrostatic potential $\phi$ is replaced with a relativistically modified potential $\phi^*$, which is given by

$$\phi^* = \phi\left(1 + \frac{e\phi}{2m_0c^2}\right). \tag{5.61}$$

Equivalently, $U$ is substituted by $U^*$. This yields the relativistic refractive index of electrons

$$\boxed{n(\boldsymbol{r}) = \sqrt{\frac{\phi^*(\boldsymbol{r})}{U^*}} - \sqrt{\frac{e}{2m_0\phi^*(\boldsymbol{r})}}\,\boldsymbol{s}\cdot\boldsymbol{A}.} \tag{5.62}$$

With the refractive index of electrons in the stationary electromagnetic field, the analogy between light optics and electron optics is in principle complete. Equation (5.62) provides a means to describe the trajectory of an electron in the electromagnetic field in a way similar to which a light ray is determined by the optical refractive index.

### 5.6.3 *Geometrical wave surfaces*

The kinetic momentum $\boldsymbol{p}$ of an electron exposed to an electromagnetic field is parallel to the corresponding ray which is described by the trajectory vector $\boldsymbol{s}$. Each point $\boldsymbol{r}$ in space can be considered to be associated with a unit vector $\boldsymbol{s}$. The multitude of electron trajectories fill a portion of space in such a way that each point of the volume belongs to an electron trajectory. Such a system of rays, or in general a system of curves, is said to form a congruence. The pencil of rays in Fig. 5.10, for instance, forms a homocentric congruence. The congruence is *homocentric* because all the rays have a common origin in point $P_0$.

For the case of an *electrostatic* field, i.e. when $\boldsymbol{A} = \boldsymbol{0}$, the rays, which are determined by the vectors $\boldsymbol{s}$, are parallel to the kinetic momentum $\boldsymbol{p}$ of the electrons. The rays cut the geometrical wave surfaces orthogonally. Hence, for the case $\boldsymbol{A} = \boldsymbol{0}$, there exists a family of surfaces whose normals are the electron trajectories. These geometrical wave surfaces are surfaces of constant point eikonal, $S =$ constant. This circumstance is essentially expressed in Eqs. (5.27) and (5.32).

The existence of a family of surfaces which are related to a congruence in such a way that the curves of the congruence are orthogonal to the surfaces imposes a strong requirement on the geometrical behavior of the curves. Indeed, the congruence is said to be a *normal* congruence. A normal congruence fulfills the condition

$$\mathrm{curl}\boldsymbol{s}(\boldsymbol{r}) \equiv \boldsymbol{0}. \tag{5.63}$$

For light rays and for electrons exposed to an electrostatic field, this condition is always met. The fact that the kinetic momentum is parallel to the trajectory vector $\boldsymbol{s}$ implies that the geometrical wave surfaces have a direct physical interpretation; the rays are orthogonal to the corresponding wave surfaces.

For electrons exposed to an electromagnetic field with $\boldsymbol{A} \neq \boldsymbol{0}$, the condition in Eq. (5.63) is usually not met. The violation of Eq. (5.63) means that the congruence associated with the electron trajectories in a electromagnetic field forms a *skew* congruence. The classification of electron trajectories in a magnetic field as a skew congruence is intuitively understandable. If we consider the propagation of a charged particle entering with a velocity $\boldsymbol{v}$ a homogeneous magnetic field $\boldsymbol{B}$ with $\boldsymbol{v} \cdot \boldsymbol{B} \neq 0$ and $\boldsymbol{v} \times \boldsymbol{B} \neq \boldsymbol{0}$, i.e. when $\boldsymbol{v}$ is neither perpendicular nor parallel to $\boldsymbol{B}$, the particle will spiral following a skew trajectory (see Chapter 6). Hence, electron trajectories in an electron microscope which contains at least one magnetic lens form a skew congruence.

A skew congruence does not have the property that there exists a family of surfaces that are orthogonal to the curves that form the congruence. However, we can still consider geometrical wave surfaces with $S =$ constant. Although these surfaces are not orthogonal to the electron trajectories, i.e. orthogonal to the kinetic momentum $\boldsymbol{p}$, they are orthogonal to the canonical momentum $\widetilde{\boldsymbol{p}}$ of the electrons. Since the canonical momentum is given by $\widetilde{\boldsymbol{p}} = \boldsymbol{p} - e\boldsymbol{A}$ and since the vector potential $\boldsymbol{A}$ is not gauge-invariant, the geometrical wave surfaces associated with the propagation of electrons in electromagnetic fields do not have a direct physical interpretation. Nevertheless, the concept of the canonical momentum allows us to apply the formalism which is actually based on the description of the optical behavior of light rays in a light optical medium, to describe the behavior of electrons in electromagnetic fields.

## 5.7 Summary

In this chapter, some very fundamental electron optical concepts were introduced. Starting with the description of the interaction of charged particles with electromagnetic fields expressed by the Lorentz force, we saw that the Hamiltonian analogy allows for translating the concepts based on classical mechanics into optical terms. Keywords addressed in this chapter are: the Hamiltonian analogy; the refractive index of electrons; the (reduced) principle of least action; the principle of Maupertius; geometrical wave surfaces; the canonical momentum (as opposed to the kinetic momentum); the (point) eikonal; and the characteristic function.

# Chapter 6

# Gaussian Dioptrics

So far, our considerations about electron optics have been general in the sense that we did not impose any geometrical limitations to the propagation of the electrons within the electromagnetic fields. The electromagnetic fields, which are generated by electrodes and magnets, are in general inhomogeneous. Furthermore, since the force of a magnetic field exerted on an electron depends on the direction of the electron's trajectory, electromagnetic fields act as inhomogeneous anisotropic optical media. This circumstance is expressed in the refractive index given in Eq. (5.62). An essential aspect of electron optics thus concerns the derivation of the electromagnetic fields associated with electron lenses. Once the fields are known, one expects that electron trajectories and wave surfaces can be derived based on the principles outlined in the previous chapter, if not analytically then in principle numerically. However, in reality, deriving electron trajectories for known electromagnetic fields is not as simple as it might seem on first sight. It can lead to complex problems with even chaotic behavior (Rose, 2009a). Furthermore, the problem to solve is often reversed. For a desired electron optical system, the imaging properties of the system and thus the behavior of the electron trajectories close to the specimen and close to the image plane are known or given parameters. What needs to be solved is the set of lenses, their particular design and excitation, which enables the desired imaging properties. Therefore, approximations are necessary.

## 6.1 Geometry and Coordinate Systems

So far, we have mentioned terms like electron source, microscope column and optical axis and we have dealt with electromagnetic fields which in some way alter the trajectories of electrons. Most of the terminology used is intuitively understandable, and for this reason we have not made attempts to define these terms explicitly.

In the following, we introduce the coordinate system and some frequently used notations which allow us to easily navigate in the electron microscope. We start with the *optical axis*. The optical axis is usually referred to as the $z$-axis (see, e.g. Fig. 5.1). Electrons propagate in positive $z$-direction. The optical axis is the line of highest symmetry along which the optical elements are arranged. For a transmission electron microscope consisting exclusively of rotationally symmetrical optical elements, the optical axis is simply the line that connects the center of each of these round elements. Looking at a transmission electron microscope, the optical axis essentially corresponds to the center of the microscope column, i.e. the axis that is intuitively recognizable as the axis of symmetry, containing the idealized point source and the center of each of the round lenses (see, e.g. Fig. 5.1). Of course, mechanical misalignments between subsequent optical units in the form of a shift or a tilt can be present. However, neglecting any of these potential misalignments, a conventional transmission electron microscope has in general a *straight* optical axis,. In contrast, electron mirrors, magnetic prism or, for instance, $\Omega$-type energy filters are systems of *curved* optical axis. Here, we limit our introductory discussion about Gaussian optics to systems with a straight optical axis. Apart from the optical axis, there are two types of planes which need to be defined. Planes which are perpendicular to the optical axis shall be referred to as *radial* planes. Surfaces that contain the optical axis are referred to as *meridional* planes or *sections*. The optical axis represents the multitude of points which all the meridional planes have in common. This is illustrated in Fig. 6.1a. For instruments of curved optical axis, meridional planes can be bent, but for instruments of straight optical axis, meridional planes are plane surfaces of constant surface normal.

While the optical axis is normally referred to as the $z$-axis, the coordinates used to navigate in radial planes are less strict. They are often chosen according to the

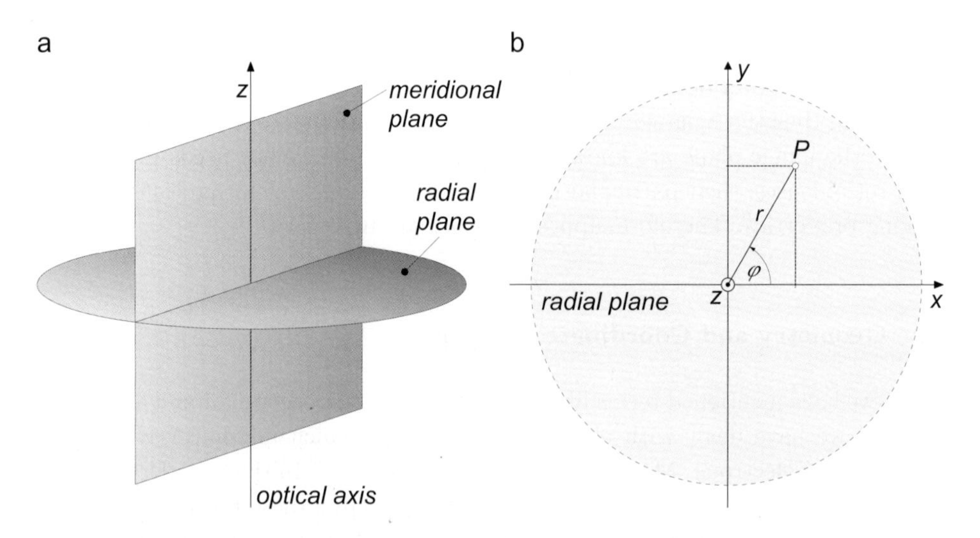


Fig. 6.1 Geometrical setup: (**a**) defines the optical axis, a radial plane and a meridional plane (or section); (**b**) illustrates a point $P$ in a radial plane described by $z =$ constant. The location of $P$ can be described by cartesian coordinates $(x, y)$, or by polar coordinates expressed by the radial distance $r$ and the azimuth angle $\varphi$.

axial symmetry of the system. The location of a point $P$ in a plane ($z$ = constant), as depicted in Fig. 6.1b, can be described by employing cartesian coordinates $x$ and $y$. Alternatively and particularly useful in the case of rotationally symmetric optical elements, one can use polar coordinates with the azimuth angle $\varphi$ and the radial distance $r$. The radius $r$ simply gives the distance between the point $P$ and the optical axis $z$. Often however, instead of using plain cartesian coordinates, a complex notation is employed. A point $P$, which has the cartesian coordinates $(x, y)$, is then described by a complex number $\mathrm{w} = x + \mathrm{i}y = r \exp \mathrm{i}\varphi$. For distinct planes, which are either planes that correspond to object and image planes or diffraction planes and their equivalent planes, we shall employ special letters to distinguish them easily. For the object plane, we employ $(x, y)$ or, in the complex notation, $\mathrm{w} = x + \mathrm{i}y$. For diffraction planes, we employ $(q_x, q_y)$ as in Fig. 2.6 or, in the complex notation, $\mathrm{w}_q = q_x + \mathrm{i}q_y$, with $|\mathrm{w}_q|$, $q_x$ and $q_y$ being spatial frequencies. Alternatively, since a diffraction plane in the back focal plane of the objective lens reflects the angular distribution of the electron scattering, we introduce a complex angle $\omega = \theta_x + \mathrm{i}\theta_y$ with[1] $\theta = |\omega| = \lambda q$ (see Fig. 6.2). The azimuth angle of the complex angle $\omega$ is $\varphi = \arctan \theta_y/\theta_x$. Similarly to the scattering angle $\theta$, we also employ the complex angle $\omega$ to navigate in the front focal plane of the objective lens. The illumination semi-angle $\alpha$ is then given by the maximum of $|\omega|$ that contributes to the illumination. The complex conjugate of a complex variable shall be denoted by a bar on top; i.e. $\overline{\omega}$ is the complex conjugate of $\omega$, or $\overline{\mathrm{w}}$ is the complex conjugate of $\mathrm{w}$.

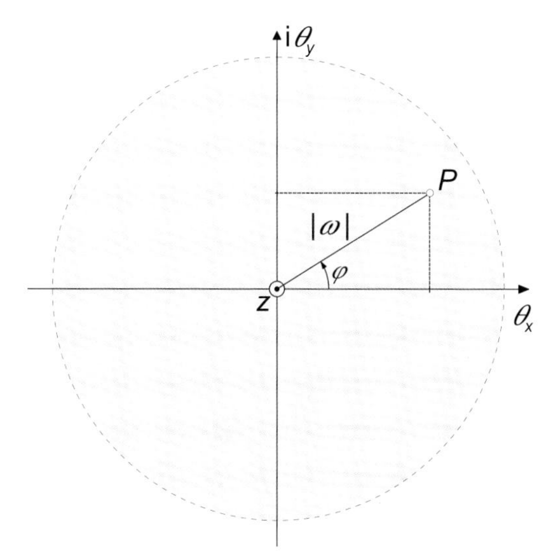


Fig. 6.2 A point $P$ in the diffraction plane. The location of the point $P$ is described by a complex angular coordinate $\omega = \theta_x + \mathrm{i}\theta_y$.

[1]This relation holds only for small $\theta$. In general, $\lambda q = 2 \sin \theta/2$.

So far, we basically defined the coordinate system according to the geometry of the electron optical instrument. The observer is expected to be outside of the system and thus describes the path of an electron inside the instrument from the coordinate system that is fixed to the instrument. This, of course, seems to be a suitable approach, particularly for electrons that do not rotate around the optical axis. However, for helical electron trajectories (see, e.g. Sec. 6.5.1), the fixed observer's coordinate system is not the most suitable coordinate system. Therefore, for electrons that follow helical trajectories, a coordinate system that follows the rotation of the electron about the optical axis is more appropriate. Instead of describing the electron trajectories in the fixed coordinate system, helical electron trajectories can better be described in a *rotating* coordinate system.

For this reason, we introduce a rotating coordinate system, which rotates about the optical axis with an angular velocity $\dot{\varphi}$. The rotating coordinate system shall be twisted by an angle $\varphi$ in respect to the fixed coordinate system. How the rotation angle of the coordinate system is related to the helical path of an electron will be shown in Sec. 6.5.2. For now, it shall suffice to state that there is a fixed coordinate system $(x, y, z)$ and a rotating coordinate system $(\hat{x}, \hat{y}, z)$, which follows the helical trajectories of the electrons.

Figure 6.3 shows two coordinate systems; the system $(x, y)$ with the unit vectors $\boldsymbol{e}_x$ and $\boldsymbol{e}_y$, which shall be the fixed one, and the rotating system $(\hat{x}, \hat{y})$ with the unit vectors $\hat{\boldsymbol{e}}_x$ and $\hat{\boldsymbol{e}}_y$. The $z$-axis shall be the same for both coordinate systems. The axes of the system $(\hat{x}, \hat{y})$ are rotated by an azimuth angle $\varphi$ in respect to the $x$- and $y$-axes of the fixed coordinate system. From Fig. 6.3 we can see that the unit vectors of the fixed coordinate system can be expressed in terms of the rotating system as

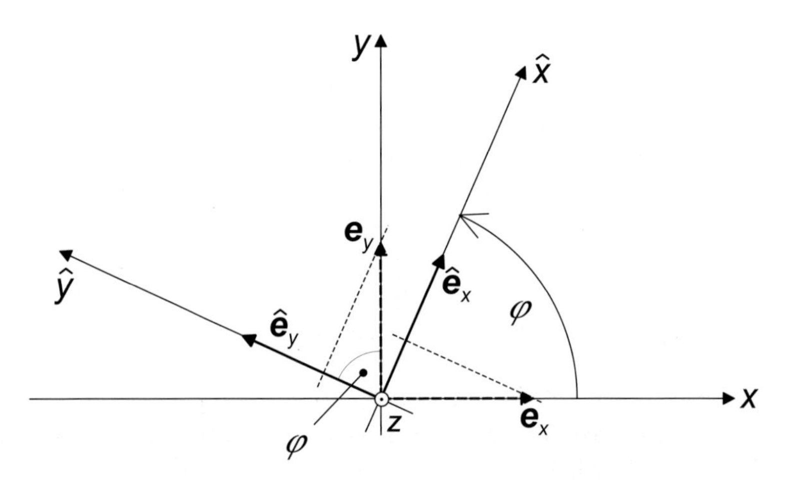


Fig. 6.3 The relation between the fixed and the rotating coordinate system.

$$\begin{aligned}\boldsymbol{e}_x &= \hat{\boldsymbol{e}}_x \cos\varphi - \hat{\boldsymbol{e}}_y \sin\varphi \\ \boldsymbol{e}_y &= \hat{\boldsymbol{e}}_x \sin\varphi + \hat{\boldsymbol{e}}_y \cos\varphi \,.\end{aligned}$$

Let us assume that there is an arbitrary point $P$ with the coordinates expressed $(\hat{x}, \hat{y})$ in the rotating coordinate system. The radial vector in the plane of $P$ pointing from the origin on the optical axis to $P$ is then given by $\overrightarrow{0P} = \hat{x}\hat{\boldsymbol{e}}_x + \hat{y}\hat{\boldsymbol{e}}_y$. Alternatively, we can express the location of point $P$ in the fixed coordinate system and express the position vector by $\overrightarrow{0P} = x\boldsymbol{e}_x + y\boldsymbol{e}_y$. Substituting the expression of the unit vectors in $x\boldsymbol{e}_x + y\boldsymbol{e}_y$ yields

$$\overrightarrow{0P} = x\boldsymbol{e}_x + y\boldsymbol{e}_y = x\left(\hat{\boldsymbol{e}}_x \cos\varphi - \hat{\boldsymbol{e}}_y \sin\varphi\right) + y\left(\hat{\boldsymbol{e}}_x \sin\varphi + \hat{\boldsymbol{e}}_y \cos\varphi\right) \,. \tag{6.1}$$

We rearrange the terms and obtain

$$\overrightarrow{0P} = \hat{\boldsymbol{e}}_x\left(x\cos\varphi + y\sin\varphi\right) + \hat{\boldsymbol{e}}_y\left(-x\sin\varphi + y\cos\varphi\right) \,.$$

Comparing this expression with $\overrightarrow{0P} = \hat{x}\hat{\boldsymbol{e}}_x + \hat{y}\hat{\boldsymbol{e}}_y$ from above, we can conclude that the rotating coordinate system $(\hat{x}, \hat{y})$ is related to the fixed cartesian coordinate system by

$$\begin{aligned}\hat{x} &= x\cos\varphi + y\sin\varphi \\ \hat{y} &= y\cos\varphi - x\sin\varphi \,.\end{aligned} \tag{6.2}$$

Furthermore, we can also express a point in the rotating coordinate system using the complex notation $u = \hat{x} + \mathrm{i}\hat{y}$. The rotating coordinate system enables us to describe the path of an electron without explicitly considering its azimuthal movement. Of course, in order to define the rotating coordinate system, knowledge about the angle $\varphi$ and thus about the helical path of the electron are still necessary.

## 6.2 Fields and Lenses

We do not draw our attention to the calculation of electromagnetic fields associated with the design of a certain electrode or magnet. For this, the reader is referred to textbooks such as those by, for example, Glaser (1952) or Hawkes and Kasper (1989a), which describe such methods in detail. One point, however, that needs our attention is the fact that electromagnetic fields have neither sharp borders nor distinct surfaces, particularly along the path of the beam. Their spatial extension is in principle infinite. How does this circumstance affect the optical behavior of the system?

Let us first consider a light optical instrument. A typical optical microscope consists of a light source, one or two condenser lenses, a (partially) transparent object, an objective lens below the object and an eyepiece. This basic setup is very similar to a transmission electron microscope. The light optical lenses, however, consist of specially shaped glass. The shape of the lens and its refractive index is chosen according to the desired path of the light rays in the optical instrument. We can even assume that the lenses act as homogeneous optical media. In contrast to electromagnetic fields, which are the optical elements in electron optics, glass lenses in light optics do have distinct surfaces. Whenever a light ray enters or leaves a lens, the direction of its path can change. This is described by Snell's law of refraction given in Eq. (5.6). For a homogeneous glass lens, the path of a ray does not change within the lens. Therefore, the path of a light ray is changed in discrete steps at surfaces of the lenses. Even for inhomogeneous optical lenses, the path of a light ray only changes within a defined length, i.e. the length that is defined by the volume and shape of the glass lens. The crucial point is that the effect of a light optical lens is confined to the physical location of the lens.

In an electron microscope, the electromagnetic fields associated with magnets and electrodes that form the optical media for the electron rays are in general inhomogeneous and anisotropic. Furthermore, electromagnetic fields do not possess sharp borders. As will be shown in a following section, the lens depicted in Fig. 5.1 creates a magnetic field that does not end abruptly. For this, the electromagnetic field of a magnetic electron lens interacts with an electron even if the electron is 'far' away from the actual physical location of the lens. Furthermore, the fields generated by individual magnetic lenses in an electron microscope would, in principle, overlap such that the series of electromagnetic lenses in an electron microscope form one single entity.

This circumstance reasons a first assumption which concerns the optical behavior of electron lenses. The spatial extension of the action of an electron lens shall be limited. Though the mathematical extension of a magnetic field is in principle infinite, the actual physical impact on electron trajectories can be regarded as being confined to a certain portion of space. Hence, if the field of a lens in a given distance to the lens is sufficiently weak, it will not significantly affect the trajectory of the electron. Therefore, in analogy to light optics, it is assumed that electromagnetic lenses only influence the trajectory of an electron if the electron is close to the physical location of the lens. If an electron is far away from a lens, the force exerted by the lens field does not significantly alter the electron's trajectory. Though electromagnetic fields do not have sharp borders, because the fields rapidly decay, their effect can be considered to be confined, too.

Of course, this approximation has its limitations. If, for instance, there are two electron optical elements in close proximity to each other, it can happen that the fields of the elements overlap. The elements can no longer be treated as separate units. This is known as crosstalk between optical elements. An example in which crosstalk can be important is the immediate neighborhood of the objective lens, which typically is the strongest but most sensitive magnetic element in an transmission electron microscope. It can happen that the rather weak field of the last condenser lens in close proximity, which can be used to switch from a broad-beam TEM mode to a probe mode (Williams and Carter, 1996), interacts with the field of the objective lens. This type of crosstalk is usually considered in the optical setup of an electron microscope. For this particular situation, the crosstalk is reflected by the fact that the nominal excitation of the objective lens differs in probe and broad-beam mode. For most other cases, however, electron lenses are treated as separate optical units of limited spatial extension.

## 6.3 The Paraxial Approximation

The statement that 'electron lenses are not perfect' is often heard in electron microscopy, referring to the fact that in particular the spherical aberration is unavoidable in rotationally symmetric, stationary electromagnetic electron lenses that are free of charges (Scherzer, 1936b). However, if we consider electromagnetic fields to be the media by which we control the trajectories of electrons, the usage of the term *imperfect* is problematic. In principle, if the electromagnetic fields are known, electron trajectories can be calculated to some precision. From a set of such trajectories, the transfer properties of the corresponding electromagnetic field and the optical unit should be derivable. We should be able to predict where a certain object point finds a corresponding (stigmatic or astigmatic) image point. In this sense, the imaging properties of electromagnetic fields are predictable. For this reason, we can only call a lens imperfect if we know *a priori* how a lens is supposed to behave, i.e. if we have expectations about how it should behave. Therefore, the statement about imperfect electron lenses requires a reference point — a reference point that defines the properties of a perfect lens, as well as the knowledge about how such a perfect lens can be approximated by using an imperfect lens. The paraxial approximation, i.e. Gaussian optics, provides us with this reference point. It is the deviation from this approximation that leaves us with imperfect electron lenses that do not fulfill our demands on the optical system in detail.

*Gaussian optics* deals with rays or electron trajectories which run in close proximity to the optical axis and whose inclination angle in respect to the optical axis is small. Such rays are called paraxial rays. *Paraxial rays* are confined to a narrow cylinder which runs along the optical axis. If we have an optical system with an optical axis of length $l$, the radius $r$ of this cylinder shall be sufficiently small such that in calculations only linear terms of $r/l$ have to be considered. Higher order terms of $r/l$ shall be negligible. In addition, the inclination angle $\gamma$ of a paraxial ray shall be small; sufficiently small such that the approximations $\sin\gamma \approx \gamma$ and $\cos\gamma \approx 1$ can be employed. Hence, a paraxial ray is a 'flat' ray which runs close to the optical axis. Such paraxial rays are the basis of Gaussian optics.

Furthermore, we limit the treatment of Gaussian optics to the case of dioptrics. *Dioptrics* deals with optical phenomena which arise through refraction. This is in contrast to *catoptrics*, which deals with optical phenomena which arise through reflection. A single electron mirror, for instance, represents a catoptric system. More generally, a dioptric imaging instrument is an instrument whose image plane moves in the same direction as the object plane when the object position is changed along the optical axis. For a catoptric imaging system, the image moves in the opposite direction.

## 6.4 Path Equation of an Electrostatic Field

Dealing with electromagnetic fields as lenses, what basically needs to be known about these fields is how they affect the trajectories of the electrons. In the present case, we would like to know how the fields affect paraxial electron trajectories. For this we aim at deriving a *path equation* for paraxial electrons. A path equation is a differential equation which defines the geometrical curves along which the electrons travel through the field, provided, of course, suitable boundary conditions are available, such as the position, direction and momentum of an electron entering the field. A solution of a path equation is the mathematical description of an electron trajectory through a given electromagnetic field.

Let us first consider an electrostatic field which is rotationally symmetric about the optical axis $z$. The field $\boldsymbol{E}(r, z)$, as depicted in Fig. 6.4, has two components, one in direction of the optical axis $E_z$ and one in radial direction $E_r$. In general, both components depend on the position $z$ along the axis and on the radial distance $r$. Because of the continuity of the field, the radial component has to vanish on the optical axis, i.e. $E_r(r = 0) = 0$. We assume that there is a plane at $z = z_0$ for which the radial component vanishes, i.e. $E_r(r, z_0) = 0$. Hence, in the plane $z_0$, the electrostatic field is parallel to the optical axis with $|\boldsymbol{E}(r, z_0)| = E_z(r)$. Such a rotationally symmetric electrostatic field is for instance realized in a so-called *einzel lens* or *unipotential* lens, whose electrostatic potential has a symmetry plane in $z = z_0$ (Glaser, 1952; Lenková, 2009). Retarding and accelerating rotationally symmetric electrostatic lenses are being used, too. The corresponding electrostatic potential does not have a symmetry plane perpendicular to the optical axis. A simple rotationally symmetric electrostatic lens can be realized by employing an annular electrode aligned in respect to the optical axis.

What we would like to know is the trajectory of an electron that enters such a field on a plane that corresponds to a section, i.e. a trajectory that lies on a plane which contains the optical axis, as illustrated in Fig. 6.1. Let us consider a narrow cylindric volume segment with one flat surface at the plane $z = z_0$, i.e. where $E_r = 0$, and another one at $z = z_0 + dz$. The volume of this segment shall be $\pi r^2 dz$, as illustrated in a cross-section in Fig. 6.4. The flux of the $\boldsymbol{E}$-field that goes inside this volume has to be equal to the flux that leaves the volume segment. The inward flux through the area $\pi r^2$ at the plane $z = z_0$ is

$$f_{\text{in}} = \int \boldsymbol{E}(r, z_0) \cdot d\boldsymbol{S} = E_z(z_0)\pi r^2, \tag{6.3}$$

with $d\boldsymbol{S}$ the vector area of the base surface of the cylinder in direction of the normal surface (in positive $z$-direction). Similarly, the outward flux through the top surface

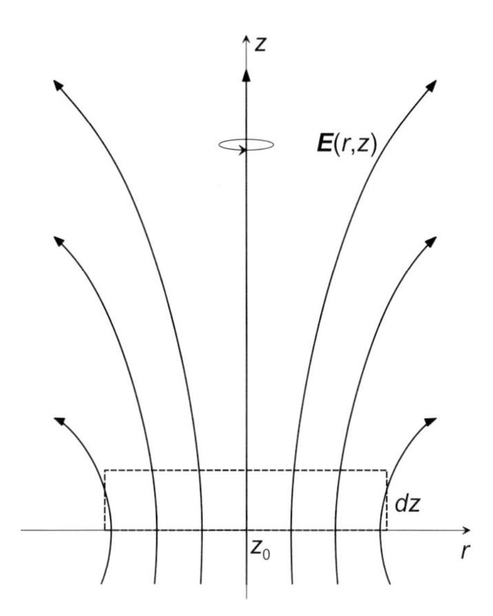


Fig. 6.4 A rotationally symmetric electrostatic field. Considering a cylindric volume element with $\pi r^2 dz$ free of charges, the inward flux trough the area $\pi r^2$ at $z_0$ (indicated by the dashed line) shall be equal to the sum of the outward fluxes through the surface shell $2\pi dz$ and the top surface $\pi r^2$ at $z_0 + dz$.

of the cylinder $\pi r^2$ at $z = z_0 + dz$ is

$$E_z(z_0 + dz)\pi r^2. \tag{6.4}$$

The outward flux through the cylinder's surface shell of area $2\pi r dz$ is

$$E_r 2r\pi dz. \tag{6.5}$$

The total outward flux $f_{\text{out}}$ is thus $E_z(z_0 + dz)\pi r^2 + E_r 2r\pi dz$. For a field which is free of sources in the respective volume segment, the inward flux $f_{\text{in}}$ has to be equal to the total of the outward flux $f_{\text{out}}$. From this condition we obtain

$$2\pi r dz E_r + \pi r^2 (E_z(z_0 + dz) - E_z(z_0)) = 0, \tag{6.6}$$

which for the case that $dz \to 0$ implies that

$$E_z(z_0 + dz) - E_z(z_0) = \frac{\partial E_z(z_0)}{\partial z} dz. \tag{6.7}$$

This leads us to

$$E_r = -\frac{r}{2} \frac{\partial E_z(0, z)}{\partial z}, \tag{6.8}$$

where $E_z(0, z)$ is the the component of $E_z(r, z)$ for $r = 0$, i.e. $E_z(r = 0, z)$. Alternatively, employing the relation between the electrostatic potential $\phi$ and the electrostatic field $\boldsymbol{E}$ given in Eq. (5.40), this can written as

$$E_r = \frac{r}{2} \frac{\partial^2 \phi(r = 0, z)}{\partial z^2} = \frac{r}{2} \frac{\partial^2 \Phi(z)}{\partial z^2} = \frac{r}{2} \Phi''(z), \tag{6.9}$$

where $\Phi(z)$ is the electrostatic potential $\phi$ along the optical axis $z$ with $r = 0$, i.e. $\Phi(z) = \phi(0, z)$, and $\Phi''$ is the second derivative of $\Phi(z)$ in respect to $z$; $\Phi'' = \partial^2\Phi/\partial z^2$. Hence, in Eq. (6.9) the electrostatic potential $\phi$ is approximated by its component along the optical axis $\Phi(z)$. This approximation is reasoned by the fact that we deal exclusively with paraxial trajectories that are close to the optical axis. For the same reason, we shall simply denote $E_z(0, z)$ by $E_z$. Hence, for small $r$, the radial component of the electromagnetic field $E_r$ can be expressed by the axial component $E_z$ for $r = 0$. Furthermore, the radial component $E_r$ can then be expressed as a function of the electrostatic potential $\Phi(z)$,

$$E_r = -\frac{r}{2}\frac{\partial E_z}{\partial z} = \frac{r}{2}\Phi''. \tag{6.10}$$

For clarity, we omitted to write out explicitly the $z$-dependence of $E_r$, $E_z$ and $\Phi$ in the above equation. With the definition of the electrostatic force in Eq. (5.5), we can write that the radial force $F_r$ exerted by a rotationally symmetric electrostatic field on an electron in the vicinity of the optical axis is given by

$$F_r = m_0\ddot{r} = -\frac{e}{2}\Phi''(z)r, \tag{6.11}$$

where $\ddot{r}$ means the second derivative of $r$ in respect to time $t$, i.e. $d^2r/dt^2$. The important point about Eq. (6.11) is that the force $F_r$ increases linearly with $r$. The force that drives a paraxial electron towards the optical axis increases with the electron's off-axial distance. This dependence is the basis of the imaging properties of a rotationally symmetric electrostatic field.

As an example, let us assume there is a point source located on the optical axis which emits electrons in positive $z$-direction. The electrons are emitted into a finite angular range. After a certain distance along the optical axis, the electrons enter a rotationally symmetric electrostatic field. The electron distribution in this entrance plane is determined by the emission angle of the electrons in the source point. The steeper the emission angle, the larger the radial distance to the optical axis. Therefore, the steeper the angle at which an electron is emitted, the stronger the force exerted by the rotationally symmetric electrostatic field that bends its trajectory back towards the optical axis. This means that for the paraxial case, the source point on the optical axis is imaged into a corresponding image point on the optical axis. The linear dependence between radial distance $r$ and $F_r$ holds only for trajectories that are close to the optical axis, otherwise the radial field $E_r$ cannot be expressed simply as a function of the axial field $E_z(r = 0)$.

Employing the (non-relativistic) relation of the conservation of energy, we can write that the kinetic energy of an electron is equal to its potential energy

$$\frac{m_0v^2(r, z)}{2} = e\phi(r, z). \tag{6.12}$$

The velocity of the electron in the direction of the optical axis is $v_z = dz/dt$. The velocity in radial direction can be approximated by $v_r = v_z dr/dz$. With $v^2 = v_z^2 + v_r^2$

we obtain, in agreement with Scherzer (1933),

$$\frac{m_0}{2}\left[1+\left(\frac{dr}{dz}\right)^2\right]\left(\frac{dz}{dt}\right)^2 = e\phi(r,z). \tag{6.13}$$

Since we deal with paraxial rays, we can replace the electrostatic potential $\phi(r,z)$ with the potential along the optical axis $\Phi(z)$. Furthermore, $(dr/dz)^2$, which is the square of the tangent of the inclination angle of the electron trajectory in respect to the optical axis, can be approximated by $(dr/dz)^2 \approx 0$. Incorporating this paraxial approximation into the equation of the conservation of energy yields

$$v_z = \frac{dz}{dt} = \sqrt{\frac{2e\Phi(z)}{m_0}}. \tag{6.14}$$

With this we can expand the radial force $F_r = m_0\ddot{r}$ in Eq. (6.11) as follows

$$\begin{aligned} m_0\ddot{r} &= m_0\frac{d}{dt}\left(\frac{dr}{dt}\right) = m_0\frac{dz}{dt}\frac{d}{dz}\left(\frac{dz}{dt}\frac{dr}{dz}\right) = m_0 v_z\frac{d}{dz}\left(v_z\frac{dr}{dz}\right) \\ &= m_0\sqrt{\frac{2e\Phi}{m_0}}\frac{d}{dz}\left(\sqrt{\frac{2e\Phi}{m_0}}\frac{dr}{dz}\right) = 2e\sqrt{\Phi}\frac{d}{dz}\left(\sqrt{\Phi}\frac{dr}{dz}\right), \end{aligned} \tag{6.15}$$

where again for readability we omitted to note explicitly the $z$-dependence of $\Phi$. The above expression for $m_0\ddot{r}$ can be incorporated into Eq. (6.11), from which we then obtain

$$\sqrt{\Phi}\frac{d}{dz}\left(\sqrt{\Phi}\frac{dr}{dz}\right) + \frac{1}{4}\Phi'' r = 0. \tag{6.16}$$

Employing the product rule for the derivative $d/dz$ and the derivative $d\sqrt{\Phi}/dz = \Phi'/(2\sqrt{\Phi})$ yields the following differential equation (see, e.g. Glaser, 1952, or Busch, 1926):

$$\boxed{\frac{d^2r}{dz^2} + \frac{\Phi'}{2\Phi}\frac{dr}{dz} + \frac{\Phi''}{\Phi}\frac{r}{4} = 0.} \tag{6.17}$$

This linear homogeneous differential equation of second order is the path equation we are looking for. Its solutions are of the form $r(z)$, i.e. they provide a measure for the radial distance $r(z)$ as a function of the axial coordinate $z$. These solutions describe the geometrical curves of paraxial electrons in a rotationally symmetric electrostatic field. Equation (6.17) reveals that within the paraxial approximation, the electron trajectory depends only on the $z$-component of the electrostatic field on the optical axis, i.e. $\Phi(z) = \Phi(0,z)$. Consequences of the characteristics of the differential equation in Eq. (6.17) on the amount of solutions that are required to fully characterize the optical properties of the rotationally symmetric electrostatic field will be discussed in Sec. 6.7.1.

## 6.5 Path Equation of a Stationary Magnetic Field

In analogy with the previous section, where we discussed the influence of an inhomogeneous rotationally symmetric electrostatic field on the propagation of electrons, here we examine the case of a magnetic field of equivalent symmetry. A rotationally symmetric magnetic field can be realized by a magnetic coil, as depicted in Fig. 5.2. The rotationally symmetric magnetic field generated by a magnetic coil is in general inhomogeneous. The magnetic induction $\boldsymbol{B}$ consists of a transverse (or radial) component $B_r(r,z)$ and a longitudinal (or axial) component $B_z(r,z)$, which both depend on the radial distance $r$ and on the distance $z$ along the axis of the coil. For the special case that the length of the coil exceeds by far the diameter of the coil, the coil is said to be *long* and the corresponding magnetic field is considered to be homogeneous. For a long coil, the radial component $B_r$ as well as the dependence of $B_z$ on $z$ and $r$ are negligible. Hence, a long coil forms approximately a homogeneous magnetic field (Glaser, 1952).

### 6.5.1 *The homogeneous rotationally symmetric magnetic field*

Before we draw our attention to the case of an inhomogeneous rotationally symmetric magnetic field, we discuss three special cases that are based upon the simpler, i.e. homogeneous magnetic field. We assume that there is a magnetic field present with the only non-vanishing field component $B_z$. The spatial extension of the field shall be limited. Yet, within the domain of the field $B_z$ is constant.

#### 6.5.1.1 *Three preliminary cases*

(i) Propagation parallel to the magnetic field vector

An electron entering a homogeneous magnetic field parallel to the field vector $\boldsymbol{B}$ is not affected by the presence of the magnetic field. If the only field component is $B_z$ and the electron travels along the $z$-direction with a velocity $v_z$, the electron proceeds along the $z$-direction without noticing the magnetic field. This directly follows from the cross product between velocity $\boldsymbol{v}$ and $\boldsymbol{B}$ in the Lorentz force given in Eq. (5.1), which we rewrite for $\boldsymbol{E} = \boldsymbol{0}$ and an electron of charge $-e$ as

$$\boldsymbol{F}_{\mathrm{L}} = -e\,\boldsymbol{v} \times \boldsymbol{B}.$$

(ii) Propagation perpendicular to the magnetic field vector

The second case concerns an electron entering the field $B_z$ with a velocity $v_n$, along a direction which is perpendicular to the $z$-axis. Such an electron experiences an acceleration which is normal to its trajectory and normal to the $z$-direction. Though the velocity of the electron remains constant, its direction and thus its momentum do change, but stay in a plane normal to $z$. Provided that the spatial extension of the field is large enough such that the electron stays within the field, the electron

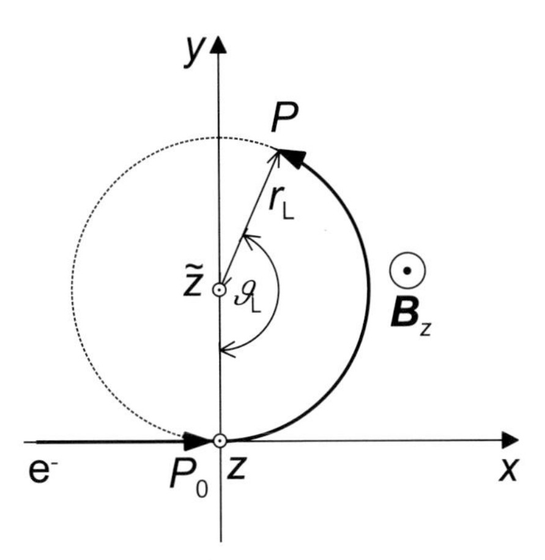


Fig. 6.5 An electron of velocity $v_n$ and charge $q = -e$ travelling in a direction perpendicular to the $z$-direction enters a homogeneous magnetic field $B_z$ at point $P_0$. The magnetic field is directed in positive $z$-direction. As long as the electron stays within the magnetic field, the electron follows a trajectory which describes a circle of radius $r_L$.

follows a trajectory described by a circle. This is illustrated in Fig. 6.5. The circle lies in a plane $z$ = constant. The radius $r_L$ of the circle is the Larmor radius, or the radius of gyration. It is given by

$$r_L = \frac{m_0}{e}\frac{v_n}{B_z}. \tag{6.18}$$

The time $\tau$ needed for one revolution of distance $2\pi r_L$ is

$$\tau = \frac{2\pi m_0}{eB_z}. \tag{6.19}$$

Hence, while the radius of the circle depends on the velocity of the electron, $\tau$ is independent from $v_n$. We can also characterize the path along the circle in terms of the angle (see Fig. 6.5). The angle $\vartheta_L$, i.e. the angle of rotation around the $z$-axis, is the angle of the Larmor rotation.

(iii) Propagation under finite inclination angle

The third and last preliminary case is the one of an electron entering a homogeneous magnetic field $B_z$ under an inclination angle $\gamma$ with respect to the $z$-axis (here, $\gamma$ refers to an angle). The electron's velocity is $v$ consisting of a longitudinal component $v_z = v\cos\gamma$ along the $z$-direction, and a transverse component $v_n = v\sin\gamma$ perpendicular to $z$. The influence of the magnetic field on the electron's trajectory can be discussed by separating the propagation of the electron into a transverse and a longitudinal motion. The path of the electron is then given by the superposition of the motion in transverse and longitudinal directions. As in the first preliminary case discussed above, the electron's propagation in $z$-direction is not affected by the magnetic field. Hence, the electron continues its travel along the longitudinal direction with the velocity $v_z$. After a time $t$, the electron covered a distance $z = tv\cos\gamma$.

In analogy to the second preliminary case, the transverse velocity $v_n$ of the electron's initial momentum leads to a circular movement around an axis $\tilde{z}$ which runs parallel to the $z$-axis. The time needed for a full revolution around this axis is $\tau = 2\pi m_0/eB_z$.

Superimposing the circular motion with the unaltered propagation along the $z$-direction, it can be seen that the path of the electron describes a helix about an axis $\tilde{z}$ parallel to the $z$-direction. This is illustrated in Fig. 6.6. The radius of the helix projected onto a plane $z =$ constant depends on the electron's transverse velocity and is given by

$$r_{\mathrm{L}} = \frac{m_0 v_n}{eB_z} = \frac{m_0 v}{eB_z}\sin\gamma. \tag{6.20}$$

Expectedly, this is equivalent to Eq. (6.18), which describes the radius of the Larmor rotation.

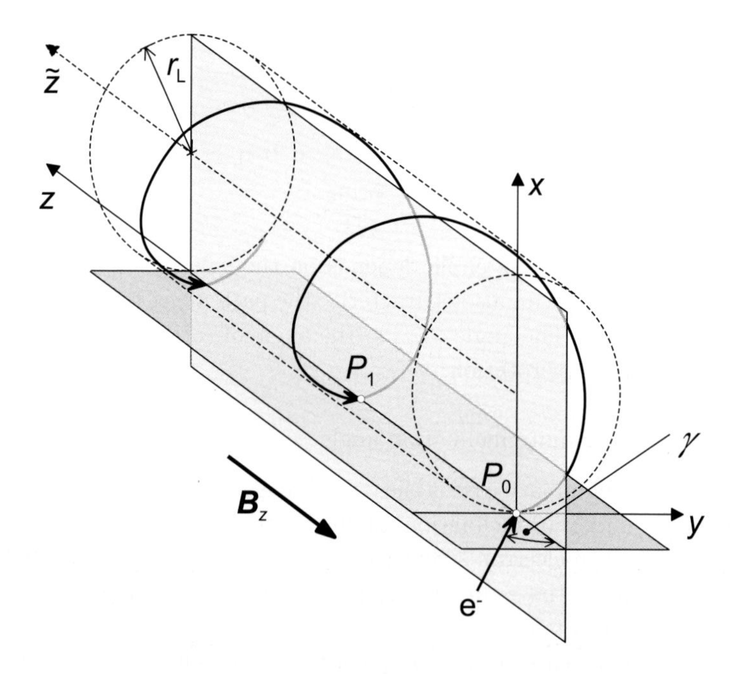


Fig. 6.6 An electron with a transverse velocity $v_n$ and a longitudinal velocity $v_z$ enters at point $P_0$ a homogeneous magnetic field $B_z$ under an inclination angle $\gamma$ given by $\tan\gamma = v_n/v_z$. While the propagation in longitudinal direction is unaffected by the magnetic field, the propagation in transversal direction leads to a radial acceleration. The path of the electron within the magnetic field describes a helix with an axis $\tilde{z}$ in longitudinal direction.

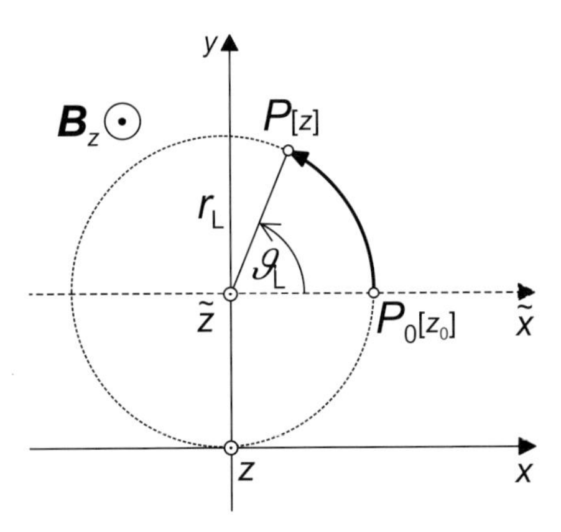


Fig. 6.7 Larmor rotation. The path of an electron from $P_0$ to $P$ along a helix projected onto a plane perpendicular to the $z$-direction. The labels $[z]$ and $[z_0]$ shall indicate that the points $P_0$ and $P$ have different $z$-coordinates. The axis of the helix denoted by $\tilde{z}$ runs parallel to the $z$-axis.

### 6.5.1.2 *Larmor precession*

Finally, we would like to deduce an expression for the angle of the Larmor rotation, i.e. the twist angle $\vartheta_L$ around the $\tilde{z}$-axis. Figure 6.7 illustrates the situation. Let us assume there is point $P_0$ in $z_0$ from which the electron starts its path on a helix. The position of $P_0$ is described in polar coordinates with respect to the $\tilde{z}$-axis as $P_0 = (r_L, \varphi_0 = 0, z_0)$, see Fig. 6.7. After a time $t$, the electron reaches a point $P$ on the helix which is described in polar coordinates with respect to the $\tilde{z}$-axis as $P = (r_L, \varphi, z)$. The distance between $P_0$ and $P$ along the $z$-direction is $\Delta z = z - z_0$. The difference between the two azimuth angles of $P_0$ and $P$ corresponds to the angle of the Larmor rotation, i.e. $\vartheta_L = \varphi - \varphi_0$.

As above, the motion of the electron can be split into a transverse component and a longitudinal component. Let us first consider the longitudinal component. Unaffected by the transverse motion, the electron travels with a velocity $v_z$ in the direction of the $z$-axis. Starting in point $P_0$, we can write that the time needed to reach point $P$ along the $z$-direction is

$$t_z = \frac{\Delta z}{v_z}. \tag{6.21}$$

Alternatively, we can consider the transverse motion, i.e. the precession about the $\tilde{z}$-axis. If we consider only this transverse motion, we can write that the time needed to reach point $P$ from $P_0$ along the perimeter of the projected helix is given by the length of the segment of the circle divided by the transverse velocity $v_n$ (see Fig. 6.7)

$$t_\varphi = \frac{\vartheta_L r_L}{v_n}. \tag{6.22}$$

Independent from analyzing either the transverse or the longitudinal component, as a matter of fact the electron must reach the point $P$ at the same time. Therefore,

$t_z \equiv t_\varphi$, and thus

$$\frac{\Delta z}{v_z} = \frac{\vartheta_{\mathrm{L}} r_{\mathrm{L}}}{v_n}. \tag{6.23}$$

With Eq. (6.20) for $r_{\mathrm{L}}$, we can deduce the relation of the angle of the Larmor rotation

$$\vartheta_{\mathrm{L}} = \frac{eB_z}{m_0 v_z} \Delta z. \tag{6.24}$$

This result can be generalized for the case that $B_z$ is not constant but a function of $z$, i.e. $B_z = B_z(z)$ (Glaser, 1952)

$$\boxed{\vartheta_{\mathrm{L}} = \int_{z_0}^{z} \frac{eB_z}{m_0 v_z} dz.} \tag{6.25}$$

This is the angle of Larmor rotation, which describes the rotation of an electron around the axis of its helical path.

#### 6.5.1.3 *Properties of a homogeneous magnetic field*

The point $P_0$ in Fig. 6.6 reflects the position where an electron enters a homogeneous magnetic field $B_z$ under an inclination angle $\gamma$. The entrance or inclination angle $\gamma$ is determined by $\tan\gamma = v_n/v_z$. Without loss of generality, we can set the origin of the coordinate system in the point $P_0$. The electron, which enters the field in $P_0$ and thus crosses the $z$-axis in this point, continues on a path described by a helix. The axis of this helix denoted by $\tilde{z}$ is parallel to the $z$-axis (see Fig. 6.6). After a distance $\ell_z$ the electron reaches a point $P_1$, which, similarly to point $P_0$, lies on the $z$-axis. Figure 6.6 illustrates that when the electron reaches $P_1$, it makes a full $2\pi$ (or 360°) revolution around the axis of its helix. The distance $\ell_z$, which connects the origin $P_0$ with the point $P_1$ along the $z$-axis, is (Busch, 1926)

$$\ell_z = \tau v \cos\gamma = \frac{2\pi m_0 v}{eB_z} \cos\gamma. \tag{6.26}$$

Let us now consider another electron, which also enters the field $B_z$ in the point $P_0$, but whose inclination angle is $\gamma_1$. The helix along which the second electron travels is different from the first helix. In particular, the axis of the second helix, though still parallel to the $z$-axis, will be different from the axis of the first helix. Finally there will be a point $P_1'$ on the $z$-axis which the second electron reaches after a full $2\pi$ revolution around the axis of its helix. The point $P_1'$, which is located in a distance $\ell_{z,1} = 2\pi m_0 v/(eB_z)\ \cos\gamma_1$ from the entrance point $P_0$, will be different from $P_1$ because $\gamma \neq \gamma_1$ and thus $\ell_z \neq \ell_{z,1}$. In general, trajectories of different inclination angles $\gamma_i$ lead to different points $P_1^i$. With increasing $\gamma$, the distance $\ell_z$ between $P_0$ and $P_1$ decreases. This is illustrated in Fig. 6.8.

This finding can be further exploited. We can consider the $z$-axis to be the optical axis, and we can picture the magnetic field forming some kind of transfer

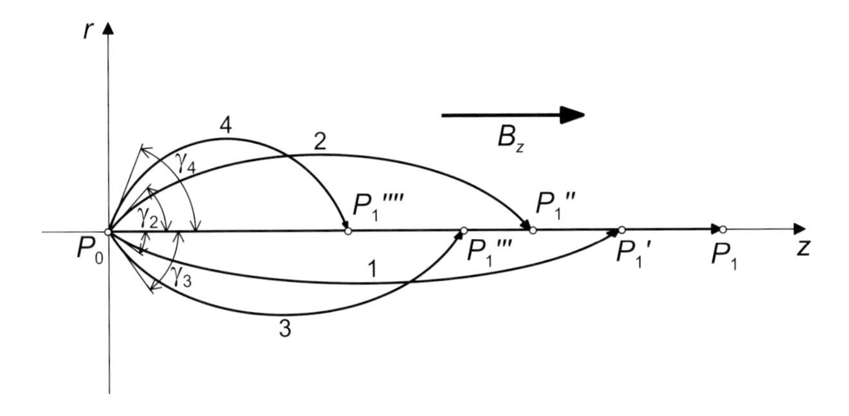


Fig. 6.8 Electrons enter a homogeneous magnetic field $B_z$ through the point $P_0$. Because the inclination angles $\gamma_i$ of the incident electron trajectories are different, each electron continues its path in the field on an individual helix. These helices are depicted in projection by the curves 0, 1, 2, 3 and 4. An electron with vanishing inclination angle $\gamma \rightarrow 0$ follows a hypothetical helix of vanishing radius and intersects the axis again in point $P_1$. Electrons that enter under finite inclination angle $\gamma_i$ (depicted by the curves 1, 2, 3 and 4) intersect the axis at a distance which is a factor $\cos\gamma_i$ smaller than the electron which enters under an angle $\gamma \rightarrow 0$. The points of intersection of the trajectories 1, 2, 3 and 4 are labelled $P_1', P_1'', P_1'''$ and $P_1''''$.

system, for instance as depicted in Fig. 5.2. Let us assume that there is a point-like electron source $P_0$ on the optical axis in front of the homogeneous magnetic field $B_z$. We can employ an annular aperture of small opening aligned on the optical axis. This narrow annular aperture shall act as a filter that only allows for passing rays which are emitted exactly under an angle $\gamma_1$. Furthermore, if we assume that the source emits electrons of equal energy, there is a point $P_1'$ behind the aperture where all the electrons emitted under the angle $\gamma_1$ intersect. The point $P_1'$ is located in a distance $\ell_{z,1} = 2\pi m_0 v/(eB_z)\cos\gamma_1$ from the source point $P_0$. Considering the annular aperture to be variable, for each opening angle $\gamma_i$ of the annular aperture there is a characteristic point $P_1^i$ in a distance $\ell_{z,i}$, which for a known magnetic field reflects the velocity of the electron $v$, the ratio $e/m_0$ and the emission angle $\gamma_i$. Figure 6.9 illustrates this situation for the case of two narrow annular aperture openings. Though this looks like a rather trivial experiment, it has a practical meaning. The electron's velocity $v$ can be determined from the known acceleration potential. Provided the magnetic field is known from the magnetic coil used, the only unknowns are the ratio $e/m_0$ and $\ell_z$. Therefore, by measuring the distance $\ell_z$, one can deduce the ratio $e/m_0$, i.e. the specific charge of the electron. Alternatively, one can change the acceleration potential to focus the electrons in a certain distance $\ell_z$. Hence, as an alternative to J. J. Thomson's well known experiment to measure the specific charge of particles, this method was proposed and applied by Hans Busch (1922), who could show a measuring precision for the specific charge of an electron which is better than 1%. It was one of the steps that

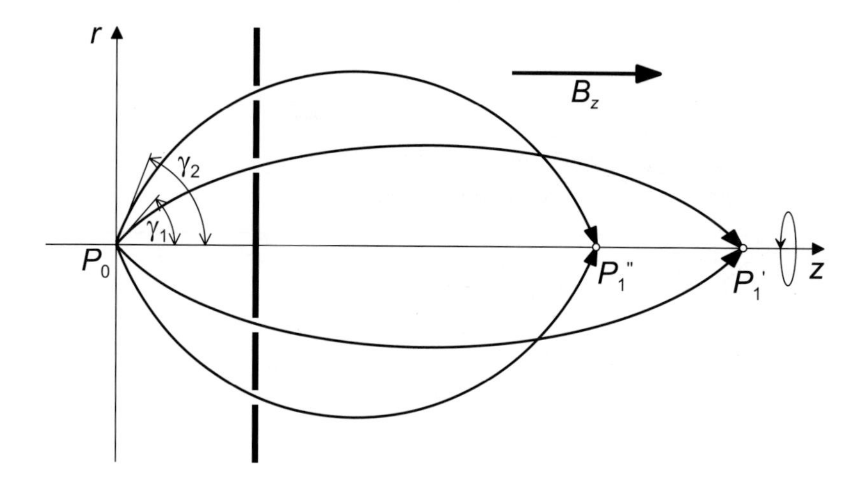


Fig. 6.9 An annular aperture with two ring openings filters electrons that are emitted into a homogeneous magnetic field $B_z$ under two different angles, $\gamma_1$ and $\gamma_2$. Rays that are emitted with angle $\gamma_1$ and pass the inner aperture opening form a stigmatic image of the point $P_0$ in point $P_1'$. Rays that are emitted under angle $\gamma_2$ and pass the outer aperture opening form a stigmatic image of the point $P_0$ in point $P_1''$. Since the image points $P_1'$ and $P_1''$ are different, the trajectories are not paraxial trajectories.

led Busch to the discovery of the focusing power of electromagnetic fields employed in electron-optical instruments (Busch, 1926).

#### 6.5.1.4 *Application of the paraxial approximation*

Within the paraxial approximation, where we deal with electrons near the optical axis following flat trajectories (i.e. $\gamma$ is small), we can approximate the cosine function simply by $\cos\gamma \approx 1$. This approximation, which in fact reflects the cosine function substituted by the first term of its Taylor-series expansion[2], is correct as long as $\gamma^2/2$ is negligible compared to 1. With this, we obtain the surprisingly simple result that $\ell_{z,1} = \ell_{z,2} = ... = \ell_{z,i}$. Hence, under the paraxial approximation, the point $P_1'$ of a first electron emitted under an angle $\gamma_1$ coincides with the point $P_1''$ of a second electron emitted under an angle $\gamma_2$. Since $P_1'$ and $P_1''$ are identical, and since any other electron entering the field through $P_0$ with small $\gamma$ will go through this exact point as well, we simply denote it by $P_1$. This is illustrated in a simplified side representation in Fig. 6.10. Comparing Fig. 6.8 with Fig. 6.10 clearly reveals that the paraxial approximation is only applicable if $\gamma$ is small. Furthermore, it is clear that none of the rays 1–4 illustrated in Fig. 6.8 are adequately described on the basis of the paraxial approximation. However, Figs. 6.8 and 6.10 amplify the effect of the simplification which is made by employing the paraxial approximation.

Considering the fact that within the paraxial approximation there is a point $P_1$ which reflects the entrance or source point $P_0$, we can identify the point $P_1$ as an

[2]The Taylor series of the cosine function is given by $\cos\gamma = 1 - \frac{1}{2}\gamma^2 + \frac{1}{4!}\gamma^4 - \frac{1}{6!}\gamma^6 + \frac{1}{8!}\gamma^8 + O(\gamma^{10})$.

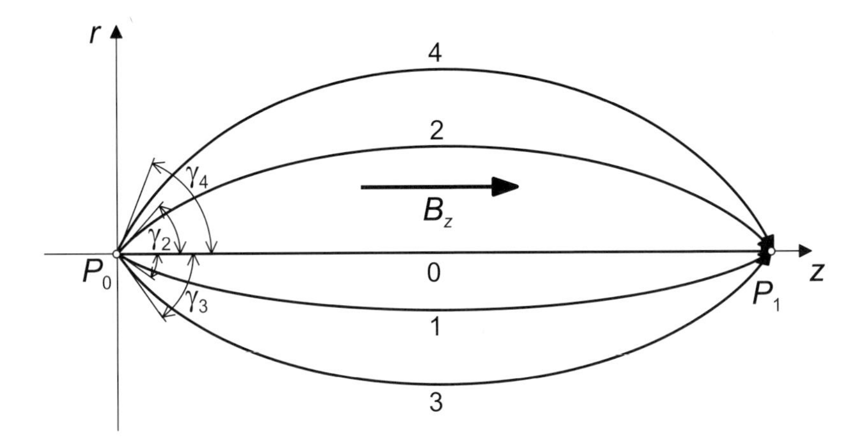


Fig. 6.10 Electrons enter a homogeneous magnetic field $B_z$ through the point $P_0$. Because the inclination angles $\gamma_i$ of the incident electron trajectories are different, each electron continues its path in the field on an individual helix. These helices are depicted in projection by the curves 0, 1, 2, 3 and 4. In contrast to Fig. 6.8, under the assumption of paraxial rays ($\cos\gamma_i = 1$), all of the trajectories intersect the axis in a single point $P_1$.

image point of point $P_0$, and call $P_0$ the object point. Hence, we find that under the assumption of paraxial rays, a source point $P_0$ can be related with an image point $P_1$ in a distance $\ell_z$ along the $z$-direction. This is valid as long as we deal with paraxial rays. From this we can conclude that the homogeneous rotationally symmetric magnetic fields acts as a lens which images the plane $z_0$ into an image plane $z_1$. This is a simple (paraxial) imaging system.

#### 6.5.1.5 *Deviation from the paraxial approximation*

As a next step in the discussion of the paraxial approximation of a homogeneous magnetic field, we would like to draw our attention briefly to the situation where deviations from the paraxial trajectories are unavoidable. What happens if we have to deal with electron trajectories which are not properly described by the paraxial approximation? Such trajectories can be further away from the optical axis and/or their inclination angle $\gamma$ might violate the approximation $\cos\gamma = 1$.

If latter deviation from the paraxial path of a ray has to be taken into account, it becomes necessary to expand the rather profound approximation $\cos\gamma \approx 1$. Having started with the substitution of the cosine function with the first term of its Taylor series, it seems obvious that in a second step, the limitations imposed by the paraxial approximation can be overcome by taking into account the next higher order term of the series expansion of the cosine function. Hence, we approximate the cosine function by $\cos\gamma \approx 1 - \gamma^2/2$. This approach is valid as long as the next following term of the Taylor series, i.e. $\gamma^4/4!$ is negligible compared to $1-\gamma^2/2$. In immediate analogy to the paraxial analysis described above, the path of an electron entering the field $B_z$ under an inclination angle $\gamma$ in point $P_0$ returns back to the $z$-axis in a point $P_1$, which is at a distance $\ell_z$ from $P_0$ (see Fig. 6.6). With our expanded approximation, we obtain that the distance $\ell_z$ is given by

$$\ell_z = \frac{2\pi m_0 v}{eB_z}\left(1 - \frac{1}{2}\gamma^2\right). \tag{6.27}$$

While in the paraxial approximation $\ell_z$ is independent from the inclination angle $\gamma$, the expanded approximation reveals that the distance $\ell_z$ depends on the square of the entrance angle $\gamma$. The larger $\gamma$, the smaller the distance $\ell_z$. This is closer to the realistic situation depicted in Fig. 6.8.

An electron of vanishingly small inclination angle ($\gamma \to 0$) entering the magnetic field $B_z$ in point $P_0$ would be transferred to a point $P_1$ on the axis which is at a distance $\ell_z(\gamma = 0) = 2\pi m_0 v/(eB_z)$ from point $P_0$. A trajectory of $\gamma \to 0$ is indeed a trajectory for which the paraxial approximation provides the true description of its path. We can use the image distance $\ell_z$ of a ray with $\gamma = 0$ as a reference distance and denote it with $\ell_0$. For a given point $P_0$ with the $z$-coordinate $z_0 = 0$, the distance $\ell_0$ shall define the location of the image plane by $z_1 = \ell_0$. With this and the approximation $\cos\gamma \approx 1 - \gamma^2/2$, an electron trajectory of finite inclination angle $\gamma$ leads to an image point $P_1'$ in a distance $\ell_z = \ell_0(1 - \gamma^2/2)$. The distance between the image plane at $\ell_0$ and the point $P_1'$, i.e. the longitudinal deviation of $P_1'$ from the image plane, is (see Fig. 6.11)

$$\Delta_S = \ell_z - \ell_0 = -\frac{1}{2}\ell_0\gamma^2. \tag{6.28}$$

The distance $\Delta_S$ is called the longitudinal spherical aberration. Let us assume that there is an ensemble of electrons entering the field in point $P_0$ under the same inclination angle $\gamma$ in respect to the optical axis but under different azimuth angles $\varphi$. In front of the magnetic field, this ensemble of rays forms a cone of half angle $\gamma$ with its vertex in $P_0$. Under the approximation we made above, the homogeneous

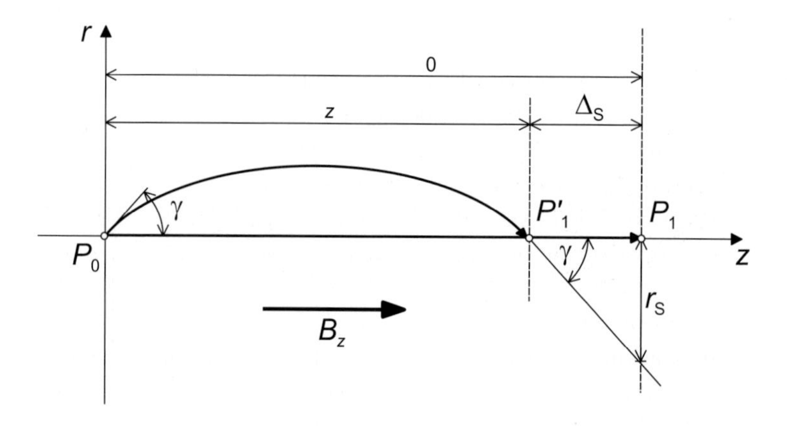


Fig. 6.11 Deviation from the paraxial approximation. Spherical aberration.

magnetic field would transfer these rays to a point $P_1'$ which represents an image of the vertex $P_0$ and lies at a distance $\ell_0(1-\gamma^2)$ from point $P_0$ on the optical axis. Though there is an image point $P_1$ in a distance $\ell_0(1-\gamma^2/2)$, the trajectories would generate an image of the point $P_0$ in the paraxial (or Gaussian) image plane $\ell_0$, which represents a circle of radius

$$r_S = \Delta_S \tan\gamma \approx \Delta_S\,\gamma = -\frac{1}{2}\ell_0\gamma^3. \tag{6.29}$$

The value $-\ell_0/2$ is a universal constant for the magnetic (or electrostatic) field under consideration which characterizes the deviation from the paraxial trajectory for the case in which the second-order term of the cosine function is considered in the path derivation. This parameter is known as the constant of third-order spherical aberration. It is a third-order aberration because its impact $r_S$ on the image increases with the third power of $\gamma$ (see Eq. (6.29)). The disk of radius $r_S$ in the plane $z_1$ is the image aberration due to the spherical aberration. Figure 6.12 illustrates the impact of the spherical aberration on a pencil of rays emerging in point $P_0$ entering a homogeneous magnetic field $B_z$. The helical path of a ray with $\gamma \to 0$ intersects the optical axis in $P_0$ and $P_1$. The distance between $P_0$ and $P_1$ is $\ell_0$. With increasing inclination angle $\gamma$ of the initial path starting in $P_0$, the point of intersection with the optical axis moves closer towards $P_0$. This is expressed by the distance $\ell_z = \ell_0(1-\gamma^2/2)$, which shrinks with increasing $\gamma$. Instead of intersecting in a single point in the plane of $P_1$, the rays with finite $\gamma$ form a circle in this plane. Hence, it is only within the paraxial approximation that $\Delta_S$ and $r_S$ are both equal zero, and that the image point $P_1$ unambiguously reflects the object point $P_0$. For any expansion of the path derivation beyond the paraxial approximation, the point $P_0$ does not have a unique point that would correspond to its image. There is no stigmatic image point reflecting the object point $P_0$.

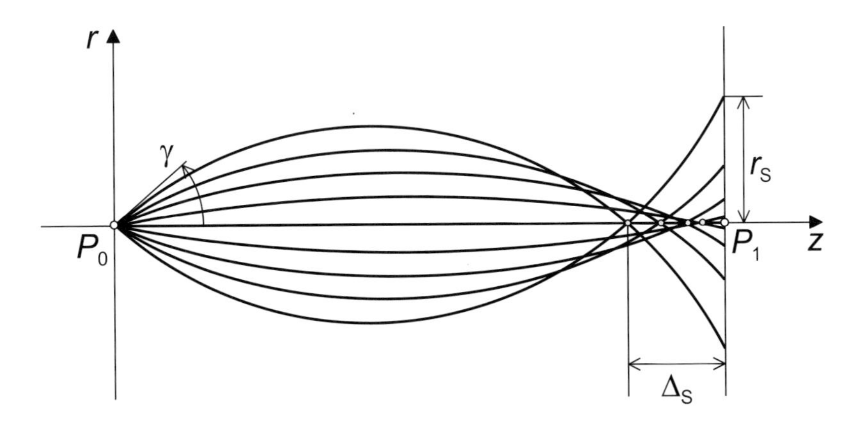


Fig. 6.12 Qualitative illustration of the spherical aberration.

In summary, the discussion of the three preliminary cases of an electron travelling in a homogeneous magnetic field and the resulting imaging properties revealed that the homogeneous rotationally symmetric magnetic field generated by a long magnetic coil can act as a lens that transfers information from an object plane to an image plane. Because the electron trajectories are helices, the magnification of a homogeneous rotationally symmetric magnetic field is equal one. The field neither magnifies nor demagnifies the object. Furthermore, under the assumption of Gaussian optics, this transfer is ideal; for an entrance point $P_0$ there exists a point $P_1$ which is a stigmatic image of point $P_0$. The presence of a stigmatic image requires that the trajectories connecting $P_0$ with $P_1$ are paraxial rays. Furthermore, we saw that by including the next higher-order term of the series expansion of the cosine function, deviations from the paraxial behavior occur, and that the presence of a stigmatic image is no longer warranted.

Although the present subsection did not follow a strategy that aimed explicitly at deriving the path equation of electrons in the homogeneous rotationally symmetric magnetic fields, it gave a qualitative description of the corresponding electron trajectories. In the following subsections, we expand our analysis to the more general case of an inhomogeneous rotationally symmetric magnetic field and derive the path equation similarly to the case of electrons in the rotationally symmetric electrostatic field in Sec. 6.4. It will be shown that apart from the axial field component $B_z$, there is a radial field component $B_r$ present, which needs to be considered in the derivation of the path equation. Inhomogeneous rotationally symmetric magnetic fields are generated by round magnetic electron lenses, as predominately used in transmission electron microscopes.

### 6.5.2 *The rotationally symmetric magnetic field*

#### 6.5.2.1 *General properties*

On the basis of the imaging properties of a homogeneous magnetic field, we would like to expand our analysis of electron trajectories and discuss the effect of an inhomogeneous but rotationally symmetric magnetic field on the propagation of electrons. Such fields can be generated by magnetic coils as used in *round* magnetic electron lenses, i.e. lenses that have an axis of rotation along the optical axis. As an example, Fig. 5.1 illustrates a simple magnetic lens which consists of a magnetic coil surrounded by a yoke. The current that runs through the coil generates a rotationally symmetric magnetic field as, for example, depicted in Fig. 5.2. The yoke is made of a ferromagnetic material to enhance the field strength on the optical axis.

The field enhancement that can be achieved depends on the permeability of the yoke and its design. A strong field concentration can be achieved by shaping the ferromagnetic material in so-called pole shoes or pole pieces (see Fig. 6.13) (Glaser, 1952; Tsuno, 2009). A magnetic lens, whose yoke ends in two pole shoes

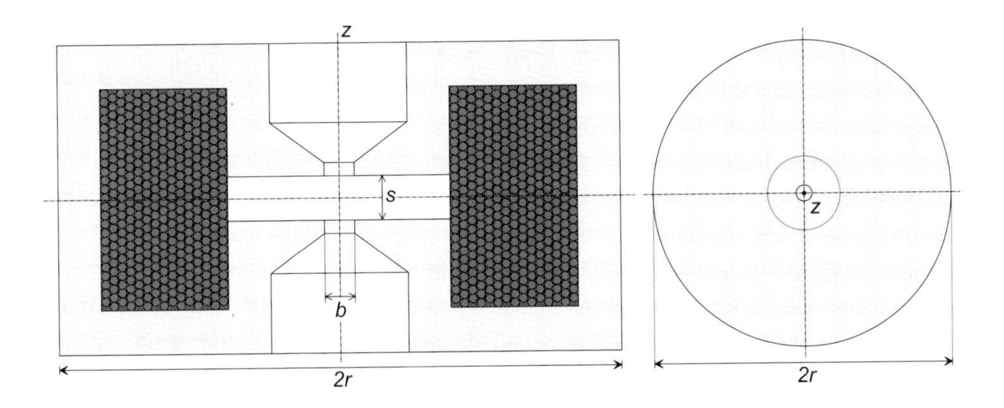


Fig. 6.13 A magnetic lens with ferromagnetic casting ends in two pole shoes which are at a distance $s$ to each other. The pole shoes concentrate the field and thus enhance the focusing strength of the lens. Left: side view cross-section. Right: top view at a scale 1:2 compared to the side view.

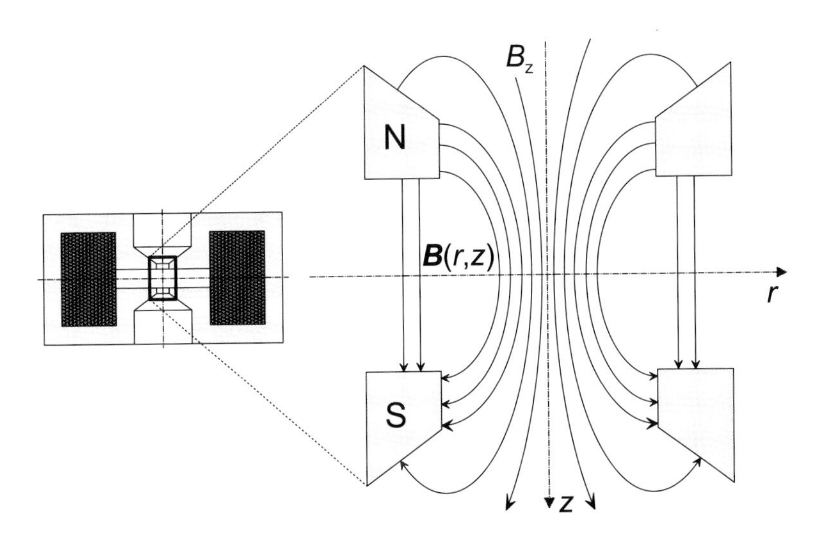


Fig. 6.14 Section through the rotationally symmetric magnetic field associated with a pole-shoe lens. The field has two components; an axial or longitudinal component $B_z(r, z)$ and a radial or transverse component $B_r(r, z)$.

is called a pole-shoe lens. The distance $s$ between the pole shoes and the diameter of the bore $b$ define the basic shape of the lens and thus the main characteristics of the corresponding magnetic field. The ratio $b/s$ is known as the shape parameter of the lens. The magnetic field, as illustrated in Fig. 6.14, depends on the design of the lens, i.e. on its geometry and in particular on the shape parameter, as well as on the ferromagnetic material of the yoke and the pole shoes, the number of windings of the coil and, of course, on the electric current that runs through the coil.

The strength of the lens reflected by its focal length can be adjusted by changing this current. The upper limit is, for instance, given by the magnetic saturation of the ferromagnetic pole shoes. The fact that the strength of an electron lens is adjustable reveals a distinct difference between light and electron optics. In general, the strength of a light optical lens made of glass is given by its design and cannot be changed. While in a magnetic lens an object can be focused by changing the current through the coil, i.e. the strength of the field, in light optics the only way to focus an object is to change the distance between object and lens.

A rotationally symmetric magnetic field, which is generated by a (short) magnetic lens, has two field components: an axial or longitudinal component $B_z(r, z)$ and a radial or transverse component $B_r(r, z)$. In general, both components depend on the distance $r$ from the optical axis, as well as on the position $z$ along the axis. Figure 6.14 qualitatively depicts the magnetic field associated with the lens shown in Fig. 6.13. Having the two field components $B_z$ and $B_r$, the magnetic field vectors $\boldsymbol{B}$ must lie in meridional planes.

The most important field component of a round magnetic lens is the axial component $B_z(r, z)$ *on* the optical axis, i.e. $B_z(0, z) = B_z(z)$. This will be shown in the next subsection when we derive the corresponding path equation. The fact that $B_z(0, z) = B_z(z)$ is of crucial importance is particularly true for paraxial trajectories which run close to the optical axis. Within the paraxial approximation, both the axial $B_z$ and the radial $B_r$ field component, can be expressed by the axial component $B_z(0, z)$ on the optical axis. This is similar to the case of a rotationally symmetric electrostatic field whose axial electrostatic potential $\Phi$ is the crucial quantity that defines the paraxial trajectories.

Figure 6.15 qualitatively depicts the axial field component $B_z$ on the optical axis $r = 0$ for four different cases. As discussed in the previous subsection, for the long magnetic coil $B_r = 0$ and $B_z =$ constant, i.e. the axial field component is constant within the coil. This situation corresponds to the dashed-dotted line in Fig. 6.15. Short magnetic lenses on the other hand show a maximum of the magnetic induction $B_z$ in the center of the coil. The height of the maximum and the decay of the field along the optical axis depend on the design of the lens. While the dotted curve in Fig. 6.15 schematically illustrates the case of a magnetic coil without yoke and pole pieces (Fig. 5.2), the dashed curve depicts the axial field component for a lens with iron casting (Fig. 5.1). The full line in Fig. 6.15 qualitatively illustrates the field concentration that can be achieved by employing a yoke which is shaped into pole shoes as, for instance, illustrated in Fig. 6.13. Provided that the current through the coils as well as the amount of windings is unchanged, there are general rules or trends which can be used to compare or forecast the strength of different lenses. The shorter the lens, the higher the field concentration on a small section on the optical axis, i.e. the higher the maximum of $B_z$. The smaller the bore of a pole shoe lens or the smaller the distance between the pole shoes, the larger the maximum of $B_z$ and thus the steeper the decay of the field along $B_z$. The decay

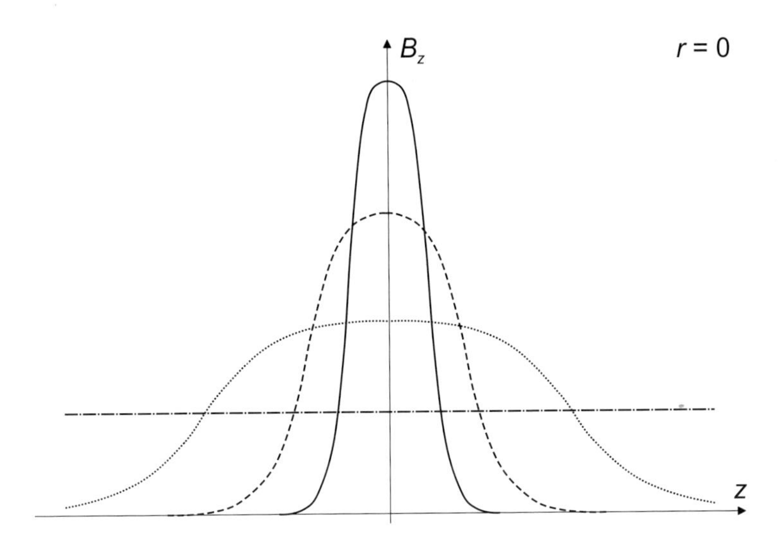


Fig. 6.15 Field profiles $B_z$. The axial magnetic induction $B_z(r = 0)$ for three different types of lenses. The dashed-dotted curve illustrates $B_z$ for a long magnetic coil. Within the long coil $B_z$ =constant. The dotted curve reflects $B_z$ of a short magnetic coil. The dashed curve shows the case of a short magnetic coil with yoke and the full line illustrates the case of a short magnetic coil with yoke and pole shoes.

of the field is expressed by the axial field gradient $dB_z/dz = B'_z$. As will be shown below, the axial field gradient $B'_z$ of $B_z$ along the optical axis essentially determines the radial field component $B_r$ within the paraxial approximation. As such, $B'_z$ is an important parameter. Furthermore, the larger the current that runs through the coils, the larger the maximum of $B_z$ and the steeper its gradient. This is reflected in the increased focusing power of a lens when the current through the coil is increased. This adjustability can be used to focus the image of an object in a given plane.

#### 6.5.2.2 *The path equation*

In the following, we derive the path equation for a rotationally symmetric magnetic field in a very similar way as was done in Sec. 6.4 for a rotationally symmetric electrostatic field. We closely follow the treatise of W. Glaser (1952).

We start with the derivation of a relation between $B_z$ and $B_r$, which allows us to express the radial component $B_r$ of the magnetic induction in terms of the axial component $B_z$. For this we employ the strategy which we have already developed for the rotationally symmetric electrostatic field in Sec. 6.4 and which led us to Eq. (6.9). Under the assumption of a source-free field, the inward flux of the field $\boldsymbol{B}$ into a small cylindrical volume element has to be equal to the outward flux. Since this strategy is identical to the case of the electrostatic field illustrated in Fig. 6.4, we do not repeat this procedure in detail and simply give the result. The radial component $B_r$ of the magnetic field close to the optical axis can be expressed as a function of the longitudinal component along the optical axis $B_z(0, z)$

$$B_r(r, z) = -\frac{r}{2}\frac{\partial B_z(0, z)}{\partial z}. \tag{6.30}$$

Under the assumption that the radial deviation of the paraxial rays from the optical axis is small, we simply denote $B_z(0, z)$ with $B_z$, keeping in mind however that $B_z$ is the component of $B_z(r, z)$ for $r = 0$. The relation in Eq. (6.30) holds under the same approximations that were made to deduce Eq. (6.8) for the rotationally symmetric electrostatic field.

In a first step towards the path equation we derive the equation of motion. While the solutions of the path equation are time-independent geometrical curves, which represent the rays, the solutions of the equation of motion represent electron trajectories which tell us at which point in time $t$ an electron passes a certain location $\boldsymbol{r}$ (or *vice versa*). Because of the rotational symmetry, we switch to cylindrical coordinates. Therefore, we also express the velocity of the electron in cylindrical coordinates $(r, \varphi, z)$: radial distance $r$, azimuth angle $\varphi$ and axial distance $z$. The components of the electron's velocity are then

$$v_r = \dot{r}\,, \quad v_\varphi = r\dot{\varphi} \quad \text{and} \quad v_z = \dot{z}, \tag{6.31}$$

with $v_r$ the radial velocity, $\dot{\varphi}$ the angular velocity, $v_\varphi$ the velocity in azimuthal (or angular) direction and $v_z$ the velocity component along the optical axis.

In order to derive the equations of motion for the three spatial components, we need to find out which forces act on an electron within a rotationally symmetric magnetic field. Similar to the velocity, the total magnetic force exerted on an electron can be divided into three components: a radial component $F_r$, a force in azimuthal (or angular) direction $F_\varphi$ and a force acting along the optical axis $F_z$. For generality, we assume that the angular velocity is finite, i.e. $\dot{\varphi} \neq 0$. The non-vanishing azimuthal velocity makes it necessary to take into account the centrifugal force. Hence, apart from the radial magnetic force $F_r$, the centrifugal force acts in radial direction as well. This leads us to the following equation of motion in radial direction

$$m_0\ddot{r} = K_r + m_0 r\dot{\varphi}^2. \tag{6.32}$$

The last term in Eq. (6.32) expresses the centrifugal force, which increases with the square of the angular velocity $\dot{\varphi}$.

Considering the force $F_\varphi$ in azimuthal direction, the law of the conservation of the angular momentum provides us with a suitable condition. The change of the angular momentum $d/dt\,(m_0 r v_\varphi)$ is equal to the angular momentum $rK_\varphi$, which is due to the angular force component $K_\varphi$ acting in direction of $\varphi$. This can be written as

$$\frac{d}{dt}\left(m_0 r v_\varphi\right) = rF_\varphi. \tag{6.33}$$

The relation for the $z$-component is simply

$$m_0\ddot{z} = F_z. \tag{6.34}$$

Equations (6.32), (6.33) and (6.34) provide us with three very general relations for the individual components of the force acting on an electron in a magnetic field. Now, however, we need to derive actual expressions for the magnetic force components $F_r$, $F_\varphi$ and $F_z$ in a rotationally symmetric magnetic field. For this we can employ the Lorentz force given in Eq. (5.1), with $\boldsymbol{E} = \boldsymbol{0}$. Substituting the velocity components from Eq. (6.31) in Eq. (5.1) yields three components of the Lorentz force $\boldsymbol{F}_\mathrm{L} = (F_r, F_\varphi, F_z)$, caused by the the magnetic induction $\boldsymbol{B} = (B_r, B_\varphi, B_z)$

$$\begin{pmatrix} F_r \\ F_\varphi \\ F_z \end{pmatrix} = q \begin{pmatrix} v_\varphi B_z - v_z B_\varphi \\ v_z B_r - v_r B_z \\ v_r B_\varphi - v_\varphi B_r \end{pmatrix}. \tag{6.35}$$

For the rotationally symmetric field under consideration, $B_\varphi = 0$. This allows us to reduce the above equation to

$$\begin{pmatrix} F_r \\ F_\varphi \\ F_z \end{pmatrix} = \begin{pmatrix} -ev_\varphi B_z \\ ev_r B_z - ev_z B_r \\ ev_\varphi B_r \end{pmatrix}. \tag{6.36}$$

Applying the relations in Eqs. (6.32), (6.33) and (6.34), as well as the expressions for the velocity in Eq. (6.31), we obtain the following equations of motion

$$m_0 \ddot{r} = -eB_z r\dot{\varphi} + m_0 r\dot{\varphi}^2 \tag{6.37}$$

$$\frac{d}{dt}\left(m_0 r^2 \dot{\varphi}\right) = er\dot{r}B_z - er\dot{z}B_r \tag{6.38}$$

$$m_0 \ddot{z} = er\dot{\varphi}B_r. \tag{6.39}$$

Equation (6.37) reveals that the radial magnetic force $F_r$ and the centrifugal force act in opposite directions. On the basis of the equations of motion for the individual components, including the expressions for the magnetic force components, we can advance by using the approximation for the field component $B_r$ in Eq. (6.30). The substitution of $B_r$ according to Eq. (6.30) yields for Eqs. (6.38) and (6.39)

$$\frac{d}{dt}\left(m_0 r^2 \dot{\varphi}\right) = eB_z r\dot{r} + \frac{e}{2} r^2 \frac{\partial B_z}{\partial z} \dot{z} = \frac{d}{dt}\left(\frac{e}{2} r^2 B_z\right) \tag{6.40}$$

$$m_0 \ddot{z} = -\frac{e}{2}\frac{\partial B_z}{\partial z} r^2 \dot{\varphi}. \tag{6.41}$$

Applying the product rule to obtain the time derivative of the last term in Eq. (6.40) shows that the result is identical to the sum in the second term of Eq. (6.40). Furthermore, the left-hand side as well as the right-hand side of Eq. (6.40) can directly be integrated over time, leading to

$$m_0 r^2 \dot{\varphi} = \frac{e}{2} r^2 B_z + \text{constant}. \tag{6.42}$$

We now make the assumption that in front of the magnetic field, i.e. in the field-free domain, the angular velocity $v_\varphi = 0$. Furthermore, we shall deal with meridional rays only. Meridional rays are rays whose initial direction in the field-free domain

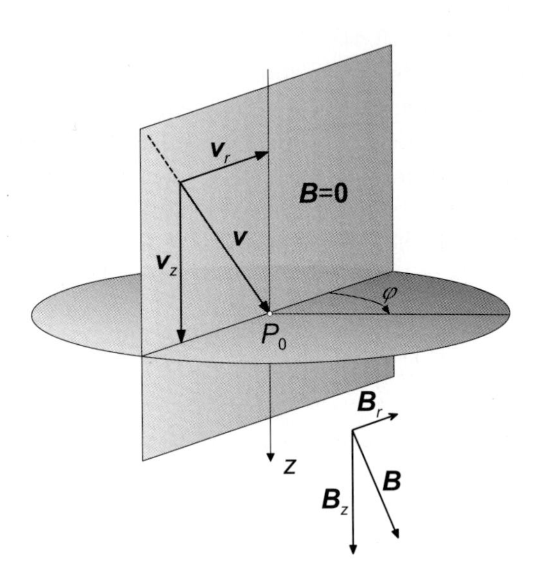


Fig. 6.16 Meridional ray. An electron in a field-free domain ($\boldsymbol{B} = \boldsymbol{0}$) enters a rotationally symmetric magnetic field in point $P_0$. The electron's velocity in front of the field has two components; a radial component $v_r$ and an axial component $v_z$. The angular velocity $\dot{\varphi}$ and thus $v_\varphi = r\dot{\varphi}$ are equal zero implying that the electron moves in a meridional plane.

lie in a flat plane which contains the origin $P_0$ and the optical axis (see Fig. 6.16). Hence, meridional rays lie in meridional planes or sections. An electron associated with a meridional path travels in radial direction with a velocity $v_r$ and in axial direction with a velocity $v_z$ (see Fig. 6.16). Since a finite angular velocity means that the electron has a velocity component normal to a meridional plane, the angular velocity of an electron following a meridional trajectory has to vanish; $\dot{\varphi} = 0$. With the restriction to meridional rays, we can conclude that the constant in Eq. (6.42) has to be equal zero. If were not, the angular velocity in the field-free domain would be proportional to this constant, and the corresponding path would not lie in a meridional plane. Setting the constant in Eq. (6.42) equal zero leads us to

$$\dot{\varphi} = \frac{eB_z}{2m_0}. \tag{6.43}$$

This relation shows that within the domain of the field $B_z$, the angular velocity $\dot{\varphi}$ is finite. The finite angular velocity $\dot{\varphi}$ is the source of the precession as, for example, illustrated by the helix in Fig. 6.6. This kind of precession in a magnetic field is the Larmor precession leading to the Larmor rotation. The important point about Eq. (6.43) is that from the equation of motion in angular direction (Eq. (6.40)), we obtain a relation for the angular velocity $\dot{\varphi}$ which we can employ in the equation of motion Eq. (6.37) in radial direction. We thus substitute the expression for $\dot{\varphi}$ from Eq. (6.43) in Eq. (6.37) and obtain that the total force acting in radial direction is

$$m_0\ddot{r} = -\frac{e^2}{4m_0}B_z^2 r. \tag{6.44}$$

This is the sum of the forces due to the magnetic field plus the centrifugal force. Equation (6.44) and the equivalent relation, which we obtained for the radial force in a rotationally symmetric electrostatic field given in Eq. (6.11), are very similar. In analogy to Eq. (6.11), Eq. (6.44) reveals that the force in a rotationally symmetric magnetic field driving an electron towards the optical axis increases linearly with its distance $r$ to the optical axis. Equation (6.44) is the equation of motion in radial direction, which describes how the radial distance $r$ changes with time as a paraxial electron travels through a rotationally symmetric magnetic field.

We now draw our attention briefly to the movement along the optical axis. We employ the expression in Eq. (6.43) and substitute $\dot{\varphi}$ in Eq. (6.41). From the substitution we obtain

$$\ddot{z} = -\left(\frac{e}{2m_0}\right)^2 r^2 B_z \frac{\partial B_z}{\partial z}. \tag{6.45}$$

This is the equation of motion in $z$-direction. The acceleration $\ddot{z}$ in $z$-direction vanishes on the optical axis, but it increases with increasing distance to the optical axis. In fact, it increases with the square of the radial distance $r$. From the discussion of the homogeneous rotationally symmetric field in the previous section, we saw that the electron does not experience an acceleration in $z$-direction, i.e. $v_z = \dot{z} =$ constant, regardless of the distance to the optical axis. This, of course, is not truly valid for an inhomogeneous rotationally symmetric magnetic field, as seen from Eqs. (6.41) and (6.45). However, within the paraxial approximation, terms that depend on the radial distance in second-order or higher ($\propto r^n$ with $n \geq 2$) are negligible. We thus conclude that we can set $\ddot{z} = 0$, because the axial force depends on $r^2$. This result is indeed equivalent to the case of the homogeneous rotationally symmetric magnetic field. However, it has to be pointed out that while for the homogeneous rotationally symmetric field $\ddot{z} = 0$ is exact, it is an approximation for the case of the inhomogeneous rotationally symmetric magnetic field, where it is strictly valid only on the optical axis, i.e. for $r \to 0$. The velocity of a paraxial electron along the optical axis $v_z$ is thus constant.

The last step of our analysis concerns the derivation of the radial path equation, i.e. the derivation of a time-independent description of the path of the electron in a rotationally symmetric magnetic field. For this we need to expand the equation of motion given in Eq. (6.44). This step is very similar to the case of the electrostatic field. We apply the law of the conservation of energy (non-relativistic) which tells us that the kinetic energy after the acceleration to the nominal electron energy is equal to the potential energy of the electron at the source point. This is expressed by

$$\frac{m_0 v^2}{2} = eU, \tag{6.46}$$

with $U$ the acceleration voltage of the microscope which essentially expresses the axial electrostatic potential $\Phi$ at the height of the specimen. For paraxial, i.e. flat electron trajectories, $v \approx v_z = \dot{z}$ and hence

$$v_z = \frac{dz}{dt} = \sqrt{\frac{2eU}{m_0}}. \tag{6.47}$$

Employing this relation, we can expand $\ddot{r}$ by

$$\ddot{r} = \frac{d}{dt}\left(\frac{dr}{dt}\right) = \frac{dz}{dt}\frac{d}{dz}\left(\frac{dz}{dt}\frac{dr}{dz}\right) = \sqrt{\frac{2eU}{m_0}}\frac{d}{dz}\left(\sqrt{\frac{2eU}{m_0}}\frac{dr}{dz}\right) = \frac{2eU}{m_0}\frac{d^2r}{dz^2}. \tag{6.48}$$

This relation allows us to substitute $\ddot{r}$ in Eq. (6.44). The substitution yields the equation of motion

$$\boxed{\frac{d^2r}{dz^2} + \frac{e}{8m_0U}B_z^2 r = 0.} \tag{6.49}$$

This equation is equivalent to the result obtained by Hans Busch (1926), who uncovered that a rotationally symmetric magnetic field affects electron trajectories in the same way as an optical lens influences the path of light rays. Rotationally symmetric magnetic fields can thus be used as electron lenses. A few years later, Otto Scherzer (1933) discussed this relation, considering in addition deviations from the ideal focusing behavior described by Eq. (6.49), i.e. deviations from the paraxial approximation.

Comparing the path equation of a rotationally symmetric electrostatic field given in Eq. (6.17) with the one for a rotationally symmetric magnetic field given in Eq. (6.49) reveals that both relations are linear homogeneous differential equation of second order. Each solution of these equations represents an electron trajectory within the paraxial approximation. A solution of Eq. (6.49) describes the radial distance $r$ of an electron as a function of its position along the optical axis. Equation (6.49) does not consider the change of the azimuthal position which changes due to the Larmor precession.

As can be seen from Eq. (6.44), the force exerted by a rotationally symmetric magnetic field on a charged particle is proportional to $B_z^2$. This implies that by changing the direction of the magnetic field, i.e. the polarity of the magnetic lens, the action of the lens is unaltered. However, what changes is the Larmor rotation, which according to Eq. (6.25) is proportional to $B_z$.

### 6.5.3 *The rotationally symmetric electromagnetic field*

#### 6.5.3.1 *Strategy*

Although the cases of the rotationally symmetric magnetic field discussed previously and the equivalent electrostatic field already provided us with essential information about the rather general structure of the path equation within the paraxial approximation (see Secs. 6.4 and 6.5.1), for completeness we treat the more general case of a rotationally symmetric electromagnetic field. Such a field consists of a rotationally symmetric electrostatic field and a rotationally symmetric magnetic field. We basically employ the same approximations which formed the fundament of the treatment of the electron propagation within the Gaussian optics in the previous subsections. The only field component that is of relevance is the one along the optical axis. However, in contrast to the treatment of the magnetic field, we will take into account the Larmor precession and describe the path of the electron in a rotating cartesian coordinate system, which considers the helical or spiral path of the electrons within the electromagnetic field. In order to do this, we need to describe a point in a plane by two independent variables; just the radial distance $r$ is no longer sufficient. Yet, we start the treatment on the basis of a fixed cartesian coordinate system and transform the raw equations of motion to the rotating coordinated system in a second step. The fixed cartesian coordinate system is determined by the $z$-axis reflecting the optical axis, and the axes $x$ and $y$, which describe the coordinates of a point in a radial plane orthogonal to the optical axis (see Fig. 6.1). Once the equations of motion are translated into the rotating coordinate system we will eliminate the time dependence. This step will allow us to come up with the path equations for $\hat{x}$ and $\hat{y}$, i.e. for the coordinates of the rotating coordinate system (see Fig. 6.3).

#### 6.5.3.2 *Equations of motion in the fixed cartesian coordinate system*

We start with the Lorentz force in Eq. (5.1), which allows us to write the equations of motion in $x$- and $y$-directions for $\boldsymbol{B} = (B_x, B_y, B_z)$ and $\boldsymbol{E} = (E_x, E_y, E_z)$ as

$$m_0 \ddot{x} = qE_x + q\left(\dot{y}B_z - \dot{z}B_y\right) \tag{6.50}$$

$$m_0 \ddot{y} = qE_y + q\left(\dot{z}B_x - \dot{x}B_z\right). \tag{6.51}$$

In analogy to the previous sections, $\dot{x} = dx/dt = v_x$ is the velocity of the electron of charge $q = -e$ in $x$-direction, and $\ddot{x} = d^2x/dt^2 = a_x$ its respective acceleration. The equivalent holds for $\dot{y}$ and $\ddot{y}$ in $y$-direction.

So far, no assumptions have been made about the symmetry of the fields. The electrostatic field components $E_x$, $E_y$ and $E_z$ as well as $B_x$, $B_y$ and $B_z$ still depend on $x$, $y$ and $z$. However, we now apply the paraxial approximation for the electrostatic potential $\phi$ and the magnetic field components $B_x$ and $B_y$, which are expressed in Eqs. (6.9) and (6.30). Though we developed these relations in the previous sections for the radial field components $r$, they can be written equivalently in cartesian coordinates for the magnetic field components as

$$B_x = -\frac{x}{2}\frac{\partial B_z}{\partial z}, \qquad B_y = -\frac{y}{2}\frac{\partial B_z}{\partial z}, \tag{6.52}$$

and for the electrostatic field components as

$$E_x = -\frac{x}{2}\frac{\partial E_z}{\partial z} = \frac{x}{2}\frac{\partial^2 \phi}{\partial z^2} = \frac{x}{2}\Phi'', \qquad E_y = -\frac{y}{2}\frac{\partial E_z}{\partial z} = \frac{y}{2}\frac{\partial^2 \phi}{\partial z^2} = \frac{y}{2}\Phi''. \tag{6.53}$$

We recall that $B_z$ and $\Phi$ are the components of the magnetic induction and of the electrostatic potential on the optical axis, i.e. $B_z = B_z(r = 0, z)$ and $\Phi = \Phi(r = 0, z)$. Substituting these relations in the above equations of motion, dividing them by $m_0$, replacing $q$ with $-e$ and writing $\partial B_z/\partial z$ as $B_z'$ yields the paraxial equation of motion for the $x$-direction

$$\ddot{x} = -\frac{e}{m_0}\frac{x}{2}\Phi'' - \frac{e}{m_0}\left(\dot{z}\frac{y}{2}B_z' + \dot{y}B_z\right), \tag{6.54}$$

and the equivalent expression for the $y$-direction

$$\ddot{y} = -\frac{e}{m_0}\frac{y}{2}\Phi'' + \frac{e}{m_0}\left(\dot{z}\frac{x}{2}B_z' + \dot{x}B_z\right). \tag{6.55}$$

Eqs. (6.54) and (6.55) are two equations of motion in the fixed cartesian coordinate system. They can be simplified by transforming them into a rotating coordinate system which considers the Larmor rotation of the electrons within the electromagnetic field. The rotating coordinate system shall rotate around the optical axis with an angular velocity $\dot{\varphi}$.

#### 6.5.3.3 *Larmor precession*

In order to find a relation between the fixed and the rotating coordinate systems, we need to find out how the rotating coordinate system is related to the Larmor rotation of an electron. Figure 6.17 illustrates the problem that needs to be solved. An electron travels on a helix, similar to the one in Fig. 6.6, from a point $A$ to a point $B$. Figure 6.17 shows the projection of the helix onto a plane normal to the $z$-axis. The axis of the helix runs parallel to the $z$-axis and is denoted by $\tilde{z}$. The helix intersects the $z$-axis in $O$. Without loss of generality, the electron shall start in point $A$ on the $x$-axis of the fixed coordinate system, which at this point in time ($t = 0$) coincides with the $\hat{x}$-axis of the rotating coordinate system. When the electron reaches point $B$, it is rotated by an angle $\vartheta_L$ around the $\tilde{z}$-axis. Since the

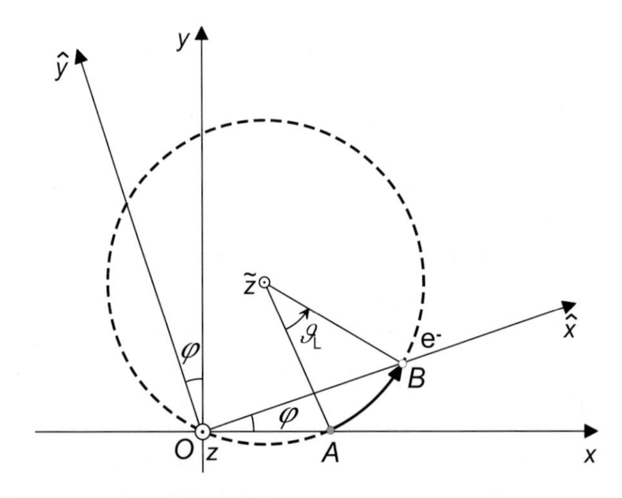


Fig. 6.17 Relation between the angle of rotation $\varphi$ of the rotating coordinate system and the angle of the Larmor rotation $\vartheta_L$.

rotating coordinate system follows the electron, the electron stays on the $\hat{x}$-axis of the rotating coordinate system. Hence, the rotating coordinate system follows the electron's path on the helix around $\tilde{z}$. While the electron rotates around the $\tilde{z}$-axis, the rotating coordinate system twists around the $z$-axis. This explains why the Larmor angle $\vartheta_{\rm L}$ and the rotation angle $\varphi$ of the $(\hat{x}, \hat{y})$-coordinate system cannot be equal. Obviously, the problem to solve is: what is the relation between the angle $\varphi$ and the Larmor angle $\vartheta_{\rm L}$? In the following paragraph we show on the basis of Fig. 6.18 that the angle of the rotating coordinate system $\varphi$ is half the angle of the Larmor rotation, i.e. $\varphi = \vartheta_{\rm L}/2$.

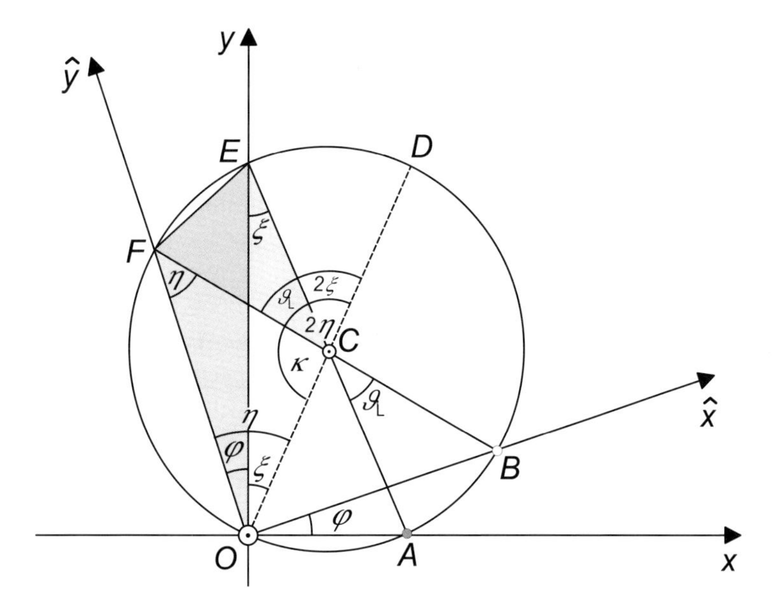


Fig. 6.18 Derivation of the relation between the rotation angle $\varphi$ of the rotating coordinate system and the angle of the Larmor rotation $\varphi = \vartheta_{\rm L}/2$.

We first redraw the situation depicted in Fig. 6.17. Figure 6.18 shows that the angle between the line connecting point $C$ with $E$ and the line connecting point $C$ with $F$, i.e. $\measuredangle ECF$ is equal to $\vartheta_{\rm L}$. The two shaded triangles in Fig. 6.18 enable us to identify the problem. Indeed, it is the well known problem of finding the relation between the central angle and the angle on the perimeter. We notice that the angles $\measuredangle COF$ and $\measuredangle OFC$ are identical; these are the angles of an equal-sided triangle. We denote this angle by $\eta$. For the same reason, the angles $\measuredangle OEC$ and $\measuredangle COE$ are identical, which we denote by $\xi$. Because the angles $\measuredangle COF = \measuredangle OFC = \eta$, the angle $\measuredangle FCO = \kappa$ is equal $180° - 2\eta$ (the sum of the angles of a triangle is equal $180°$). Hence, the angle $\measuredangle FCD$ is equal $2\eta$. Inspecting the other shaded triangle (ECF), we see that the central angle $\measuredangle ECO (= \kappa + \vartheta_{\rm L})$ is equal $180° - 2\xi$. Therefore, the angle

$\measuredangle DCE$ is equal $2\xi$. Finally, inspecting the angle $\measuredangle DCF$ reveals that $\vartheta_{\mathrm{L}} = 2\eta - 2\xi$. Furthermore, from the angle $\measuredangle COF$ we can read that $\varphi = \eta - \xi$. Comparing these two relations leads us to $\vartheta_{\mathrm{L}} = 2\varphi$.

The relation between $\vartheta_{\mathrm{L}}$ and $\varphi$ provides us with a purely geometrical relation between these two angles. However, what is actually needed is an expression for the angular velocity $\dot{\varphi}$. For this, we have to find an expression for the Larmor rotation of a paraxial electron in an inhomogeneous rotationally symmetric magnetic field. Let us start by rewriting Eq. (6.25), which expresses the Larmor rotation $\vartheta_{\mathrm{L}}$ caused by a homogeneous magnetic field of an electron with a non-vanishing normal velocity component $v_n$

$$\vartheta_{\mathrm{L}} = \int_{z_0}^{z} \frac{eB_z}{m_0 v_z} dz.$$

This relation describes the Larmor rotation as an integral of the magnetic field vector $B_z$ along the optical axis $z$. In general, the Larmor rotation needs to be considered if the electron has a non-vanishing velocity component perpendicular to a magnetic field component. Hence, for the case of an inhomogeneous rotationally symmetric magnetic field with $B_r$ and $B_z$, one would need to consider that the axial field component $B_z$ leads to a Larmor rotation if $v_x$ or $v_y$ are not equal zero. This is equivalent to the case of the homogeneous magnetic field where $B_z$ leads to the Larmor rotation because $v_n \neq 0$. However, in addition one has to consider that the radial magnetic field component $B_r$, which is perpendicular to the velocity of the electron $v_z$, leads to an additional term that contributes to the total Larmor rotation of the electron propagating in the inhomogeneous rotationally symmetric magnetic field.

Here again, the paraxial approximation allows us to make a simplification. For flat electron trajectories, the total velocity $v$ of the electron can be approximated by the axial velocity of the electron, i.e. $v \approx v_z = dz/dt$. This, of course, is only valid if the inclination angle with respect to the optical axis of the electron's trajectory is sufficiently small, i.e. if $v_n = \sqrt{v_x^2 + v_y^2} \ll v_z$. With this approximation, we essentially obtain the same expression for the Larmor rotation as for the homogeneous magnetic field. However, what we need to know is in fact the angular velocity $\dot{\varphi}$ of the rotating coordinate system and thus the angular velocity of the electron in the magnetic field, i.e. $\dot{\vartheta}_{\mathrm{L}}$. Hence, instead of expressing the Larmor rotation as a function of the distance along the optical axis (see Eq. (6.25)), we need to derive its dependence on time $t$. For this, we substitute $v_z$ with $dz/dt$ in Eq. (6.25). This yields

$$\vartheta_{\mathrm{L}} = \frac{e}{m_0} \int_{z_0}^{z} B_z \frac{dt}{dz} dz. \tag{6.56}$$

We now can exchange the integration variable and obtain

$$\vartheta_{\mathrm{L}} = \frac{e}{m_0} \int_{t_0}^{t} B_z dt, \tag{6.57}$$

which enables us to write the angular velocity of the Larmor rotation as

$$\dot{\vartheta}_{\mathrm{L}} = \frac{d\vartheta_{\mathrm{L}}}{dt} = \frac{eB_z}{m_0}. \tag{6.58}$$

The angular velocity $\dot{\varphi}$ of the rotating coordinate system is half this value, i.e.

$$\dot{\varphi} = \frac{\dot{\vartheta}_{\mathrm{L}}}{2} = \frac{e}{2m_0}B_z. \tag{6.59}$$

Furthermore, since $B_z = B_z(z)$, the corresponding angular acceleration $\ddot{\varphi}$ is

$$\ddot{\varphi} = \frac{\ddot{\vartheta}_{\mathrm{L}}}{2} = \frac{e}{2m_0}\frac{dB_z(z)}{dt} = \frac{e}{2m_0}\frac{dB_z}{dz}\frac{dz}{dt} = \frac{e}{2m_0}B_z'\dot{z}. \tag{6.60}$$

6.5.3.4 *The path equation in the rotating coordinate system*

With the description of the angular velocity $\dot{\varphi}$ of the rotating coordinate system, we are almost ready to transform the equations of motion given in Eqs. (6.54) and (6.55) into the rotating coordinate system $(\hat{x}, \hat{y}, \hat{z})$. According to Sec. 6.1 and Fig. 6.3, the rotating coordinate system $(\hat{x}, \hat{y}, \hat{z})$ can be described in terms of the stationary $(x, y, z)$ as

$$\begin{aligned} \hat{x} &= x\cos\varphi + y\sin\varphi \\ \hat{y} &= -x\sin\varphi + y\cos\varphi \\ \hat{z} &= z. \end{aligned} \tag{6.61}$$

This is a static transformation between the fixed and the rotating coordinate system. What we would like to know is how the rotating coordinate system evolves with time. Hence, we need to express its time dependence. Considering the product rule, the first derivative of $\hat{x}$ with respect to time yields

$$\dot{\hat{x}} = \dot{x}\cos\varphi - x\dot{\varphi}\sin\varphi + \dot{y}\sin\varphi + y\dot{\varphi}\cos\varphi = \dot{x}\cos\varphi + \dot{y}\sin\varphi + \dot{\varphi}\hat{y}.$$

Taking the time derivative for the $\hat{y}$ component yields a similar result. The first derivative of $\hat{x}$ and $\hat{y}$ with respect to time thus leads us to

$$\begin{aligned} \dot{x}\cos\varphi + \dot{y}\sin\varphi &= \dot{\hat{x}} - \hat{y}\dot{\varphi} \\ -\dot{x}\sin\varphi + \dot{y}\cos\varphi &= \dot{\hat{y}} + \hat{x}\dot{\varphi}. \end{aligned} \tag{6.62}$$

The second derivative of $\hat{x}$ and $\hat{y}$ with respect to time provides us the information about how the angular velocity of the rotating coordinate system changes with time. We obtain

$$\ddot{x}\cos\varphi + \ddot{y}\sin\varphi = \ddot{\hat{x}} - 2\dot{\hat{y}}\dot{\varphi} - \hat{y}\ddot{\varphi} - \hat{x}\dot{\varphi}^2 \tag{6.63}$$

$$-\ddot{x}\sin\varphi + \ddot{y}\cos\varphi = \ddot{\hat{y}} + 2\dot{\hat{x}}\dot{\varphi} + \hat{x}\ddot{\varphi} - \hat{y}\dot{\varphi}^2. \tag{6.64}$$

Now we have all the tools ready to start with the transformation of the equations of motion given in Eqs. (6.54) and (6.55) into the rotating coordinate system. We can multiply Eq. (6.54) with $\cos\varphi$ and Eq. (6.55) with $\sin\varphi$ and add them up. We obtain

$$\ddot{x}\cos\varphi + \ddot{y}\sin\varphi = \frac{e}{2m_0}\{-\Phi'' x\cos\varphi - (B_z'\dot{z}y\cos\varphi + 2B_z\dot{y}\cos\varphi)\}$$
$$+ \frac{e}{2m_0}\{-\Phi'' y\sin\varphi + (B_z'\dot{z}x\sin\varphi + 2B_z\dot{x}\sin\varphi)\}. \quad (6.65)$$

Employing the relations in Eqs. (6.61) and (6.62), the above equation can be summarized, yielding

$$\ddot{x}\cos\varphi + \ddot{y}\sin\varphi = \frac{e}{2m_0}\left(-\hat{x}\Phi'' - \hat{y}\dot{z}B_z' - 2B_z\{\dot{\hat{y}} + \hat{x}\dot{\varphi}\}\right). \quad (6.66)$$

Comparing Eq. (6.63) with Eq. (6.66) reveals that the right-hand side of Eq. (6.66) has to be equal to the right-hand side of Eq. (6.63), i.e.

$$\frac{e}{2m_0}\left(-\hat{x}\Phi'' - \hat{y}\dot{z}B_z' - 2B_z\{\dot{\hat{y}} + \hat{x}\dot{\varphi}\}\right) \equiv \ddot{\hat{x}} - 2\dot{\hat{y}}\dot{\varphi} - \hat{y}\ddot{\varphi} - \hat{x}\dot{\varphi}^2. \quad (6.67)$$

This can be rearranged, yielding

$$\ddot{\hat{x}} = \hat{x}\left(\dot{\varphi}^2 - \frac{e}{m_0}B_z\dot{\varphi} - \frac{e}{2m_0}\Phi''\right) + \dot{\hat{y}}\left(2\dot{\varphi} - \frac{e}{m_0}B_z\right) + \hat{y}\left(\ddot{\varphi} - \frac{e}{2m_0}\dot{z}B_z'\right). \quad (6.68)$$

Finally, we can replace $\dot{\varphi}$ and $\ddot{\varphi}$ from Eqs. (6.59) and (6.60) to obtain

$$\ddot{\hat{x}} = -\frac{e}{2m_0}\left(\Phi'' + \frac{e}{2m_0}B_z^2\right)\hat{x}. \quad (6.69)$$

This is the equation of motion in $\hat{x}$-direction, i.e. the equation of motion in the rotating coordinate system. In order to derive the corresponding equation for the $\hat{y}$ component, one can multiply Eq. (6.54) with $-\sin\varphi$ and Eq. (6.55) with $\cos\varphi$ and add the resulting relations. Then, Eqs. (6.61) and (6.62) can be employed to derive an expression that can be set equal to the right-hand side of Eq. (6.64). As in the treatment above for the $\hat{x}$ component, this can be simplified by substituting $\dot{\varphi}$ and $\ddot{\varphi}$ from Eqs. (6.59) and (6.60). One finally obtains

$$\ddot{\hat{y}} = -\frac{e}{2m_0}\left(\Phi'' + \frac{e}{2m_0}B_z^2\right)\hat{y}. \quad (6.70)$$

Equations (6.69) and (6.70) reveal again that within the paraxial approximation, the electromagnetic force ($m_0\ddot{\hat{x}}$ and $m_0\ddot{\hat{y}}$), which drives the electron towards the optical axis, increases linearly with the electron's distance to the optical axis.

In the last step, we need to eliminate the time dependence in Eqs. (6.69) and (6.70) to obtain the geometrical description of the electron trajectories, i.e. to obtain the path equations for $\hat{x}$ and $\hat{y}$. This can be done by making use of the conservation of energy as we did in the previous subsections.

$$\frac{m_0 v^2}{2} = \frac{m_0}{2}\left(\dot{x}^2 + \dot{y}^2 + \dot{z}^2\right) = e\phi(x, y, z) \quad (6.71)$$

For paraxial trajectories, the velocity in $z$-direction is much larger than the velocity in radial direction, i.e. $\dot{x}^2 \ll \dot{z}^2$ and $\dot{y}^2 \ll \dot{z}^2$, and for the potential $\phi$ we can use the potential along the optical axis $\Phi$. This leads us to

$$\frac{m_0}{2}\dot{z}^2 = e\Phi \quad \text{and further to} \quad \dot{z} = \frac{dz}{dt} = \sqrt{\frac{2e\Phi}{m_0}}. \tag{6.72}$$

Following the same procedure we applied in Eqs. (6.15) and (6.48), the relation for $dz/dt$ allows us to eliminate the time dependence in the equations of motion in Eqs. (6.69) and (6.70). With this we obtain the two path equations

$$\begin{aligned} \hat{x}'' + \frac{\Phi'}{2\Phi}\hat{x}' + \left(\frac{\Phi''}{4\Phi} + \frac{e}{8\Phi m_0}B_z^2\right)\hat{x} = 0 \\ \hat{y}'' + \frac{\Phi'}{2\Phi}\hat{y}' + \left(\frac{\Phi''}{4\Phi} + \frac{e}{8\Phi m_0}B_z^2\right)\hat{y} = 0. \end{aligned} \tag{6.73}$$

Since these two equations are independent from each other, i.e. there is no $\hat{y}$ term in the equation of the $\hat{x}$ component and *vice versa*, we can employ a single complex coordinate which expresses the $\hat{x}$ and the $\hat{y}$ component. For this, we use the complex coordinate $u = \hat{x} + \mathrm{i}\hat{y}$, as introduced in Sec. 6.1. This leads us to the path equation of an electron in a rotationally symmetric electromagnetic field

$$\boxed{u'' + \frac{\Phi'}{2\Phi}u' + \left(\frac{\Phi''}{4\Phi} + \frac{e}{8\Phi m_0}B_z^2\right)u = 0} \tag{6.74}$$

This relation holds for the non-relativistic case. If the speed of the electron is comparable to the speed of light in vacuum, relativistic correction is necessary. This correction, however, does not change the basic character of the path equation, which then can be written as (Scherzer, 1936b; Rose, 2009a)

$$u'' + \frac{\gamma}{2}\frac{\Phi'}{\Phi^*}u' + \left(\frac{\gamma}{4}\frac{\Phi''}{\Phi^*} + \frac{e}{8\Phi^* m_0}B_z^2\right)u = 0, \tag{6.75}$$

with $\gamma$ the relativistic factor from Eq. (5.4) for an electron travelling along the optical axis (i.e., $v_x = v_y = 0$). Furthermore, $\Phi^*$ is the relativistically modified electrostatic potential $\Phi$ as introduced in Eq. (5.61).

## 6.6 Series Expansion of the Fields

In order to derive the paraxial path equations given in Eqs. (6.17), (6.49) and (6.74), the assumption was made that the the respective electromagnetic fields can be described reasonably accurately by their field or potential components on the optical axis where $r = 0$. Hence, as long as the rays stay close to the optical axis where $r$ is small, i.e. within the approximation of Gaussian optics, this is justifiable. Strictly speaking, however, the usage of the axial field component is valid only for one ray — the ray that follows the optical axis. For any other ray, deviations from the Gaussian behavior have to be expected. In praxis, it turns out that the paraxial approximation is of sufficient precision to describe rays that are within a paraxial domain; that is, about one-fifth of the total inner diameter of a lens (Rose, 2009a). For the lens depicted in Fig. 5.1, this would correspond to trajectories that are within one-fifth of the bore of the lens, which represent the surfaces of the poles. This estimate shows that the approximation is reasonable and does not restrict the treatment to rays that have an unrealistically small spatial expansion.

In order to investigate deviations from the ideal Gaussian optics and to expand the understanding to rays that are further away from the optical axis, it is, however, necessary to consider a more advanced description of the electromagnetic fields. Of course, one could try to derive analytical or numerical solutions for the entire electromagnetic field associated with a certain optical unit. From there, one could try to derive the actual trajectories which are in agreement with the real fields. This can, however, easily lead to problems of high complexity. Realizing that the paraxial approximation might not be of sufficient accuracy, instead of deriving the real fields valid for any arbitrary trajectory without restrictions, we can consider an extended paraxial domain. We just widen the narrow paraxial cylinder along the optical axis and thus end up with a somewhat larger volume wherein the geometrical restrictions imposed on the trajectories are less strict.

The approach of employing an extended paraxial domain makes it possible to describe the electromagnetic fields associated with electron lenses by power series. The fields are expanded in power series of the radial distance $r$ from the optical axis, still, however, making use of the field (or potential) on the optical axis. Within the paraxial approximation, only the first term of such a series is considered and we saw that for $r \to 0$, the electrostatic potential $\phi(r, z) \to \Phi(z)$ and the magnetic field $B(r, z) \to B_z(z)$, with $\Phi$ the electrostatic potential on the optical axis and $B_z$ the magnetic induction on the optical axis. Indeed, the series expansion of electromagnetic fields within an extended paraxial domain has already been employed at the very beginning of electron optics (Busch, 1926).

For the case of a rotationally symmetric magnetic lens it can be shown (Glaser, 1952) that the magnetic induction $B_z$ parallel to the optical axis is given by

$$B_z(z, r) = B_z - \frac{r^2}{4} B_z^{(2)}(z) + \frac{r^4}{64} B_z^{(4)}(z) + O(r^6), \tag{6.76}$$

where $B_z = B_z(z)$ is the component of the magnetic field on the optical axis. The superscript in brackets denotes the degree of the derivative with respect to $z$, i.e. $B_z^{(n)} = \partial^n B_z / \partial z^n$. Similarly, the radial component of the magnetic field can be described by the following series

$$B_r(z, r) = -\frac{r}{2} B_z^{(1)}(z) + \frac{r^3}{16} B_z^{(3)}(z) - \frac{r^5}{384} B_z^{(5)}(z) + O(r^6). \tag{6.77}$$

For the rotationally symmetric electrostatic field, equivalent expressions can be deduced (Glaser, 1952). The series expansion is described as a polynomial of the radial distance $r$, with the individual terms containing the derivatives to varies degree of the electrostatic potential on the optical axis $\Phi$. The electric field component in $z$-direction can be described by

$$\phi(z,r) = \Phi - \frac{r^2}{4}\Phi^{(2)} + \frac{r^4}{64}\Phi^{(4)} + O(r^6), \tag{6.78}$$

where we omitted to write out explicitly the $z$-dependence of the electrostatic potential on the optical axis $\Phi = \Phi(z)$. The residual error, expressed by $O(r^6)$, is of the order of $r^6$. The important point about these series expansions is that the fields are still described approximately but we can now do this to any desirable degree within an extended paraxial domain.

Describing the electromagnetic fields as power series of the radial distance $r$ opens the way to consider only a certain amount of terms up to a desired order of $r$. If only terms of low order are considered in the derivation of the path equations, one can basically restrict the treatment to trajectories that are close to the optical axis where $r$ is small. Within the Gaussian optics, only terms of the series expansions are considered that are either constant or linear in $r$. From this one obtains the paraxial path equations. Of course, if one goes beyond the paraxial approximation and includes higher-order terms of the fields in the derivation of the path equations, the corresponding relations become more complex.

## 6.7 Imaging Within the Paraxial Approximation

### 6.7.1 *Theorem of optical imaging*

In the previous subsections we derived path equations for a rotationally symmetric electrostatic field, a magnetic field and an electromagnetic field. They all have an axis of rotational symmetry along the optical axis $z$. Comparing the path equations in Eqs. (6.17), (6.49) and (6.74) reveals that they have a very similar structure, which can be written as follows

$$u'' + Au' + Bu = 0, \tag{6.79}$$

with $u(z)$ the desired solution, $u' = du/dz$, $u'' = d^2u/dz^2$ and $A$ and $B$ are constants or, in our case, functions of $z$, i.e. $A = A(z)$ and $B = B(z)$. This is a linear homogeneous differential equation of second order. In general, a solution $u = u(z)$ of a differential equation is a function — a function which can represent the path of a paraxial electron. The function $u(z)$ describes the position $u = \hat{x} + i\hat{y}$ of the trajectory as a function of $z$. The above differential equation is called *linear* because the derivatives appear only to the power of one, i.e. there are no terms like $u'^2$. Furthermore, the equation is *homogeneous* because the right-hand side of the equation is equal to zero; there are no terms that do not depend on $u''$, $u'$ or $u$. Finally, it is a *second-order* differential equation because the highest derivative is of second order, i.e. $u'' = d^2u/dz^2$. All of the path equations we derived have this structure (see Eqs. (6.17), (6.49) and (6.74)). However, it is important to note that not every mathematical solution of the path equation is necessarily a paraxial trajectory. The mathematical expression of the path equation does not imply the paraxial conditions *per se*. For instance, we can obtain a solution for the paraxial path equation which would correspond to a trajectory far away from the optical axis, or which has a large inclination angle. Although such solutions are correct solutions of the path equation, they do not fulfill the conditions of paraxial rays.

It is not our goal to find explicit solutions to the path equations mentioned above; for this the reader is referred to mathematical textbooks. We would like to discuss the general characteristics of the solutions of the path equations. What we would like to point out is that a linear homogeneous differential equation of second order always has two linearly independent solutions, which are called particular solutions. Moreover, any solution of the differential equation can be represented as a linear combination of two particular solutions.

Let us assume that the functions $u_1(z)$ and $u_2(z)$ are two linearly independent, i.e. particular, solutions of Eq. (6.79) or of any of the path equations we derived above. The fact that $u_1$ and $u_2$ are linearly independent simply means that they cannot be transformed to each other by a multiplication. The particular solutions can be derived from a general solution by applying appropriate boundary conditions. Any other function $u(z)$, which also solves the differential equation, can then be expressed as a linear combination of the particular solutions

$$u(z) = c_1 u_1(z) + c_2 u_2(z).$$

The constants $c_1$ and $c_2$ are characteristic for each trajectory $u(z)$. The two-dimensional parameter space $(c_1, c_2)$ defines all mathematically correct solutions of the differential equation, but not necessarily all solutions that correspond to paraxial trajectories.

Since we deal with path equations of electrons in electromagnetic fields, two particular solutions $u_1$ and $u_2$ of the path equations in Eqs. (6.17), (6.49) or (6.74) are referred to as *fundamental rays*. They are called *fundamental* because any other trajectory can be derived from them. An actual set of particular solutions depends on the choice of the boundary conditions. This, in principle, leaves us with a huge variety of possible fundamental rays, as long as they are linearly independent. Yet, there are fundamental rays that are better than others; better in a sense that they provide a direct insight and overview of all feasible trajectories. Following the notation of H. Rose (2009a), we can set the following boundary conditions for a ray $u_\alpha(z)$ and a ray $u_\pi(z)$ (see, e.g. Scherzer, 1936b)

$$\begin{aligned} u_\alpha(z_0) &= 0, \qquad u'_\alpha(z_0) = \frac{du_\alpha}{dz}(z_0) = 1, \\ u_\pi(z_0) &= 1, \qquad u'_\pi(z_0) = \frac{du_\pi}{dz}(z_0) = 0. \end{aligned} \tag{6.80}$$

These are two rays where $u_\alpha$ starts in the plane $z_0$ on the optical axis with a slope of 1 (i.e. at an inclination angle $\gamma = \pi/4 = 45°$), and $u_\pi$ starts in an off-axial distance of 1 in a plane at $z_0$ with vanishing slope. The solution $u_\alpha$ is referred to as the *axial ray* and the solution $u_\pi$ is called the *principal ray*. They are illustrated in Fig. 6.19.

It is clear that the condition of the axial ray $u'_\alpha(z_0) = 1$ does not fulfill the paraxial condition of small inclination angles $\gamma$ (Glaser, 1952). However, as pointed out above, the individual solutions do not need to fulfill the conditions of the paraxial approximation. This is also true for the fundamental rays. It has to be considered that by a suitable choice of the constants $c_1$ and $c_2$, any desired paraxial ray can be deduced from these two linearly independent solutions. It is the choice of $c_1$ and $c_2$ which shall reflect the paraxial approximation.

Let us assume $u(z)$ is a paraxial ray determined by the fundamental rays $u_\alpha$ and $u_\pi$ and the constants $c_1$ and $c_2$ such that

$$u(z) = c_1 u_\pi(z) + c_2 u_\alpha(z). \tag{6.81}$$

The paraxial ray $u(z)$ shall run through a point $P_0$ in the plane of $z_0$, which is at a distance $u_0$ from the optical axis (see Fig. 6.20). According Eq. (6.80), we can write that the trajectory $u(z)$ in the plane $z_0$ is given by

$$u(z_0) = c_1 u_\pi(z_0) + c_2 u_\alpha(z_0) = c_1. \tag{6.82}$$

Hence, the value of the constant $c_1$ simply reflects the off-axial distance of the point $P_0$, which we can denote by $u_0$. On the other hand, the value of the constant $c_2$ represents the slope of the trajectory $u$ in point $P_0$. This can be seen from

$$u'(z_0) = c_1 u'_\pi(z_0) + c_2 u'_\alpha(z_0) = c_2, \tag{6.83}$$

where again we employed Eq. (6.80). Since the constant $c_2$ reflects the initial slope of a trajectory through $P_0$, we can denote it by $\gamma$. With the substitutions for $c_1$ and $c_2$ we can write that an arbitrary trajectory through point $P_0$ is determined by

$$u(z) = u_0 u_\pi(z) + \gamma u_\alpha(z). \tag{6.84}$$

This is essentially the same relation as the one in Eq. (6.81), with the only difference as being that the constants now have a direct interpretation. From this we can picture that there is a pencil of rays emerging in point $P_0$. Each ray enters the electromagnetic field under a different angle $\gamma$. Yet, since all of them go through $P_0$, the value $u_0$ is the same for all the rays forming the pencil. On the other hand, because all of the rays have a different $\gamma$, they follow different trajectories. This is expressed in Eq. (6.84).

Let us assume that the electromagnetic field is strong enough to bend the axial ray $u_\alpha$ back to the optical axis, such that apart from the intersection with the optical axis in the plane $z_0$ (see Fig. 6.19), there is another intersection with the optical axis in a plane $z_1$, i.e. $u_\alpha(z_1) = 0$. What are the consequences of the condition $u_\alpha(z_1) = 0$ for the pencil of rays that emerges in point $P_0$?

The trajectories $u(z)$ through point $P_0$ are defined by Eq. (6.84). We thus obtain for the plane $z_1$

$$u(z_1) = u_0 u_\pi(z_1) + \gamma u_\alpha(z_1) = u_0 u_\pi(z_1) = \text{constant}. \tag{6.85}$$

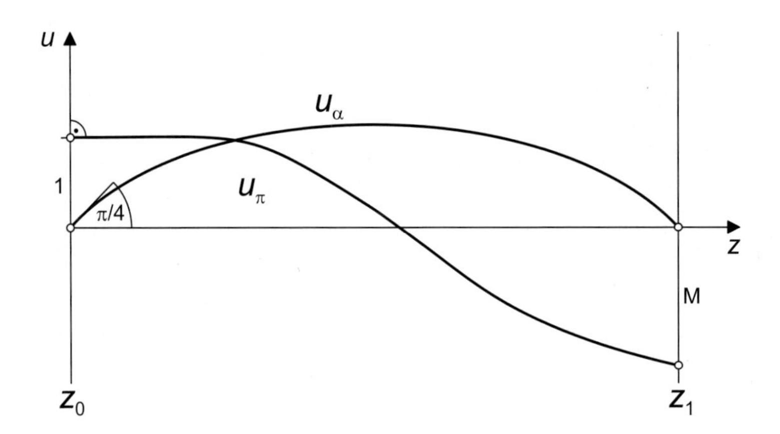


Fig. 6.19 Fundamental rays. The axial ray $u_\alpha$ originates at $z = z_0$ on the optical axis with a slope of $\pi/4$, and the principal ray $u_\pi$ originates with vanishing slope at a distance $u(z_0) = 1$ from the optical axis.

Since $u_\pi(z_1)$ is simply a constant value as well as $u_0$, all the rays emerging under varying angles $\gamma$ in point $P_0$ in the plane $z_0$ intersect in the point $P_1$ located in the plane $z_1$. Regardless of the position of $P_0$ in $z_0$, there is a corresponding point $P_1$ in the plane at $z_1$. This is a special characteristic of the plane $z_1$. Although the contribution of $u_0 u_\pi(z)$ is identical for all rays emerging in point $P_0$ for any value of $z$, the contribution of $\gamma u_\alpha(z)$ varies with the initial inclination angle $\gamma$ of a particular ray. Hence, it is only in the plane $z_1$ that the rays intersect again in a single point.

Let us imagine that there is a small radial plane around the optical axis, and that the points in this segment emit bundles of electrons. The requirement that the plane is small simply means that we deal with paraxial rays. For each of the points in this plane, the condition depicted in Fig. 6.20 is fulfilled. We did not make any assumption about the position of the point $P_0$, except that $u_0$ as well as the initial inclination angle $\gamma$ of the rays that emerge in point $P_0$ shall be small. Hence, each point in $z_0$ is transferred to a point in $z_1$. This simply means that a plane in $z_0$ around the optical axis is transferred to a corresponding plane at $z = z_1$.

This result enables us to apply optical terminology to characterize the electron transfer of a rotationally symmetric electromagnetic field. Point $P_0$ can be called an object point and the plane $z_0$ is the object plane. Point $P_1$ is the image point, which is conjugate to $P_0$, and lies in the image plane denoted by $z_1$. From this we can say that the electromagnetic field images the plane $z_0$ into the plane at $z = z_1$. Furthermore, since the distance to the optical axis of the object point $P_0$ is $u_0$ and the distance to the optical axis of the image point $P_1$ is $u_0 u_\pi(z_1)$, we can say that the ratio $u_0 u_\pi(z_1)/u_0 = u_\pi(z_1)$ provides a means to describe the lateral relation between object plane and image plane. Hence, the value of the principal ray in the

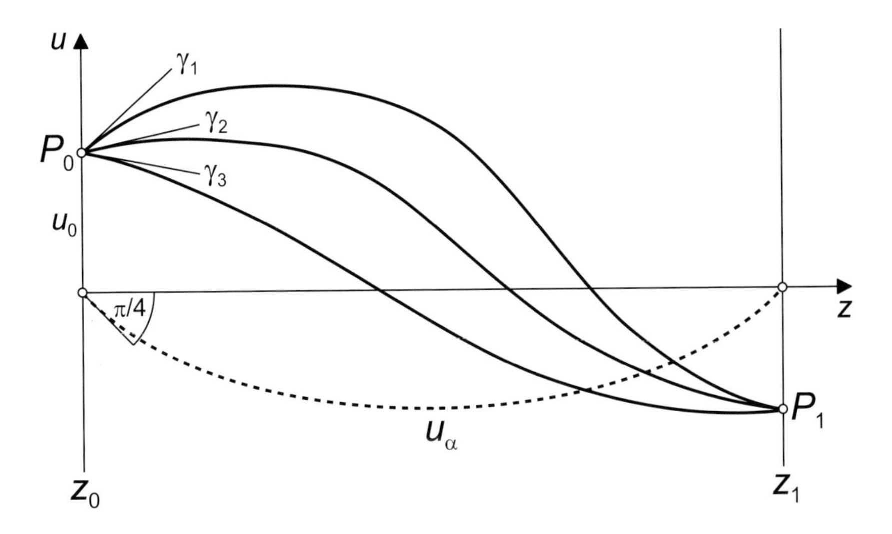


Fig. 6.20 Object point and image point. The axial ray $u_\alpha$ intersects the optical axis at two positions along the axis, in the planes located in $z_0$ and $z_1$.

image plane, i.e.

$$M = u_\pi(z_1), \tag{6.86}$$

is called the *lateral* magnification $M$. Since there is only one principal ray $u_\pi$ with a unique value $u_\pi(z_1)$, the magnification is independent from the position of the object point $P_0$.

On the assumption that the force driving an electron back to the optical axis increases linearly with the electron's distance to the optical axis, H. Busch (1926) came to exactly the same result; the rays of a paraxial pencil emerging from a single point which is located on (or near) the optical axis, entering an electromagnetic field strong enough to bend the electron trajectories back to the optical axis, intersect again in a single point located on (or near) the optical axis. This finding from 1926 reasoned the application of rotationally symmetric electromagnetic fields as lenses for electrons in analogy to glass lenses that act on light rays. A rotationally symmetric electromagnetic field focuses electrons. This leads us to the theorem of optical imaging (Glaser, 1952):

**Theorem of optical imaging**
Paraxial rays, which emerge from a set of points in a plane, recombine in a corresponding set of points such that the set of corresponding points forms a (magnified or demagnified) image of the original set of points.

It was essentially the theorem of optical imaging applicable to electrons in rotationally symmetric electromagnetic fields which motivated Knoll and Ruska (1932) to build the first electron microscope.

The theorem of optical imaging summarizes the imaging effect of rotationally symmetric electromagnetic fields for the case of paraxial electrons. A point in the object plane is imaged in a point in the image plane. Since all trajectories going through the object point $P_0$ intersect in a *single* point $P_1$ in the image plane, the point $P_1$ forms a *stigmatic* image of $P_0$. This leads us to the very fundamental result summarized in Busch's theorem (see, e.g. Rose, 2009a):

> **Busch's theorem**
> Within the paraxial approximation, a rotationally symmetric electromagnetic field acts as a double convex, i.e. converging lens, which forms a distortion-free stigmatic image.

### 6.7.2 *Generalized theorem of Lippich*

In Sec. 6.5.2 we restricted the treatment to meridional rays, i.e. rays that lie in meridional planes and, provided that they do not run parallel to the optical axis, thus intersect the optical axis at some value of $z$. We excluded rays that do not lie in sections and thus do not intersect the optical axis. Such rays are called *skew*, see Fig. 6.21. Since we excluded skew trajectories from the derivation of the path equations, one might expect that the equations given in Eqs. (6.17), (6.49) and (6.74) are not valid for such rays but apply only to meridional rays. However, there is a straightforward principle which goes back to F. Lippich and is called, according to Glaser (1952), the generalized theorem of Lippich. It can be formulated as

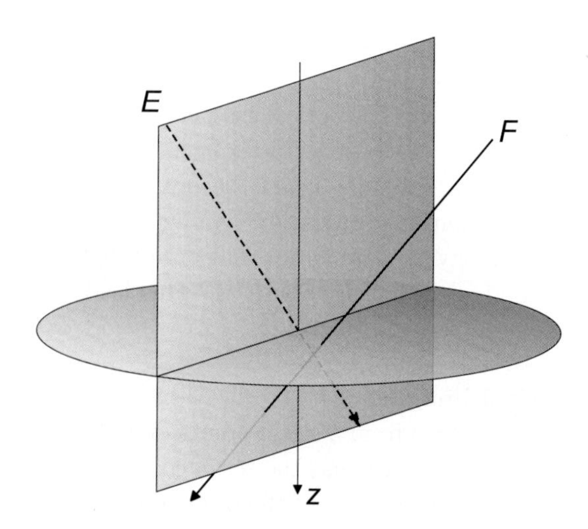


Fig. 6.21 Skew ray vs. meridional ray. The dashed line, $E$, corresponds to a meridional ray as illustrated in Fig. 6.16. The line $F$ corresponds to a skew ray or skew trajectory. A skew ray does not cross the optical axis $z$ and does not lie in a meridional plane.

follows:

**Generalized theorem of Lippich**
An arbitrary skew ray $F$ can be considered to be composed of two projected rays $F^{\hat{x}}$ and $F^{\hat{y}}$ which run in the meridional planes $\hat{y} = 0$ and $\hat{x} = 0$, defined by the rotating coordinate system. The rays $F^{\hat{x}}$ and $F^{\hat{y}}$ are the top and side views of the skew ray $F$, respectively.

In the path equations we derived above and which are given in Eqs. (6.17), (6.49) and (6.74), the coordinates $\hat{x}$ and $\hat{y}$ are treated separately. There are no mixed terms like $\hat{x}\hat{y}$. This, for instance, enabled us to straightforwardly summarize the equations for $\hat{x}$ and $\hat{y}$ in a single one employing the complex notation $u = \hat{x} + \mathrm{i}\hat{y}$, see Eq. (6.74). We can thus conclude that a skew ray can be treated as a superposition of two projections that represent meridional rays. Furthermore, since the meridional rays are fully describable by, for example, Eq. (6.74), we can state that the path equations we developed for meridional rays are adequate to describe skew rays as well. Hence, a skew ray can be described by the abovementioned equations simply by considering two projections on meridional planes. This is depicted in Fig. 6.22.

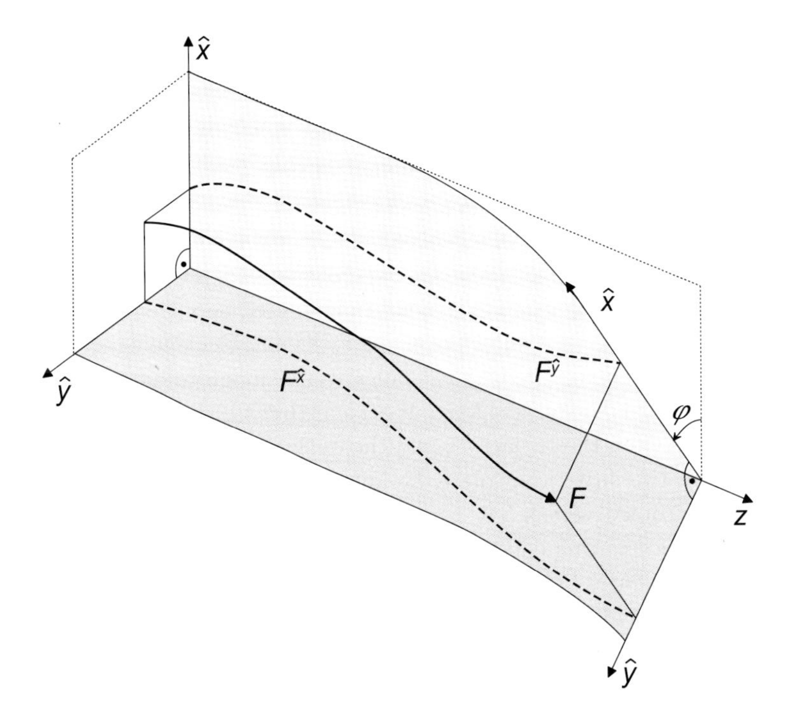


Fig. 6.22 Generalized theorem of Lippich. A skew ray $F$ can be considered to be composed of two projections in a rotating coordinate system $(\hat{x}, \hat{y}, z)$. The projection of $F$ onto the plane $\hat{x} = 0$, i.e. the plane formed by the axes $\hat{y}$ and $z$, gives the meridional ray $F^{\hat{x}}$, and the projection of $F$ onto the plane $\hat{y} = 0$, i.e. the plane formed by the axes $\hat{x}$ and $z$, gives the meridional ray $F^{\hat{y}}$.

As a consequence of the generalized theorem of Lippich, we can restrict the optical discussion to meridional rays without loss of generality.

### 6.7.3 *Real image and virtual image*

The above discussion about the paraxial imaging properties of a rotationally symmetric electromagnetic field showed that the field acts as a convex lens. The electromagnetic field exerts a radial force which bends an electron's trajectory towards the optical axis. This is illustrated for two fundamental rays in Fig. 6.19. Figure 6.19 shows the trajectory of an electron entering an electromagnetic field parallel to the optical axis and the trajectory of an electron entering the field on the optical axis with finite inclination angle. In both cases, the rotationally symmetric electromagnetic field drives the electron from its original direction of propagation towards the optical axis. Within the paraxial approximation, the radial force increases linearly with the electron's distance to the optical axis. Furthermore, the radial force increases with the square of the magnetic field $B_z$ and linearly with the second derivative of the electrostatic potential on the optical axis $\Phi''$ which, within the paraxial approximation, is proportional to the radial field component $E_r$ (see, e.g. Eqs. (6.11), (6.44) and (6.69)). Hence, with increasing field strength an electron is accelerated more strongly towards the optical axis and thus finds its way to the axis within a shorter axial distance $\Delta z$.

So far, we have simply implied that the radial force exerted by the electromagnetic field inflects the trajectories towards the optical axis, and thus assumed that the trajectories either intersect the optical axis within the field or intersect the optical axis at some point behind the domain of the field (see, e.g. Fig. 6.19). This of course is only correct provided that the force exerted by the electromagnetic field is sufficiently strong or, alternatively, if the field is long enough.

However, what happens if an electron enters a weak electromagnetic field under a given inclination angle and the field is not strong enough to inflect the trajectory of the electron towards the optical axis? It is clear that for this case there is no point of intersection between the trajectory and the optical axis in or behind the domain of the field. Does this situation violate the imaging properties of an electromagnetic lens? This is a problem which has already been addressed by Busch (1926), when he investigated the focusing effect of rotationally symmetric electromagnetic fields.

Figure 6.23 illustrates two trajectories: A and B. Both trajectories enter the field in the axial point $P_0$ under an inclination angle $\gamma$. Let us assume that the electrons associated with the two rays experience different electromagnetic fields. While ray A represents the interaction with a strong field, the electron of trajectory B shall interact with a weak field. The gray shaded area in Fig. 6.23 indicates the spatial extension of the rotationally symmetric electromagnetic field along the optical axis. If the slope of a trajectory changes its sign within the field, the trajectory intersects the optical axis behind the entrance point $P_0$, either within the field or beyond.

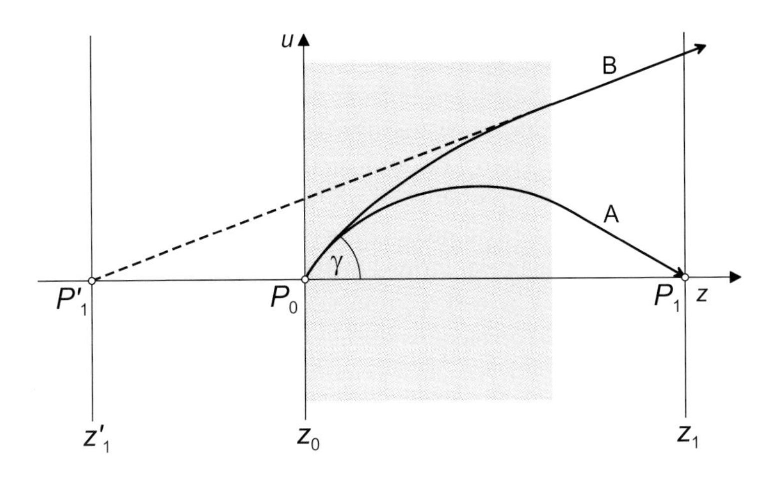


Fig. 6.23 Real image $P_1$ and virtual image $P_1'$.

This is essentially the condition for the formation of an image of point $P_0$. For ray A, the field is sufficiently strong to cause an inflection of the trajectory. This results in a point of intersection of the trajectory with the optical axis. The point of intersection is point $P_1$, which is an image of point $P_0$. Since the slope of the trajectory does not change outside the field (i.e. no radial acceleration beyond the field), the trajectory is a straight line in the field-free domain. Hence, for the case that the point of intersection with the optical axis lies outside the domain of the field, $P_1$ can be constructed simply by drawing the asymptote to the ray at the point it leaves the field.

Let us assume point $P_0$ is a source point such that there is an entire pencil of rays entering the field from $P_0$. Since the field is strong, all the rays of the pencil intersect the optical axis behind point $P_0$. Regardless of the (small) initial inclination angle $\gamma$, there is only one point of intersection with the optical axis for the entire paraxial pencil. This is point $P_1$. All the rays that emerge in the source point $P_0$ intersect the axis in point $P_1$. Hence, from a position further down the optical axis, it seems that all the rays emerge from the point $P_1$. According to the theorem of optical imaging, point $P_1$ is an image of the source point $P_0$. For the case that there is a real point of intersection between the individual trajectories and the optical axis, it can be said that $P_1$ is a *real* image of the entrance point $P_0$.

For ray B, which represents the interaction with a weaker field, the situation is different. The field is not strong enough to inflect the trajectory. This simply means that the electron, though affected by the presence of the field, follows a trajectory of increasing radial distance. The inclination angle of the ray which leaves the field is smaller than $\gamma$, but an inflection point is not present. Consequently, ray B does not intersect the optical axis behind point $P_0$. However, we can apply the same procedure as we did for ray A. We simply draw the asymptote to the ray from the point it leaves the field. This is indicated by the dashed line in Fig. 6.23. Instead of intersecting the optical axis behind point $P_0$, the asymptote to ray B intersects the optical axis in point $P_1'$, which lies in front of point $P_0$. This is not a real point of intersection between the trajectory and the optical axis. However, point $P_1'$ and point $P_1$ are still equivalent. From a position along the optical axis beyond the field, it seems that ray B emerges from point $P_1'$, though it actually emerges from point $P_0$. Since the point $P_1'$ is not a real point, $P_1'$ is called a *virtual* image of point $P_0$.

A virtual image is, for instance, formed in the case of an electron source where the electrons are accelerated by a short rotationally symmetric electrostatic field. Though the electrostatic field accelerates the electrons in direction along the optical axis, the electrons also experience an acceleration towards the optical axis (see Sec. 6.4). If the field is nearly homogeneous, the radial force is weak and does not lead to an inflection of the trajectories. The individual trajectories of the emitted electrons do not intersect along the optical axis (see Fig. 6.24). However, the rays form a virtual image of the source point which lies inside the cathode. Hence, looking at the emitted electrons after being accelerated by the nearly homogeneous electrostatic field, the electrons seem to emerge from a point $P_1'$, which is the virtual image of the actual source point. For a real electron source, there is not a single point emitting electrons but a small area. The emitted electrons are focused to a cross-over, which is then further transferred by the condenser lenses towards the specimen plane. For field emission electron sources, this first cross-over lies within the tip and is thus virtual (Swanson and Schwind, 1997).

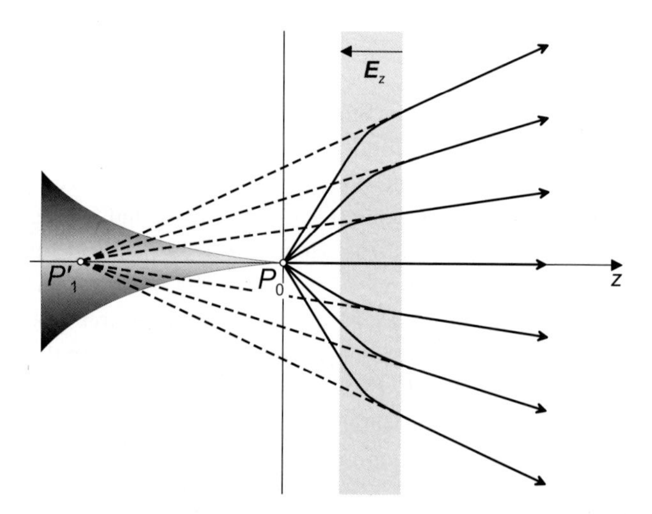


Fig. 6.24 A cathode emits electrons in point $P_0$. The electrons are accelerated in a short, nearly homogeneous electrostatic field. The asymptotes of the trajectories in front of the field intersect in the source point $P_0$. The asymptotes of the rays behind the field intersect in the virtual image point $P_1'$ which lies within the cathode.

### 6.7.4 *Asymptotic cardinal elements*

#### 6.7.4.1 *Real vs. asymptotic focal point*

Although the spatial extension of the electromagnetic field employed in an electron lens is in principle infinite, as the field is highly concentrated to a small area (see Fig. 6.15), the main effect of a lens field on the propagation of electrons can be considered to be confined to a certain domain. This was explained in detail in Sec. 6.2. We can imagine that the domain wherein the field effectively affects the trajectories of the electrons is limited by two boundary planes perpendicular to the optical axis. These planes shall define the spatial extension of the electromagnetic lens. The important point about this concept is that the trajectory of an electron is altered only within the domain of the electromagnetic field. Hence, within the virtual boundary planes of the lens the rays are in general curved lines, but beyond the domain of the lens the rays are straight lines. This is illustrated for the case of a principal ray $u_\pi$ in Fig. 6.25. The principal ray $u_\pi$ is defined by the boundary conditions given in Eq. (6.80). The ray is parallel to the optical axis in front of the field. Within the field, which is indicated by the gray shaded area in Fig. 6.25, it is a curved line, and behind the field it is a straight line again.

Figure 6.26 shows the effect of the electromagnetic field in Fig. 6.25 on a set of rays which all run parallel to the optical axis in front of the field. One of the rays shall be the principal ray $u_\pi$ defined by the boundary conditions in Eq. (6.80), and the two other rays are given by $c_1 u_\pi$ with $c_1$ an arbitrary constant. The principal ray intersects the optical axis in a point $F_1^*$. Since the two other rays are given by simple multiplication of $u_\pi$, and since $u_\pi(F_1^*) = 0$, all the rays that run parallel to the optical axis in front of the field intersect the optical axis in point $F_1^*$. Since point $F_1^*$ is a real point of intersection with the optical axis, and since all the rays

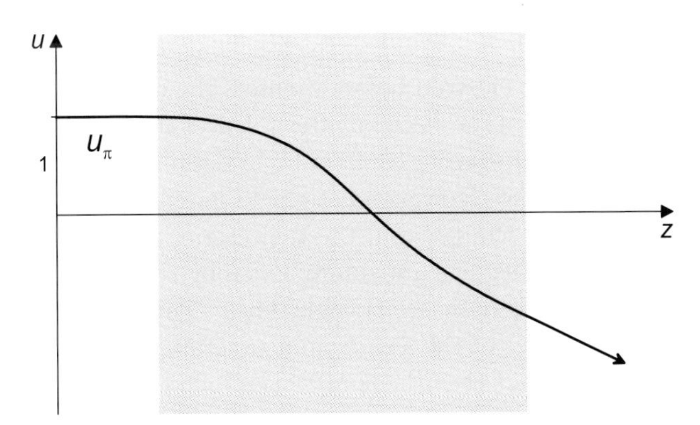


Fig. 6.25 Principal ray $u_\pi$. The ray is defined by the boundary conditions given in Eq. (6.80). The ray is a straight line parallel to the optical axis in front and behind the electromagnetic field, which is indicated by the gray shaded area.

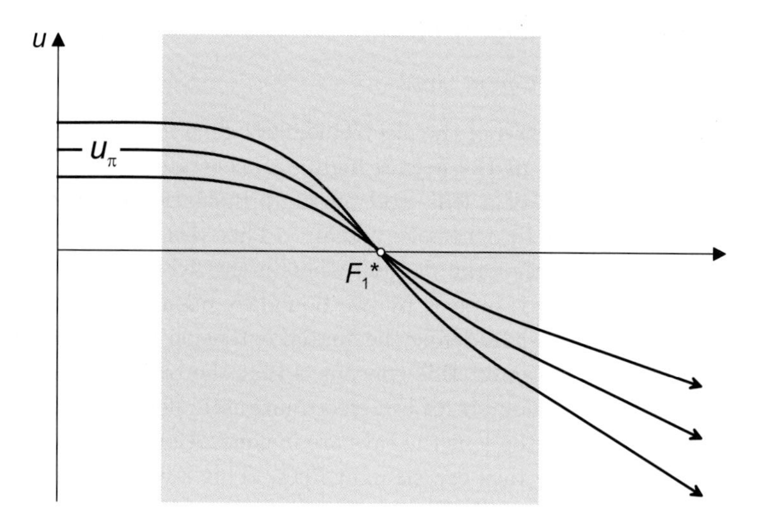


Fig. 6.26 Real focal point $F_1^*$ and the principal ray $u_\pi$.

that run parallel to the optical in front of the field intersect the axis in $F_1^*$, point $F_1^*$ is called the *real* focal point of the lens associated with the electromagnetic field. The real focal point is a real cardinal element. A *cardinal element* characterizes a certain aspect of the lens' functionality. The usage of *real* cardinal elements is certainly legitimate. However, if we consider that the lens is some kind of a black box whose function it is to translate the incoming rays into a set of outgoing rays, we do not need to know in detail the rays inside the domain of the lens. From a very practical point of view, the knowledge of the location of the real focal point is not necessary to describe the basic functionality and characteristic of the lens. We can make use of the fact that the effect of the electromagnetic field is considered to be confined to the domain of the lens.

The confinement of the lens action has the following very useful consequence: if we know the direction of an electron before it enters the domain of the electromagnetic field and the direction of the electron after it leaves the field, we basically know the effect of the lens without having to describe in detail the trajectory within the field. Furthermore, since the electromagnetic field is only of relevance within the domain of the lens, outside the domain, the trajectories follow straight lines. The essential characteristics of the lens are simply given by the directions of the initial and final trajectories, which both are straight lines. This concept is employed to determine the characteristics of the lens from a practical point of view. The main idea is to make use of the fact that the trajectories follow straight lines outside the domain of the field, and thus to determine the characteristics of the lens based on the asymptotes of the trajectories extrapolated into the domain of the lens. The set of cardinal elements determined in this way are then called *asymptotic* cardinal elements.

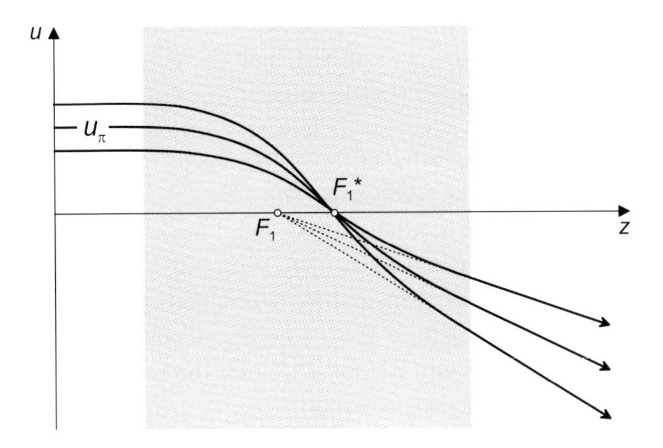


Fig. 6.27 Real $F_1^*$ and asymptotic focal point $F_1$.

Figure 6.27 illustrates the situation already depicted in Fig. 6.26. However, while in Fig. 6.26 the location of the real focal point is illustrated, whose location is determined by the curvature of the rays within the lens domain, Fig. 6.27 shows how the *asymptotic* focal point is constructed, based upon the asymptotes of the outgoing trajectories $c_1 u_\pi$. In principle, only the principal ray $u_\pi$ or any of the rays $c_1 u_\pi$ is necessary to geometrically determine the location of the asymptotic focal point $F_1$. As illustrated in Fig. 6.27, the asymptotes of all the rays $c_1 u_\pi$ intersect the optical axis in the same asymptotic focal point. Hence, similar to the real focal point, the location of the asymptotic focal point is a characteristic measure defining the focusing effect of the lens field.

### 6.7.4.2 *Object and image focal points*

So far, we have drawn our attention to one particular fundamental ray; the principal ray which in front of the lens is parallel to the optical axis. Hence, it is the ray which according to Eq. (6.80) can be defined in a more general way as

$$u_\pi(z \to -\infty) = 1, \qquad u'_\pi(z \to -\infty) = 0, \qquad u'_\pi(z \to \infty) = \frac{1}{f_1}, \tag{6.87}$$

with the slope of the outgoing ray being some value $1/f_1$, which we specify in the next subsection. We saw that $u_\pi$ is a solution of a paraxial path equation and that it is linearly independent from the axial ray $u_\alpha$ (see Eq. (6.80)). As explained above, any two solutions which are linearly independent are sufficient to describe all possible rays within a given rotationally symmetric electromagnetic field. Hence, we have the choice of finding a corresponding linearly independent ray. Another solution to the path equation that is linearly independent from $u_\pi$ (and $u_\alpha$) is the fundamental ray $u_{\bar{\pi}}$, which is determined by the boundary conditions

$$u_{\bar{\pi}}(z \to \infty) = 1, \qquad u'_{\bar{\pi}}(z \to \infty) = 0, \qquad u'_{\bar{\pi}}(z \to -\infty) = \frac{1}{f_0}. \tag{6.88}$$

The principal ray $u_{\bar{\pi}}$ is called the *object* principal ray, and $u_\pi$ is the *image* principal ray. From the above definitions we can see that the slope of the object principal ray for $z \to \infty$ approaches a value that is equal to the reciprocal value of the image focal length $f_1$. Similarly, the image principal ray approaches a slope for $z \to -\infty$, which is equal to the reciprocal value of the object focal length $f_0$. The focal lengths $f_0$ and $f_1$ will be defined in the next subsection.

Figure 6.28 shows the rays $u_\pi$ and $u_{\bar{\pi}}$ and, in addition, for each principal ray $u_\pi$ and $u_{\bar{\pi}}$ two rays which are determined by $c_1 u_\pi$ and $c_2 u_{\bar{\pi}}$, respectively. Furthermore, Fig. 6.28 shows that the object principal ray $u_{\bar{\pi}}$ defines the object focal point $F_0$, and the image principal ray defines the image focal point $F_1$. Both focal points $F_0$ and $F_1$ in Fig. 6.28 are asymptotic focal points; the real object and image focal points are labelled $F_0^*$ and $F_1^*$, respectively.

### 6.7.4.3 *Object and image focal lengths*

From the definition of the asymptotic focal points we would like to introduce the focal lengths, which we mentioned in Eqs. (6.87) and (6.88). For this, we first have to introduce the principal planes. The (asymptotic) principal planes are planes normal to the optical axis. The image principal plane is located at the height of the intersection between the asymptote of the image principal ray $u_\pi$ for $z \to \infty$ and the line $u = 1$, which corresponds to the asymptote of the image principal ray $u_\pi$ for $z \to -\infty$. On the left-hand side of Fig. 6.29, the (asymptotic) image principal plane is denoted by $p_1$. The point of intersection between the image principal plane $p_1$ and the optical axis is the image principal point $H_1$.

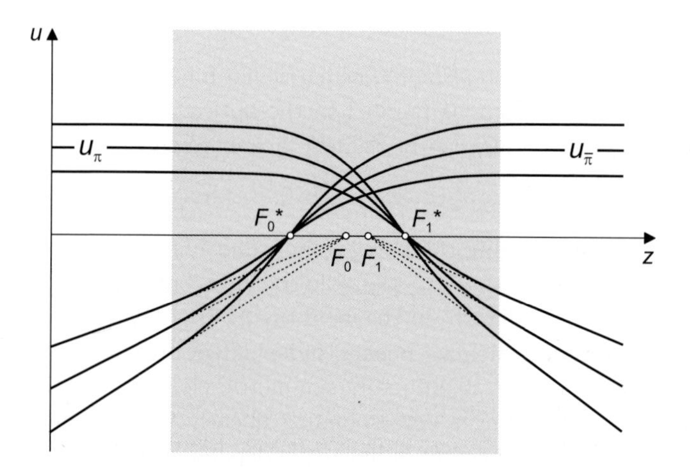


Fig. 6.28 Focal points: asymptotic object focal point $F_0$; asymptotic image focal point $F_1$; real object focal points $F_0^*$; and real image focal point $F_1^*$.

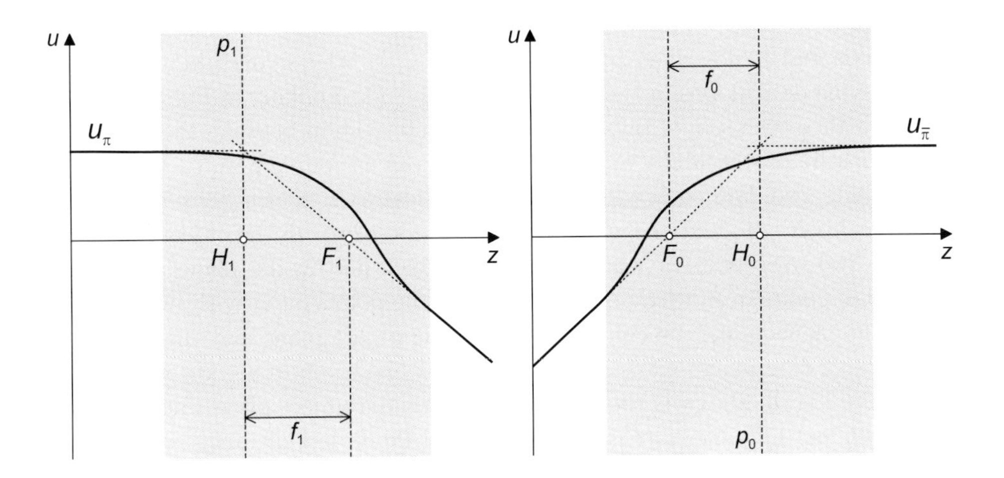


Fig. 6.29 Principal planes and focal distances. Image principal plane $p_1$, image principal point $H_1$, image focal length $f_1$, object principal plane $p_0$, object principal point $H_0$ and object focal length $f_0$.

Similarly, the (asymptotic) object principal plane $p_0$ (see right-hand side of Fig. 6.29), is located at the height of the intersection between the asymptote of the object principal ray $u_{\bar{\pi}}$ for $z \to -\infty$ and the line $u = 1$, which corresponds to the asymptote of $u_{\bar{\pi}}$ for $z \to \infty$. The point of intersection between the object principal plane $p_0$ and the optical axis $z$ is the image principal point $H_0$.

With the definition of the principal planes we can define the focal lengths. The image focal length $f_1$ is given by

$$f_1 = z(H_1) - z(F_1). \tag{6.89}$$

The image focal length is thus the distance between the image focal point $F_1$ and the image principal point $H_1$. Analogously, the object focal length $f_0$ is given by

$$f_0 = z(H_0) - z(F_0), \tag{6.90}$$

which corresponds to the distance between the object focal point $F_0$ and the principal point $H_0$. As can be seen from Fig. 6.29, for the case of a simple rotationally symmetric electromagnetic field, which acts as a convergent lens, the image focal length $f_1 < 0$ whereas the object focal length is $f_0 > 0$. For readability we omitted to mention explicitly that these quantities are asymptotic quantities.

### 6.7.4.4 *Newton's equation*

In the following we would like to derive an expression which, on the basis of the knowledge of the focal lengths, relates the object plane to the image plane. For this, let us consider two sets of rays. One of them enters the field parallel to the optical axis and is described by $c_1 u_\pi$ with $c_1 > 0$. From the above discussion (see, e.g. Fig. 6.28), it is clear that this first set of rays intersects the optical axis in the image focal point $F_1$. The second set of rays leaves the field parallel to the optical axis and is described by $c_2 u_{\bar{\pi}}$ with $c_2 < 0$. The second set of rays thus intersects the optical axis in the object focal point $F_0$. Looking at Fig. 6.30, which illustrates these two sets of rays, the points of intersection between the two sets of rays, which lie on the plane $z_0$, can be considered as source points. Hence, each source emits two rays, one parallel to the optical axis, which goes through the image focal point $F_1$, and one whose asymptote goes through the object focal point $F_1$. The two rays intersect again in a point which lies on the plane $z_1$. We can identify the points in $z_0$ with object points and the points on $z_1$ with image points. Hence, the plane $z_0$ can be considered to be the object plane and the plane $z_1$ is the corresponding image plane. What we would like to know are the distances $\zeta_0$ and $\zeta_1$ (see Fig. 6.30); $\zeta_0$ is the distance between the object plane and the object focal point, i.e. $\zeta_0 = z_0 - z(F_0)$, which is called the object distance, and $\zeta_1$ is the distance between the image plane and the image focal point, i.e. $\zeta_1 = z_1 - z(F_1)$, which is called the image distance. These measures define the location of the object plane and the image plane in respect to the focal points. While $\zeta_0$ is negative, $\zeta_1$ is positive (see Fig. 6.30).

Let us now focus on a pair of rays which intersects in the object plane $z_0$ and in the image plane $z_1$. Such a pair of rays provides, in principle, all necessary information about an object point $P_0$ and its image point $P_1$ (see Fig. 6.31). The projection of the object point $P_0$ onto the optical axis is $P_0'$, and $P_1'$ is the projection of the image point $P_1$ onto the optical axis. The (asymptotic) focal points are denoted by $F_0$ and $F_1$, and the intersection of the principal planes with the optical axis are the object principal point $H_0$ and the image principal point $H_1$. Their projections onto the asymptotes, which run parallel to the optical axis, are $H_0'$ and

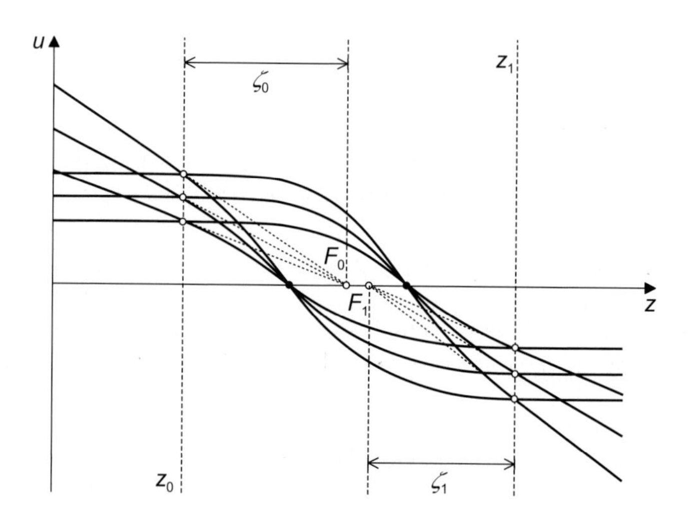


Fig. 6.30 Object plane $z_0$ and image plane $z_1$. Object distance $\zeta_0$ and image distance $\zeta_1$.

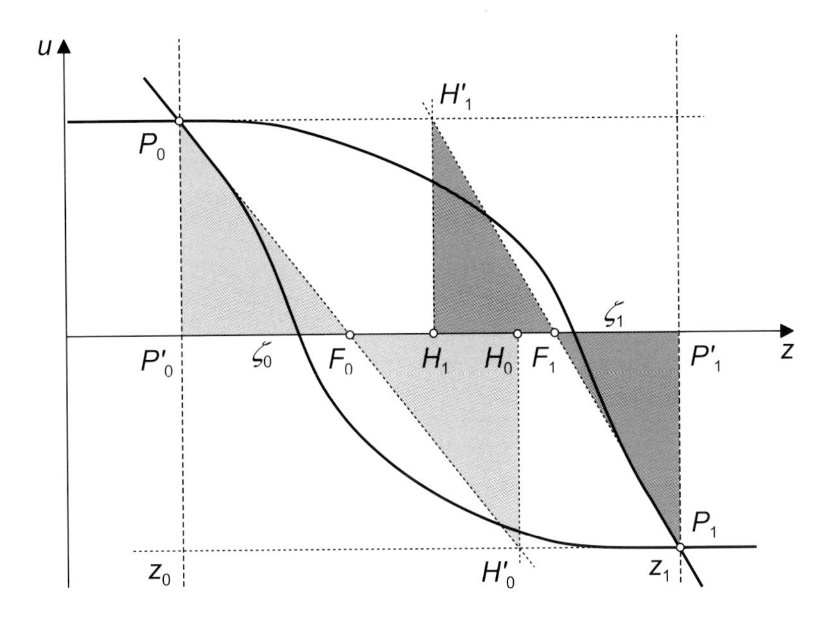


Fig. 6.31 Newton's equation. Two fundamental rays, $c_1 u_\pi$ and $c_2 u_{\bar{\pi}}$ (with $c_1 > 0$ and $c_2 < 0$), emerge in the object point $P_0$, whose projection onto the optical axis $z$ is $P_0'$. The point of intersection of the asymptote of $c_1 u_\pi$ for $z \to -\infty$ with the asymptote of $c_1 u_\pi$ for $z \to \infty$ yields point $H_1'$, whose projection onto the optical axis is the image principal point $H_1$. The image principal point defines the location of the image principal plane. The intersection of the asymptote of $c_1 u_\pi$ for $z \to \infty$ with $z$ yields the (asymptotic) image focal point $F_1$. The point of intersection of the asymptote of $c_2 u_{\bar{\pi}}$ for $z \to -\infty$ with the asymptote of $c_2 u_{\bar{\pi}}$ for $z \to \infty$ yields point $H_0'$, whose projection onto $z$ is the object principal point $H_0$. The object principal point defines the location of the object principal plane. The intersection of the asymptote of $c_2 u_{\bar{\pi}}$ for $z \to -\infty$ with the optical axis yields the (asymptotic) object focal point $F_0$. From the similarity of the triangles $P_0 F_0 P_0'$ and $H_0' F_0 H_0$, as well as of the triangles $H_1' F_1 H_1$ and $P_1 F_1 P_1'$, Newton's equation in Eq. (6.95) is derived.

$H_1'$. Basic geometrical considerations reveal the similarity between the triangles $P_0 F_0 P_0'$ and $H_0' F_0 H_0$. The same holds true for the triangles $H_1' F_1 H_1$ and $P_1 F_1 P_1'$. The distance between $F_0$ and $P_0'$ corresponds to the object distance $\zeta_0$, and the distance between $P_1'$ and $F_1$ is the image distance $\zeta_1$. The values of $\zeta_0$ and $\zeta_1$ define the location of the object and image plane, respectively, and thus are the quantities we are looking for.

From the definition of the principal rays and the definition of the lateral magnification in Eq. (6.86) we can write that the lateral magnification $M$ in Fig. 6.31 is given by

$$M = \frac{P_1' P_1}{P_0' P_0}. \tag{6.91}$$

Because of the similarity of the first two triangles mentioned above ($P_0 F_0 P_0'$ and

$H_0'F_0H_0$), we can write that

$$\frac{P_1'P_1}{P_0'P_0} = \frac{z(H_0) - z(F_0)}{z(F_0) - z(P_0')} = \frac{f_0}{\zeta_0}. \tag{6.92}$$

From the other two triangles ($H_1'F_1H_1$ and $P_1F_1P_1'$) we obtain

$$\frac{P_1'P_1}{P_0'P_0} = \frac{z(P_1') - z(F_1)}{z(F_1) - z(H_1)} = \frac{\zeta_1}{f_1}. \tag{6.93}$$

Comparing the two equations above yields

$$\frac{f_0}{\zeta_0} = \frac{\zeta_1}{f_1}, \tag{6.94}$$

and thus

$$\boxed{\zeta_0\zeta_1 = f_0 f_1.} \tag{6.95}$$

This relation is known as Newton's (lens) equation, which is identical in light optics (Born and Wolf, 2001). Newton's equation relates the object and image distances to the focal lengths of the lens.

The rather general case of having two different focal lengths, $f_0$ and $f_1$, as well as two different principal planes, can often be simplified. It can be shown that for the case of magnetic electron lenses and for electrostatic lenses, which do not cause an acceleration, the image focal length and the object focal length are identical in their absolute value, i.e. $f_0 = |f_1| = f$ (see, e.g. Hawkes and Kasper, 1989a). This is the case if the electrostatic potential in the object space $\phi_0$ is equal to the electrostatic potential $\phi_1$ in the image space, i.e. if $\phi_0 = \phi_1$. As pointed out in Sec. 6.4, this is realized in a so-called einzel lens.

A further simplification is possible if we assume the lens to be thin. A lens is said to be thin if the focal distance $f$ is clearly larger than the longitudinal spacial extension of the lens, which we shall describe by $L$. Hence, if $f \gg L$, the principal planes through point $H_0$ and $H_1$ coincide. This simply means that there is only one principal plane and only one principal point $H = H_1 = H_0$. Hence, a thin magnetic lens can be described by one focal distance $f$ and by a principle plane going through the principle point $H$, which is in a distance of $\pm f$ from the focal points $F_0$ and $F_1$.

#### 6.7.4.5 *Geometrical image construction*

Figure 6.31 illustrates how an image point can be constructed provided the object point $P_0$ as well as the locations of the (asymptotic) focal points $F_0$ and $F_1$ and the principal planes in $H_0$ and $H_1$ are known.

One draws a line parallel to the optical axis through the object point $P_0$. The point of intersection of this line with the image principal plane is point $H_1'$. From this point a line $\overline{H_1'F_1}$ is drawn through the image focal point $F_1$ and extrapolated into the image space. This essentially corresponds to the (asymptotic) image principal ray $c_1u_\pi$. In the second step, the object principal ray is drawn based on the asymptotes; a line $\overline{P_0F_0}$ is drawn from the object point $P_0$ through the object focal point $F_0$ and extrapolated to the point of intersection with the object principal plane, which goes through point $H_0$. This point of intersection is $H_0'$. From point $H_0'$ a line is drawn parallel to the optical axis. This is the construction of the asymptotic object principal ray $c_2 u_{\bar{\pi}}$. The point of intersection between the extrapolated line $\overline{H_1' F_1}$ and the line parallel to the optical axis through point $H_0'$ is the image point $P_1$.

Hence, the construction is based upon two simple rules. Firstly, a ray parallel to the optical axis becomes a ray through the object focal point behind the object principal plane. Secondly, a ray that crossed the image focal point becomes a ray parallel to the optical axis behind the image principal plane. For a thin lens, where there is only one principal plane, which is the plane of the lens, the construction is simplified by the fact that the distinction between object and image principal plane can be neglected.

#### 6.7.4.6 *Lateral and longitudinal magnification*

From Newton's equation we learn that the image distance $\zeta_1$ is given by

$$\zeta_1 = \frac{f_0 f_1}{\zeta_0}. \tag{6.96}$$

Furthermore, by taking the derivative we can determine the change of the image distance related to a change of the object distance

$$\frac{d\zeta_1}{d\zeta_0} = -\frac{f_0 f_1}{\zeta_0^2}. \tag{6.97}$$

This derivative is known as the *longitudinal* magnification $M_{\mathrm{L}} = d\zeta_1/d\zeta_0$. Employing Newton's equation given in Eq. (6.95) we can write that the longitudinal magnification is related to the lateral magnification $M$ by

$$\frac{d\zeta_1}{d\zeta_0} = -\frac{f_1}{f_0} M^2. \tag{6.98}$$

Furthermore, since $f_1 < 0$ and $f_0 > 0$, independent of the sign of the lateral magnification $M$, the longitudinal magnification $d\zeta_1/d\zeta_0$ is greater than zero. Keeping in mind that $\zeta_1 > 0$ and $\zeta_0 < 0$, we see that if the object is displaced in direction parallel to the optical axis, the image moves in the same direction. This characteristic, together with the fact that $f_0 f_1 < 0$, defines a *dioptric* optical system. Systems where $f_1 f_0 > 0$ and $d\zeta_1/d\zeta_0 < 0$, i.e. when a displacement of the object leads to a displacement of the image in opposite direction, are called *catoptric*. While dioptric systems are instruments, which are based solely on refracting elements or an even number of reflecting elements, catoptric systems are based on instruments which contain an odd number of reflecting elements. In addition, dioptric instruments with $f_0 < 0$, which thus lead to images inverted in respect to the object, are called *convergent*, while systems with $f_0 > 0$, which lead to images which are equally oriented as the corresponding objects, are called *divergent*. From this we can conclude

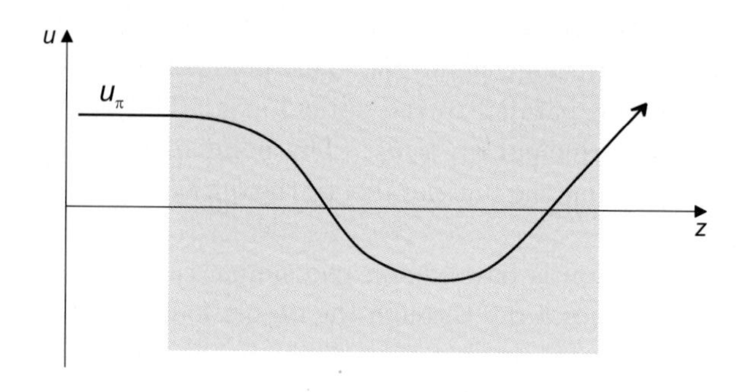


Fig. 6.32 Long lens. A long (or strong) electromagnetic lens can lead to two crossovers within the domain of the lens. The overall effect of the lens would be divergent. This behavior can, however, be explained by conceptually dividing the long lens into smaller units of convergent behavior. Consequently, this behavior is not in disagreement with the statement that electromagnetic electron lenses are always convergent lenses.

that an electron microscope consisting of (short) electromagnetic fields, which act as refracting electron lenses, are dioptric instruments of convergent electron lenses.

The restriction to thin lenses excludes the situation depicted in Fig. 6.32. Figure 6.32 illustrates the effect of a long electromagnetic lens on an electron trajectory which corresponds to a principal ray $u_\pi$. If the rotationally symmetric electromagnetic field is long, there is nothing that would stop the electron trajectory inflecting once more towards the optical axis after already having crossed it. Hence, in this case the long electromagnetic field would be described by a divergent lens. However, the effect of such a long divergent lens can be described by two short convergent lenses. As such, a long lens is not something that would add functionality to a microscope.

#### 6.7.4.7 *Telescopic system*

Let us consider a system consisting of two identical rotationally symmetric magnetic electron lenses. Both lenses shall be thin such that $f_0 = -f_1 = f$. The distance between the lenses shall be $2f$, i.e. exactly twice the focal distance of the lenses. This system, depicted in Fig. 6.33, is called a $4f$ system.

From left to right in Fig. 6.33, the first focal point $F_{1,0}$ is the object focal point of the first lens $L_1$. The principal plane of the first lens goes through point $H_1$. Since the distance between the two thin lenses is $2f$, the image focal point of the first lens $F_{1,1}$ coincides with the object focal point of the second lens. The image focal point of the second lens is $F_{2,1}$, which is located at a distance $f$ from the principal point $H_2$ of the second lens $L_2$.

Figure 6.33 reveals that by placing an object in the image (or front) focal plane of the first lens, like, for example, point $P_0$ in the plane of $F_{1,0}$, the $4f$ system forms an inverted image $P_1$ of the object in the object (or back) focal plane of the second lens which is at a distance $f$ from the principal point $H_2$. The magnification of the

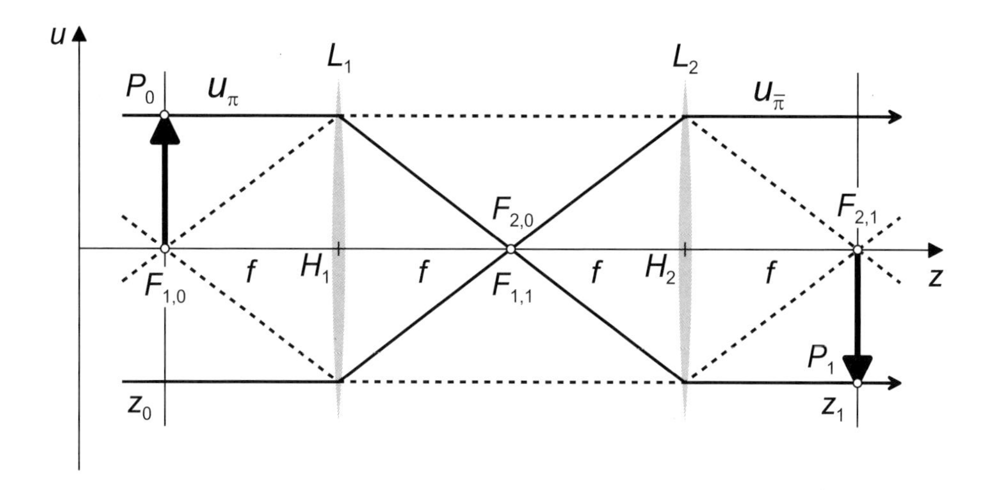


Fig. 6.33 $4f$ system. The $4f$ system consists of two thin round lenses $L_1$ and $L_2$ whose focal lengths are $f_0 = -f_1 = f$. If the lenses are placed at a distance of $2f$ from each other and an object is placed in a distance $f$ in front of the first lens, i.e. in the front focal plane $z_0$, in the back focal plane of the second lens $z_1$, which is in a distance $f$ behind $L_2$, an image is formed which shows the object inverted with a lateral magnification of $M = -1$. Since the focal planes of the entire $4f$ system are at infinity, it is a telescopic system.

transfer is $-1$. The distance between the object plane in the front focal plane and the image plane in the back focal plane is $4f$. As will be shown in Chapter 8, the $4f$ system is a crucial component of the hexapole-type aberration corrector.

So far, we have been considering the two lenses and their individual cardinal elements. However, we can also consider the $4f$ system as an entire optical unit. Doing so, we would like to know the cardinal elements of the $4f$ system, not just the one of the subunits it is made of. Straightforward geometrical considerations (see, e.g. Born and Wolf, 2001) reveal that for a system consisting of two thin lenses $L_1$ and $L_2$, which are in a distance $l = f_1 + f_2 + \Delta z$ from each other and which have the focal lengths $f_1$ and $f_2$, the focal length $\tilde{f}$ of the two-lens system is given by

$$\frac{1}{\tilde{f}} = -\frac{1}{\tilde{f}'} = -\frac{\Delta z}{f_1 f_2}, \tag{6.99}$$

where $\tilde{f}$ is the object focal lengths and $\tilde{f}' = -\tilde{f}$ is the image focal length of the two-lens system. For the $4f$ system this can be simplified (see Fig. 6.33). Since $f_1 = f_2 = f$, the focal length of the entire $4f$ system is thus

$$\tilde{f} = -\frac{f^2}{\Delta z}. \tag{6.100}$$

Since in a $4f$ system the lenses are in a distance $l = 2f$ from each other, $\Delta z \to 0$. With this, the focal length of the $4f$ system is thus $\tilde{f} \to \infty$. Hence, both focal planes of the $4f$ system are at infinity. This is characteristic for a *telescopic* system.

It is important to note that although the lenses constituting a $4f$ system have finite focal lengths (see Fig. 6.33), the focal planes of the $4f$ system are at infinity.

We would further like to emphasize a special aspect of the $4f$ system. For this we have to recall that the two principal rays $u_\pi$ and $u_{\bar{\pi}}$ are fundamental rays, which, by linear combination, can be used to derive an arbitrary ray for a given system. This is possible because $u_\pi$ and $u_{\bar{\pi}}$ are linearly independent solutions of a path equation. If we draw the object principal ray $u_\pi$ and the image principal ray $u_{\bar{\pi}}$ for the $4f$ system, we can see that they both run parallel to the optical axis in front of the first lens and behind the second lens. These two rays seem to be linearly dependent in the $4f$ system; the object principal ray is simply given by $u_\pi = -u_{\bar{\pi}}$ (see Fig. 6.33). From this we can conclude that for the $4f$ system, the principal rays $u_\pi$ and $u_{\bar{\pi}}$ are not linearly independent and thus do not allow for constructing the whole set of possible electron trajectories. Hence, we miss one linearly independent solution of the path equation. It is for this reason that in addition to the principal and axial rays, other cardinal elements are used, which further can characterize an optical system.

### 6.7.4.8 *Nodal ray and nodal points*

An alternative to the principal rays discussed above is the *nodal* ray $u_\nu$. On the basis of the principal rays of an optical system with an object focal length $f_0$ and an image focal length $f_1(< 0)$, the nodal ray $u_\nu$ is defined as

$$u_\nu = -f_1 u_\pi - f_0 u_{\bar{\pi}}. \tag{6.101}$$

From the definition of the principal rays in Eqs. (6.88) and (6.87) and their behavior for $z \to \pm\infty$, we can deduce the asymptotes of the nodal ray. For $z \to -\infty$, the nodal ray $u_\nu$ approaches $u_{\nu,-\infty} = -f_1 - (z - z(F_0))$, and for $z \to \infty$, $u_{\nu,\infty} = (z - z(F_1)) - f_0$. These asymptotes intersect the optical axis in a point $z = z(F_0) - f_1 \equiv z(N_0)$ and in a point $z = z(F_1) + f_0 \equiv z(N_1)$, respectively. These points of intersections are the object nodal point $N_0$ and the image nodal point $N_1$. They define the locations of the nodal planes. For a thin lens with $f_0 = -f_1 = f$, the nodal planes coincide with the principal plane, which also lies in a distance $f$ from the focal planes. The nodal points have the characteristics that any ray whose pre-field asymptote intersects $N_0$ under an arbitrary angle $\gamma$ asymptotically leaves the field along a line which intersects $N_1$ under the same angle $\gamma$. One says that the angular magnification is unity. This is illustrated in Fig. 6.34.

In a telescopic system with $f_{0/1} \to \pm\infty$, as for example the highly symmetric $4f$ system depicted in Fig. 6.33, the nodal points are at infinity. In this case, another cardinal ray can be used to describe the system. This ray is called the *symmetric* ray $u_\sigma$, which is defined as follows (Rose, 2009a):

$$u_\sigma = -f_1 u_\pi + f_0 u_{\bar{\pi}}. \tag{6.102}$$

The intersections of the asymptotes of the symmetric ray with the optical axis define two planes which are the so-called unit planes. The characteristics of the unit planes

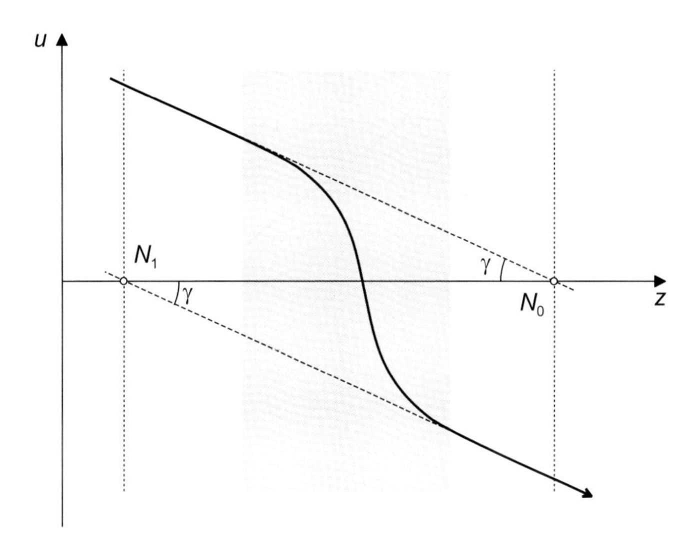


Fig. 6.34 Nodal points $N_0$ and $N_1$. A ray entering the field through the object nodal point $N_0$ under an inclination angle $\gamma$ leaves the field through the image nodal point $N_1$ under an angle $\gamma$.

are that the image unit plane contains an image of the object unit plane of lateral magnification $-1$.

With the introduction of the nodal ray and the symmetric ray, we would like to introduce a further cardinal ray, which is called the *field* ray $u_\gamma$. While the axial ray $u_\alpha$ is defined according to Eq. (6.80) by $u_\alpha(z_0) = 0$ and $u'_\alpha(z_0) = 1$, i.e. by the intersection of the optical axis in the object plane $z_0$ and its slope $u'_\alpha$ in this point, the field ray is defined according to its distance from the optical axis in the object plane and by its intersection with the optical axis in the image focal point of the objective lens post-field, which corresponds to the back focal plane $z_\mathrm{d}$. This is expressed as

$$u_\gamma(z_0) = 1, \qquad u_\gamma(z_\mathrm{d}) = 0. \tag{6.103}$$

Figure 6.35 illustrates the idea of the field ray $u_\gamma$.

### 6.7.4.9 *The TEM objective lens*

The function of the TEM objective lens, depicted schematically in Fig. 2.1, is to transfer the electron wave at the exit plane of the specimen into an image which is recorded on the fluorescence screen. A typical objective lens consists of two pole pieces which form a highly localized magnetic field. The specimen is immersed in the magnetic field formed by the two pole pieces. The focal length of the objective lens, i.e. the objective focal length $f_\mathrm{o}$, in TEM mode, is typically defined as

$$f_\mathrm{o} = -\frac{1}{u'_\pi(z_{H_1})}. \tag{6.104}$$

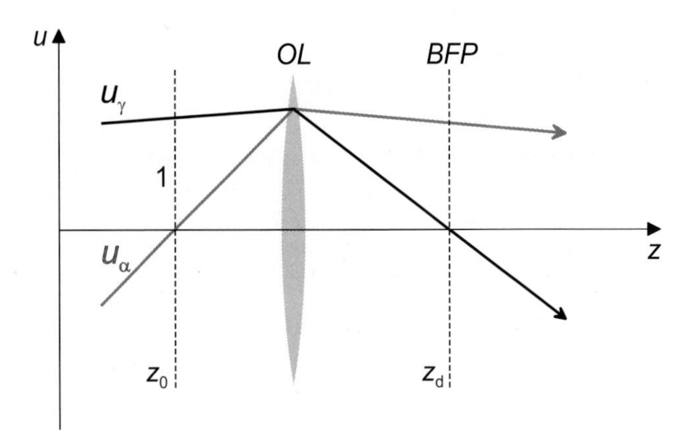


Fig. 6.35 Field ray $u_\gamma$ in comparison to the axial ray $u_\alpha$.

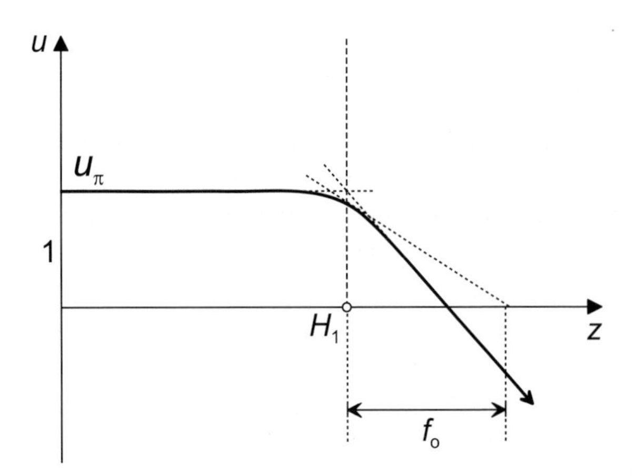


Fig. 6.36 The principal ray $u_\pi$ defines the focal length $f_o$ of the objective lens.

This is the negative reciprocal value of $u'_\pi(z_{H_1})$, i.e. the derivative of the principal ray $u_\pi$ at the image principal plane going through $H_1$. This definition can be reasoned on the basis of Fig. 6.36. We assume that the principal ray $u_\pi$ at the point of intersection with the image principal plane is approximately equal to 1, i.e. equal to the value $u_\pi \to -\infty$. The tangent to the curve $u_\pi(z_{H_1})$ in the image principal plane has the slope $u'_\pi(z_{H_1})$. The tangent thus intersects the optical axis in a distance $-1/u'_\pi(z_{H_1})$ away from the image principal point $H_1$. This value is identical to the focal length $f$ given above. Furthermore, according to its definition in Eq. (6.87), the image principal ray $u_\pi$ approaches a slope of $u'_\pi = 1/f_o$ for $z \to \infty$. Hence, this definition of the focal length $f_o$ of the objective lens is based on the approximation that the lens is short and that the change of direction of $u_\pi$ from $u'_{\pi,-\infty} = 0$ to $u'_{\pi,\infty} = 1/f_o$ occurs within a very short domain.

On the other hand, the focal distance can be defined according to the electromagnetic field of the lens. For a rotationally symmetric magnetic lens, it can be shown (see, e.g. Rose, 2009a) that the (magnetic) focal length is given by

$$\frac{1}{f_o} = \frac{e}{8m_0\Phi_0^*}\int_{-\infty}^{\infty} B_z^2(z)dz, \tag{6.105}$$

with $\Phi_0^*$ the relativistically corrected axial electrostatic potential at the height of the specimen (see, e.g. Eq. (5.61)). This relation clearly shows that with increasing field strength $B_z$ the focal distance decreases, and that with increasing electron energy, which scales with $\Phi_0^*$, the focal distance increases. The higher the electron energy, the stronger the field necessary to achieve a certain focal length.

## 6.8 Summary

In this chapter, the Gaussian dioptrics of round electromagnetic lenses was discussed. This chapter provides an insight into the electron optical treatment and description of electron lenses. The chapter started with the introduction of the relevant coordinate systems, treated the paraxial approximation and then focused onto the derivation of the path equations of rotationally symmetric electromagnetic fields. In a rather qualitative way, deviations from the paraxial approximation were also discussed. Such deviations are considered as aberrations. Although the limitation to rotationally symmetric fields, i.e. round lenses, does not directly target onto the multi-pole lenses employed in aberration correctors, the basic concepts about how the effect of an electromagnetic lens is tackled are in general also applicable to multi-pole lenses. However, for multi-pole lenses additional symmetry restrictions need to be incorporated. For this, the reader is referred to the specific literature referenced in the foregoing chapters. In the second part of the present chapter, the imaging properties of an electromagnetic lens were discussed, introducing the idea of fundamental rays and cardinal elements. This treatment is in principle identical to the discussion of light optical systems.

Keywords addressed in this chapter are: paraxial approximation; Gaussian optics; equation of motion; path equation; Larmor precession; the rotating coordinate system; theorem of optical imaging; Busch's theorem; the (asymptotic) cardinal elements; the $4f$ system; and Newton's lens equation.

# PART III
# Aberration Correction

## Chapter 7

# Aberrations

Throughout the preceding chapters, we have come across the term *aberration*. We have mentioned that electron lenses are not perfect and talked in a rather unspecified manner about so-called lens effects, referring to defocus, spherical aberration and chromatic aberration. Spherical aberration and chromatic aberration are not the only aberrations which require our attention. Though spherical aberration is known to be the resolution-limiting aberration, it is only limiting in the sense that on conventional microscopes it cannot be corrected. On the other hand, it is a routine step towards an atomic-resolution micrograph to employ a stigmator in order to correct for twofold astigmatism, which, indeed, is also an aberration. However, due to the fact that it can be manually corrected to a precision where its residual impact can be ignored compared to the effect of spherical aberration, twofold astigmatism is rarely considered in the discussion of the image formation.

In aberration-corrected microscopes, it is commonly understood that the spherical aberration can be corrected. Hence, spherical aberration is not the resolution-limiting factor in such instruments anymore. This, however, does not mean that the microscope is free of aberrations, or that the point resolution becomes a meaningless quantity, or even a quantity which is determined solely by the wavelength or the diffraction limit. There are additional aberrations whose effects can become resolution-limiting. In fact, there is always at least one aberration which limits the optical resolution of a microscope. The ideal imaging instrument does not exist. Hence, with the advent of aberration-corrected electron microscopy, a need has emerged for understanding the individual residual aberrations of an aberration-corrected instrument. Microscopists who used to deal with spherical aberration and defocus have started to adopt the optical concepts which were previously mostly relevant for experts working in electron optics.

At this point it needs to be emphasized that besides optical effects, additional factors can be resolution-limiting in real electron microscopes. It is indeed often the case that the observable resolution is not limited primarily by the actual optical behavior of the instrument but by disturbances. Factors like the overall mechanical stability of the microscope or the stability of the electromagnetic lens fields generated by the microscope's power supplies can be decisive. Instabilities are particulary important if one aims at enhancing the resolution of a given microscope up to its theoretical optical limit. Moreover, in some cases it is even the specimen which imposes an unbreakable barrier to the observable resolution. In the present chapter, however, we accentuate the discussion on purely optical effects and presume that apart from such optical effects, the instrument behaves like an ideal noise-free instrument.

## 7.1 Overview

Figures 3.4 and 3.5, redrawn and summarized in Fig. 7.1, illustrate the blurring of the focal point of a non-ideal lens due to spherical aberration (Fig. 7.1a) and due to chromatic aberration (Fig. 7.1b). The identification of these aberrations essentially led us to the conclusion that electron lenses cause unwanted artifacts which demand our attention in the optical setup of the microscope, as well as in the discussion of the imaging characteristics of a particular microscope. For instance, in the presence of spherical aberration one preferably employs a defocus such that the disk of least confusion is imaged onto the Gaussian image plane. This is explained in detail in Chapter 3.

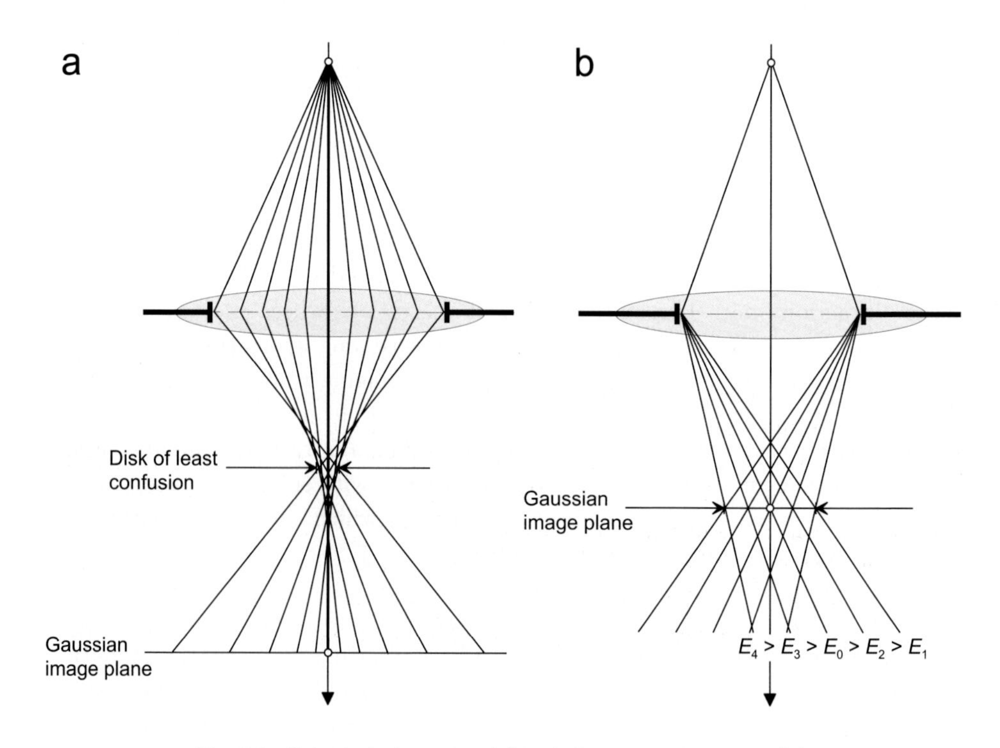


Fig. 7.1 Spherical aberration (**a**) and chromatic aberration (**b**).

Spherical and chromatic aberrations impact the broad-beam phase-contrast imaging mode as well as the formation of the electron probe in scanning probe mode. While the spherical aberration determines the resolution (see Eqs. (2.32) and (3.11)), the chromatic aberration essentially limits the information transfer of high-frequent object information (see, e.g. Eq. (2.33) and Nellist and Pennycook, 1998). In any case, if aberrations are present and the energy spread of the beam is finite, the formation of a stigmatic image point reflecting an object point is not feasible.

On the other hand, we have seen in Chapter 6 that as long as the electrons stay close enough to the optical axis, i.e. close enough such that the Gaussian dioptrics is applicable, electromagnetic lenses can provide stigmatic images. As a matter of fact, a stigmatic image, i.e. an image where each object point is transferred to the image plane as a point such that the image is a (de)magnified and undistorted representation of the object, corresponds to the imaging characteristics that we would like to have. The possibility of stigmatic images is not in contradiction to the above argumentation. However, in order to produce images of *finite* resolution, it is necessary to consider electron trajectories that are at a small but finite distance to the optical axis or which run under a finite angle to the optical axis. The consideration of such rays complicates the situation. As soon as we start dealing with such rays we have to expand the conception about idealized imaging, which is essentially based on the paraxial approximation. As a consequence, we have to include higher-order terms in the description of the various trajectory equations. This was illustrated in Chapter 6, where we exemplified the effect of including terms of higher order in the trajectory equation for the simple case of the focusing strength of a homogeneous magnetic field (see Figs. 6.10 and 6.12). The rather simple expansion of the cosine function from the paraxial approximation $\cos\gamma = 1$ to $\cos\gamma = 1 - \gamma^2/2$ enabled us to derive an expression for the constant of spherical aberration of the homogeneous rotationally symmetric magnetic field. Hence, aberrations become effective when we treat a non-ideal imaging system beyond the paraxial approximation.

The above considerations essentially illustrate that aberrations describe the deviations from the optical behavior of an ideal instrument. The reference state, which provides an ideal image, is the imaging characteristics of the Gaussian optics, which is valid only within the paraxial approximation. As a matter of course, we choose the Gaussian optics, i.e. the optical behavior describable within the paraxial approximation, as our reference state. However, as will be shown, it is particularly evident to consider the paraxial approximation as this reference state when we generalize the idea of the aberration function and the concept of the geometrical wave surface, which we have already introduced in Chapters 2, 3 and 5. Nonetheless, having a well-defined reference state which is based on the Gaussian optics enables us to conceptually describe the term *aberration*.

## 7.2 Image Aberrations

An electron ray is describable by two geometrical ray parameters; the complex coordinate in the aperture plane $\omega$ and the position of the trajectory in the object plane, which, in analogy to $\omega = \theta_x + i\theta_y$, we describe by a complex coordinate $w_o = x_o + iy_o$. Knowing $\omega$ and $w_o$, the Gaussian optics can be applied to derive the position of a ray in the Gaussian image plane. This position, also described by a complex value $w_i = x_i + iy_i$, corresponds to the Gaussian image point which reflects the object point $w_o$ in the object plane, i.e. the source point of the trajectory[1]. An aberrated ray does not intersect the Gaussian image plane in $w_i$ but in a point $w_i'$. This is the basis of the definition of the image aberrations; the *image aberration* $\Delta w_i$ is the distance between the location where the Gaussian ray intersects the Gaussian image plane and the location where the actual (aberrated) ray intersects the Gaussian image plane, i.e.

$$\Delta w_i = w_i' - w_i. \tag{7.1}$$

This is illustrated in Fig. 7.2.

Each aberration has a characteristic aberration figure which can depend on the location of the object point $w_o$ as well as on the location $\omega$ in the aperture plane where a ray intersects the aperture plane.

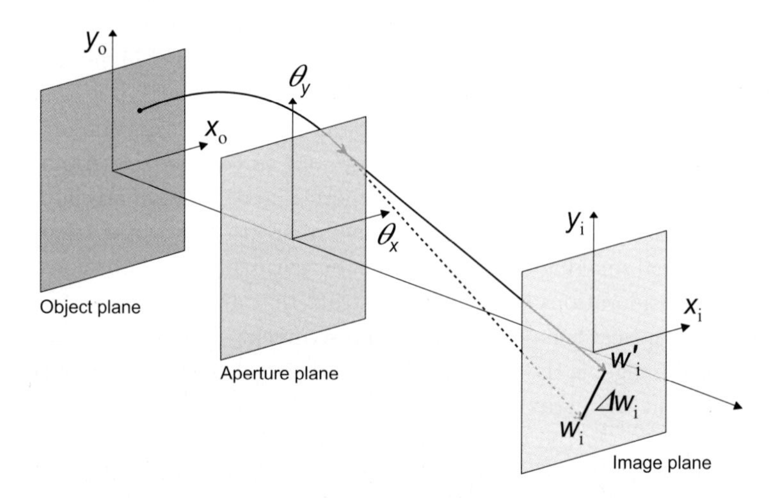


Fig. 7.2 Definition of the image aberration $\Delta w_i$. The paraxial or Gaussian ray (dashed line) intersects the Gaussian image plane in $w_i$, while the aberrated ray (full line) hits the plane in $w_i'$. The difference corresponds to the image aberration of the ray $\Delta w_i$.

[1]In the foregoing text, we do not strictly specify whether the cartesian coordinate $(x, y)$ is in the object or image plane. From the context it should be clear which plane is meant. However, whenever needed, we employ subscripts in the complex coordinate $w_o$ and $w_i$ to elucidate whether object or image plane is meant.

## 7.3 Types of Aberrations

Aberrations have different facets and they affect the imaging process in various ways, but they all have in common that at least for one set of geometrical ray parameters $(\omega, w_o)$, the image aberration $\Delta w_i \neq 0$. There are aberrations which are sensitive to the wavelength of the electrons, aberrations which depend on the scattering or illumination angle and aberrations which depend on the location of the object point within the object plane. There are aberrations which alter the coherence between the electrons in the beam and aberrations which do not influence coherency. It seems obvious therefore to distinguish aberrations according to their effect on electrons. Hence, for now we shall be concerned with the classification of aberrations.

### 7.3.1 *Axial aberrations vs. off-axial aberrations*

The aberrations we have considered so far are defocus, spherical aberration and chromatic aberration. They have a common characteristic: regardless of whether they impact a STEM probe or the imaging process in TEM, their effect increases with the inclination angle of the electron trajectories in respect to the optical axis. The impact of defocus and spherical aberration is described by the aberration function which is given for HRTEM and STEM in Eqs. (2.16) and (3.14), respectively, and which can be rewritten as

$$\chi(\theta) = \frac{1}{2}\theta^2 C_1 + \frac{1}{4}\theta^4 C_3.$$

Here, $\theta$ shall be the scattering angle in HRTEM as well as the illumination angle of a particular electron trajectory in STEM. Employing the notation of the complex angle $\omega$ for the angle $\theta$

$$\chi(\omega) = \Re\left(\frac{1}{2}\omega\overline{\omega}C_1 + \frac{1}{4}\left(\omega\overline{\omega}\right)^2 C_3\right),$$

we obtain a two-dimensional description of the effect of defocus and spherical aberration of third order, where $\omega$ and $\overline{\omega}$ are the complex angle and its complex conjugate according to Fig. 2.7. The complex angle $\omega$ is the two-dimensional position coordinate of an electron trajectory in the aperture plane. For STEM, the aperture plane is in front of the specimen conjugate to the front focal plane of the objective lens, and for HRTEM, the aperture plane is the back focal plane of the objective lens. The angle $|\omega| = |\theta_x + \mathrm{i}\theta_y| = \theta$ essentially defines the inclination angle of an electron trajectory in respect to the optical axis at the object plane, irrespective of the point in the object plane.

The behavior of the chromatic aberration is similar: the effect of the chromatic aberration increases with the angle $|\omega|$ as well. This can be seen from Eq. (3.6); the impact of the chromatic aberration increases linearly with the illumination semi-angle $\alpha$ and thus with $|\omega|$. For the case of HRTEM, it is the envelope function due to partial temporal coherence which describes the effect of the chromatic aberration.

According to Eq. (2.24), the incoherent dampening of the coherent transfer increases with the scattering angle $\theta$ or with $|\omega|$, respectively. Furthermore, we recall that the spread of defocus $\Delta C_1$ is proportional to $C_C$ (see Eq. (2.23)). Hence, the effect of the chromatic aberration $C_C$ increases with the scattering angle.

Defocus $C_1$, spherical aberration $C_3$ and chromatic aberration $C_C$ thus have in common that their effects on the STEM probe or on a HRTEM image depend on the illumination or scattering angle, i.e. on the position coordinate $\omega$ in the respective aperture plane. However, their effects do not depend on the position in the object plane $w_o = x_o + iy_o$. Indeed, all object points $w_o$ and all image points $w_i$ are identically affected by these aberrations. For $C_1$, $C_3$ and $C_C$, the relevant position variable or the relevant geometrical ray parameter is the position $\omega$ in the aperture plane. An aberration whose effect is solely a function of the position coordinate in the aperture plane (STEM or HRTEM) is called an *aperture* aberration or an *axial* aberration. Defocus $C_1$, spherical aberration $C_3$ and chromatic aberration $C_C$ are axial aberrations[2]. On the other hand, an aberration whose effect depends on the position $w_o$ in the object plane (and possibly on the position in the aperture plane as well) is called an *off-axial* aberration.

*The isoplanatic approximation*

In the previous discussion about lens effects, we have dealt with axial aberrations only and thus assumed that for all object points which are imaged onto the image plane, the same transfer function is valid. Hence, implicitly we expect the same resolution for all image points. This assumption, which is called the *isoplanatic* approximation, is of course only justifiable if all relevant aberrations are axial aberrations. In the presence of off-axial aberrations, the resolution is not equal for all image points and the isoplanatic approximation fails. This applies to TEM imaging as well as to STEM imaging.

Considering the fact that the electromagnetic fields in electron microscopes are rather inhomogeneous with a distinct radial dependence, and considering the fact that the specimen in a transmission electron microscope is immersed in the field of the objective lens, it almost seems to be a surprise that the isoplanatic approximation holds at all. Indeed, it holds so well that it is very often implicitly presumed and not even mentioned in the literature about the image formation in HRTEM. The reason for this lies in the high magnifications employed to do HRTEM imaging and atomic-resolution STEM imaging. The bore of a high-resolution objective lens is typically 2 to 3 mm in diameter (see, e.g. Fig. 6.13), while the field of view of a high-resolution micrograph is typically less than 100 nm. Hence, the field of view is four orders of magnitude smaller than the spatial extension of the electromagnetic field within the bore. It is for this reason that off-axial aberrations can often be ignored such that the entire object area imaged in a high-resolution micrograph can

[2]This statement already implies a simplification concerning the chromatic aberration, which is resolved in the following section.

be considered to be affected by a transfer function which in principle is only valid for the central point of the object plane.

Though the isoplanatic approximation holds reasonably well in high-resolution imaging, as with any approximation, it has, of course, its limitation. This limitation reasons the definition of a slightly longish term that is known as the *number of equally well resolved image points*. The maximum number of equally well resolved image points, measured along the diameter of the recorded image, defines the largest field of view for which the isoplanatic approximation holds. For a field of view which exceeds the maximum number of equally well resolved image points, the effect of off-axial aberrations becomes the resolution limiting factor, i.e. for points beyond the maximum number of equally well resolved image points.

### 7.3.2 *Chromatic aberrations vs. geometrical aberrations*

We have already come across the term 'chromatic aberration' and discussed the effect of $C_C$ for HRTEM as well as for STEM imaging. The chromatic aberration is usually referred to as an incoherent aberration, which essentially implies that it leads to a loss of information. This is, for instance, expressed by the fact that the chromatic aberration limits the information transfer in HRTEM imaging. In STEM, the chromatic aberration essentially enhances the intensity of the side lobes of the electron probe. This leads mainly to the same effect as in HRTEM imaging, i.e. it reduces the contrast of high-resolution information (see, e.g. Nellist and Pennycook, 1998). There are other incoherent aberrations which are not actual optical effects; mechanical vibrations, the finite size of the electron source, instabilities of the power supplies used to feed the electron optical elements, specimen drift or the information loss due to the point spread function of the detector. All these issues can be considered to be incoherent aberrations as well. Similarly to the chromatic aberration, these additional incoherent aberrations lead to a loss of information.

The effect of the chromatic aberration increases with decreasing electron energy and increasing energy spread of the beam. The energy spread of conventional field-emission electron microscopes is, however, sufficiently narrow such that when operating the microscope above about 100 kV, the chromatic aberration imposes a limit on the information transfer (see Eq. (2.33)) but not on the resolution. Hence, though the chromatic aberration can in principle limit the resolution of a microscope, for dedicated high-resolution microscopes this is normally not the case. For typical transmission electron microscopes, Eqs. (2.32) and (3.11) define the point resolution for HRTEM imaging and for STEM imaging, respectively. Hence, it is the spherical aberration of the objective lens which imposes a limit on the achievable resolution.

The spherical aberration is a geometrical aberration. Geometrical aberrations, which are also referred to as coherent aberrations, do not necessarily lead to an actual loss of information. They primarily affect the interpretability of the micrographs. As we have already seen in Chapters 2, 3 and 4, coherent (axial) aberrations cause a phase shift of the electron wave in the respective aperture plane, leading to a displacement of the electron trajectories from the non-aberrated trajectories. This becomes particularly clear when we consider HRTEM imaging, where the delocalization due to the oscillating phase contrast transfer function complicates the interpretability of phase contrast micrographs. The point resolution given in Eq. (2.32), which defines the smallest distance that can directly be resolved and interpreted, is typically larger than the information limit given in Eq. (2.33). Hence, since the information limit exceeds the point resolution, a typical HRTEM micrograph contains information beyond the point resolution. The information between point resolution and information limit is not lost, but because of the effect of the geometrical aberrations, namely the spherical aberration, it is not directly interpretable. Employing rather complex restoration procedures on series of micrographs each recorded under slightly different imaging conditions, the information between point resolution and information limit can be restored. Such restoration techniques thus allow for numerically extending the resolution of a microscope up to the information limit of the instrument, provided all relevant geometrical aberrations are known with sufficient precision (see Chapter 4). The possibility of restoring the aberrated image information in the presence of coherent aberrations is in contrast to incoherent aberrations, like the chromatic aberration, which lead to an actual loss of information that cannot be restored.

#### 7.3.2.1 *Chromatic aberrations*

Figure 7.1b illustrates that the chromatic aberration causes the focal point of a lens to change with the electron energy and thus with the wavelength of the electrons. We have also seen that the effect of the chromatic aberration $C_C$ is an axial effect and that it can be described in the aperture plane, respectively in the front- or back-focal plane of the objective lens depending on whether we do STEM or HRTEM. However, what we called simply the chromatic aberration $C_C$ and schematically illustrated in Fig. 7.1b is in fact only one particular aspect of the overall chromatic effects that can occur in an electron optical instrument. Though $C_C$ as illustrated in Fig. 7.1b is often the dominant chromatic effect, there is not a *single* chromatic aberration. There are chromatic aberrations.

In principle, any lens effect which depends on the *chromatic parameter* $\kappa$

$$\kappa = \frac{\delta E}{E_0} \tag{7.2}$$

is a chromatic aberration. The chromatic parameter describes the relative energy deviation of an electron in respect to the nominal electron energy $E_0$, where $\delta E$ is the deviation from $E_0$. If a certain functionality of an instrument does not depend on $\kappa$, it is an achromatic functionality or, more generally, an *achromatic* effect.

For the case in which the effect of a particular chromatic aberration depends only on the scattering (or illumination) angle $\omega$, the aberration is called an axial chromatic aberration or a chromatic aperture aberration. On the other hand, if the effect of a chromatic aberration depends on the position $w_o$ in the object plane, it is called a chromatic distortion or an off-axial chromatic aberration, as discussed in the previous section.

What we called simply the coefficient of chromatic aberration $C_C$ is indeed the primary axial chromatic aberration $C_{C\,1}$, whose impact is isotropic. Equation (3.6), rewritten as

$$\delta_C = C_{C\,1}\,\theta\,\frac{\delta E}{E_0} = C_{C\,1}\,|\omega|\,\frac{\delta E}{E_0},$$

describes the blurring of the focal point in the Gaussian image plane due to the primary axial chromatic aberration $C_{C\,1}$. While the above relation describes the radius of the circle of confusion in the Gaussian image plane (see Eq. (3.6)), the image aberration $\Delta w_i$ can be expressed as (see, e.g. Typke and Dierksen, 1995)

$$\Delta w_i = -M C_{C\,1}\,\omega\,\kappa, \tag{7.3}$$

where we substituted $\delta E/E_0$ with the chromatic parameter $\kappa$. The factor $M$ in Eq. (7.3) is the lateral magnification[3].

We see that the impact of the chromatic aberration $C_{C\,1}$ depends linearly on the geometrical ray parameter $\omega$ and on the chromatic parameter $\kappa$. Since according to Eq. (7.2) the chromatic parameter increases with decreasing nominal electron energy $E_0$, by choosing a sufficiently high microscope high tension, the effect of the chromatic blurring can be reduced such that it is not the resolution-defining quantity.

*Order, degree and rank*

The *order* $n$ of an aberration can be defined as the sum of the exponents of the geometrical ray parameters which describe the effect of the aberration in the image plane, i.e. its image aberration $\Delta w_i$ (Rose, 2009a). For STEM, the image plane, i.e. the plane where the demagnified image of the source is formed, is the specimen plane, which for HRTEM is the actual object plane. The geometrical ray parameters are the coordinate in the aperture plane $\omega$ and the coordinate in the object plane $w_o$. Since the effect of the primary axial chromatic aberration depends linearly on the geometrical ray parameter $\omega$ (see Eq. (7.3)), its order $n = 1$. Hence, $C_{C\,1}$ is a first-order aberration. The numerical subscript 1 thus indicates the order $n$ of the aberration. Furthermore, the *degree* of an aberration is equal to the exponent of

[3]The reason why the magnification $M$ needs to be incorporated in the description of the image aberration lies in the fact that the aberration coefficients are defined in terms of the object plane. This makes the aberration coefficient independent of the magnification $M$. Incorporating the magnification in Eq. (7.3) means that $\Delta w_i$ describes the (size of the) aberration figure in the image plane, as for instance defined by the fluorescence screen. On the other hand, it is also quite common to omit the magnification in the equation of the image aberrations, which means that the image aberrations are then described in a plane referred back to the object with $M = 1$ (see, e.g. Hawkes, 2009b). This is discussed in more detail in Sec. 7.4.0.1.

the chromatic parameter $\kappa$, which is used to describe the effect of the aberration in the image plane. For the case of $C_{C\,1}$, the degree is equal to 1. Finally, the *rank* of an aberration is the sum of the degree and the order of the aberration. Hence, $C_{C\,1}$ is a second-rank aberration. Besides, with the introduction of the degree of an aberration, we can simply define chromatic aberrations as aberrations whose degree is larger than zero.

The effect of the primary axial chromatic aberration is illustrated in Fig. 7.1b; it causes an object point to be imaged in the image plane as a disk because the focal point varies with electron energy. This is the dominant chromatic aberration of a round lens[4]. Apart from the axial chromatic aberration $C_{C\,1}$, in general round electron lenses possess contributions of off-axial chromatic aberrations which show a distortion character. These non-aperture chromatic aberrations, or chromatic distortions, are described by the coefficient of chromatic aberration of magnification $C_{\mathrm{D}}$ and the anisotropic chromatic distortion coefficient $C_\theta$, which, for purely electrostatic electron lenses, is zero (Hawkes and Kasper, 1989a). $C_{\mathrm{D}}$ and $C_\theta$ can be summarized in a complex quantity $A_{C\,1\,1} = C_{\mathrm{D}} + \mathrm{i}C_\theta$, which is the chromatic magnification change (Typke and Dierksen, 1995). While the first numerical index of $A_{C\,1\,1}$ refers to the order of the aberration as described above, the second numerical subscript indicates the power of $w_{\mathrm{o}}$ appearing in the relation of the corresponding image aberration $\Delta w_{\mathrm{i}}$. The second numerical subscript thus describes the dependency of the aberration on the object position $w_{\mathrm{o}}$. For axial chromatic aberrations this value is zero, and is thus omitted. The subscript C indicates a chromatic aberration.

For conventional electron microscopes consisting of round optical elements, the axial chromatic aberration $C_{C\,1}$ is the dominant chromatic aberration. The effects of other chromatic aberrations, like $A_{C\,1\,1}$, are usually negligible. This reasons the usage of the notation $C_{\mathrm{C}}$ for $C_{C\,1}$ and its colloquial naming as the chromatic aberration. However, besides the isotropic axial aberration $C_{C\,1}$, for non-round electron lenses additional axial chromatic aberrations are feasible. Examples of these are the dispersion $A_{C\,0}$ and the chromatic twofold astigmatism $A_{C\,1}$. The chromatic aberrations $C_{C\,1}$, $A_{C\,0}$ and $A_{C\,1}$ are axial chromatic aberrations and thus do not depend on $w_{\mathrm{o}}$. This reasons the omission of the second numerical subscript. While $C_{C\,1}$ leads to a change of defocus with changing electron energy (see Fig. 3.5), the dispersion $A_{C\,0}$ leads to an image shift with changing energy and $A_{C\,1}$ induces twofold astigmatism whose amount depends on the energy of the electron. Dispersion $A_{C\,0}$ is an aberration of order zero and rank one, whereas the chromatic twofold astigmatism $A_{C\,1}$ is a first-order aberration of rank two. All chromatic aberrations discussed are of first degree, i.e. they depend linearly on the chromatic parameter $\kappa$.

With this we close our short excursion into the field of chromatic aberrations. The essential point about this section is that there is not a single chromatic

[4]The term *round* lens can be misleading. What we mean by *round* lens is a lens which produces a rotationally symmetric electromagnetic field that is used to control the path of the electron beam.

aberration, as commonly presumed in the literature. In fact, the reduction of the overall chromatic effect of a round electron lens to a single quantity $C_C$ is only justifiable because there is a dominant chromatic aberration, which is the isotropic axial chromatic aberration $C_{C\,1}$. Furthermore, the impact of $C_{C\,1}$ does not vanish for object points close to the optical axis. Hence, even at very high magnifications, where in principle off-axial (chromatic) aberrations are negligible, $C_{C\,1}$ needs to be taken into account.

The topic of chromatic aberrations has many more facets than have been elucidated here briefly. Particularly in the case of non-round optical elements and in the discussion of energy filters, the proper treatment of chromatic aberrations is essential. For an extended discussion of chromatic aberrations, the reader is referred to more advanced electron-optical texts, as can be found in Hawkes and Kasper (1989a), Rose and Krahl (1995), Hawkes (2007) and Rose (2009a). In particular, for a complete derivation of all aberrations up to second rank, including geometrical aberrations, see Rose (2009a).

#### 7.3.2.2 *Geometrical aberrations*

We essentially defined the chromatic aberrations as aberrations which depend on the relative energy of the electrons and thus depend on the chromatic parameter $\kappa$. This dependency is described by the degree of an aberration which for chromatic aberrations is greater than zero. Since the rank of an aberration is the sum of its order and its degree, the rank of a chromatic aberration is thus greater than its order. *Geometrical* aberrations, on the other hand, do not depend on the relative energy of the electrons. Their impact depends solely on the geometrical ray parameters $\omega$ and $w_o$. The degree of geometrical aberrations thus vanishes and the order of a geometrical aberrations is equal to its rank.

Geometrical aberrations lead to well defined phase shifts of electron rays. The phase shifts depend on the geometrical ray parameters that define a given ray. Rays of different ray parameters can thus undergo different phase shifts. Letting these rays interfere either to form a HRTEM micrograph or a STEM probe reveals the various phase shifts in the form of an aberrated micrograph or an aberrated STEM probe. The coherence between different electron trajectories is not affected by the purely geometrical effect of these aberrations. This reasons why geometrical aberrations are also called coherent aberrations.

We have seen that the effects of defocus $C_1$ and third-order spherical aberration $C_3$ do not depend on the chromatic parameter $\kappa$. Hence, $C_1$ and $C_3$ are geometrical aberrations. Figures 7.1a and 6.12 illustrate the effect of spherical aberration; the focal point depends on the geometrical parameter $\omega$ in the aperture plane. While the focal point of the trajectory along the optical axis, i.e. the paraxial trajectory, defines the Gaussian focal point, for rays with $|\omega| \neq 0$, the effective focal distance is shorter than the one for the rays with $|\omega| \to 0$. With increasing $|\omega|$, the effective focal distance decreases. Furthermore, since according to Eqs. (2.16) and (3.14) the effects of $C_1$ and $C_3$ do not depend on the position $w_o$ in the object plane, $C_1$ and $C_3$ are a special type of geometrical aberration, namely axial aberrations. The image aberrations $\varDelta w_i$ (see Fig. 7.2) due to defocus $C_1$ and third-order spherical aberration $C_3$ can be written as

$$\begin{aligned} C_1 : \quad \varDelta w_i &= -MC_1\omega \\ C_3 : \quad \varDelta w_i &= -MC_3\overline{\omega}\omega^2. \end{aligned} \tag{7.4}$$

The above relations reveal that the effect of defocus $C_1$ in the image plane depends linearly on the geometrical ray parameter $\omega$. Hence, $C_1$ is a first-order axial aberration. On the other hand, since the effect of $C_3$ depends on $\overline{\omega}\omega^2$, its order is three. This is reflected in their respective numerical subscripts.

*Geometrical aberrations vs. parasitic aberrations*

A further distinction between actual geometrical aberrations and *parasitic* aberrations is feasible. According to Hawkes (2007), the first category, i.e. geometrical aberrations, describes the permitted aberrations of an ideal optical element of a given symmetry. An optical element of a given symmetry can only lead to a certain amount of aberrations, which are predetermined by its symmetry. Which type of aberrations are feasible for a given symmetry is discussed in detail by Hawkes and Kasper (1989b). For round electron lenses, the possible aberrations are the so-called Seidel aberrations, which are discussed in a forthcoming section of this chapter. Hence, according to Hawkes (2007), coherent aberrations which are permitted by the symmetry of the optical element represent the actual geometrical aberrations.

*Parasitic* aberrations, on the other hand, are aberrations which are induced by mechanical imperfections or, for instance, by inhomogeneities of the magnetic material the poles are made of. Hence, even though a given optical element theoretically can only cause a well defined set of symmetry-allowed aberrations, in practice it can cause a much larger variety of aberrations because no lens can ever be perfectly machined. The effect of these coherent parasitic aberrations depends on the geometrical ray parameters but not on the chromatic parameter $\kappa$.

Nonetheless, the differentiation between parasitic and symmetry-permitted aberrations can also be applied to (incoherent) chromatic aberrations. While an ideal round electron lens can only possess non-vanishing coefficients $C_{C\,1}$ and $A_{C\,1\,1}$ (up to second rank), i.e. the primary axial chromatic aberration and the chromatic magnification change, for non-ideal round electron lenses other chromatic aberration contributions could appear, like for instance the chromatic twofold astigmatism $A_{C\,1}$. Furthermore, any mechanical instability or the instability of the electromagnetic fields that form the optical units in an electron microscope can in principle be regarded as incoherent parasitic aberrations.

Another example which elucidates the difference between parasitic and symmetry permitted aberrations can be recognized in the standard alignment routine of an electron microscope. Doing high-resolution imaging in the broad-beam TEM mode or in STEM mode, it is unavoidable to correct for twofold axial astigmatism, which either directly impairs the image formation in HRTEM or the STEM probe and thus indirectly influences the respective micrographs. This astigmatism is a geometrical axial aberration which, however, is not permitted by the symmetry of a round objective lens. While the geometrical aberrations permitted by a round electron lens are discussed in the next section, axial twofold astigmatism is a direct consequence of the lens' deviation from producing a perfectly rotationally symmetric electromagnetic field. Hence, twofold axial astigmatism is a parasitic aberration which can be of importance in (conventional) atomic-resolution electron microscopy.

In the foregoing text we limit our treatise to geometrical aberrations. First, we will discuss the so-called Seidel aberrations of a round electron lens. Then, we will generalize the concepts of the geometrical wave front and the aberration function, in order to describe the optical behavior of electron optical elements of lower symmetry. This will lead us to the (residual) geometrical aberrations which have to be considered in aberration-corrected electron microscopy.

## 7.4 Geometrical Aberration of a Round Electron Lens

The number of geometrical aberrations an optical element possesses is in principle infinite. However, the amount of aberrations, or aberration coefficients, that need to be considered in the practical treatment of a particular optical system depends on the geometrical ray parameters that are of importance for the image formation. In general, the number increases with the precision by which the electron optical system needs to be described. It increases with the resolution of the optical system. For a limited regime of geometrical ray parameters, which is determined by the design of the optical instrument, one can assume that certain aberrations are of importance and need to be considered in the discussion of the image formation while others are not. Aberrations that are negligible are the ones which do not significantly affect rays within the relevant range of geometrical ray parameters $\omega$ and $w_o$.

In transmission electron microscopy, one is not interested in how a lens field acts on electrons that are scattered to large angles or on electrons which are even backscattered. In fact, one is interested only in how the lens field acts on rays which are describable by small $\omega$ and small $w_o$. Hence, for the treatment of practical (electron-) optical instruments, one can reduce the problem. The way to do this is to consider aberrations up to a certain order. This essentially means that in electron optics, one is interested in aberrations whose effect on the image (or on the electron probe) depend on first, second and third power on the ray parameters, i.e. one is interested in aberrations that are of first, second and third order. However, with increasing resolution, the regime of the ray parameters needs to be expanded such that higher-order aberrations, i.e. fourth- and fifth-order aberrations, need to be considered as well. This is also important if the symmetry of the system deviates from its ideal design (see, e.g. Rose, 1968a, 1968b).

Ludwig Seidel is commonly said to be the first person who analyzed the permitted geometrical aberrations up to third order of a round light optical lens. The work he carried out around 1855 was published in 1857 under a rather lengthy title[5]. Yet, it has to be mentioned that Joseph Maximilian Petzvals performed an equivalent analysis prior to Seidel's work. However, Petzvals' manuscript was lost even before it could be published. Hence, the primary geometrical aberrations up to third order of a round lens are commonly said to be the *Seidel aberrations*.

Seidel performed the analysis for a light optical lens, which, however, has fewer aberration coefficients than the electron optical equivalent in the presence of magnetic fields. Nonetheless, the naming has been adopted to describe the aberrations up to third order of a round electron optical lens as well.

The amount of geometrical aberration coefficients up to third order of a round light optical lens is five (Born and Wolf, 2001), whereas in the presence of magnetic fields, a round electron lens can possess up to eight (Hawkes and Kasper, 1989a). Hence, in general, eight geometrical aberration coefficients have to be considered in order to describe the optical behavior of an ideal round electron lens up to third order. The following aberrations can be distinguished; spherical aberration, coma, field astigmatism, field curvature and distortion. In the foregoing subsections, the effects of these aberrations are discussed on a qualitative basis. The reference is the Gaussian optics where a point in the object plane is imaged as a point in the image plane. The ideal, i.e. stigmatic image point in the image plane, is called the Gaussian image point. For TEM, the object plane corresponds to the plane where the specimen is located, the corresponding aperture plane is the back focal plane of the objective lens, and the image plane is the plane where the detector is located. In STEM, the object plane is essentially a plane that corresponds to the plane of the electron source, and the image plane is the plane where the electron probe is brought to focus, i.e. the plane of the specimen. Hence, it is the image of the electron source projected onto the specimen plane which is affected by the lens aberrations in STEM mode and thus ultimately limits the STEM resolution. The relevant aperture plane in STEM is the front focal plane of the objective lens. Hence, when we talk about the impact of the aberrations on the image plane, we mean in TEM mode the plane of the detector, while for STEM it is the specimen plane. The relevant aperture planes are the front focal plane for STEM and the back focal plane for HRTEM, respectively.

[5] The title of Seidel's work is *Über die Theorie der Fehler, mit welchen die durch optische Instrumente gesehenen Bilder behaftet sind, und über die mathematischen Bedingungen ihrer Aufhebung* (in Abhandlungen der naturwiss.-techn. Commission bei der Königl. Bayerischen Akademie der Wissenschaften in München, Nr. 1. (1857) 227–267), which can be translated as *On the Theory of Errors by which Images Observed by Optical Instruments are Affected and on the Mathematical Conditions on Their Annihilation.*

#### 7.4.0.1 *Spherical aberration*

We have already discussed the impact of the spherical aberration and illustrated its effect in Fig. 7.1a. The aberration figure in the image plane associated with spherical aberration is a disk which is centered at the Gaussian image point. In fact, each set of rays that passes the aperture plane under a given angle $|\omega|$ forms a circle in the Gaussian image plane. The radius $r_\mathrm{S}$ of the circle that is formed by rays that pass the aperture plane in $|\omega|$ is

$$r_\mathrm{S} = MC_3|\omega|^3, \tag{7.5}$$

while the actual image aberration is described by

$$\Delta \mathrm{w_i} = -MC_3\overline{\omega}\omega^2. \tag{7.6}$$

If an aperture of opening $|\omega_\mathrm{a}| = \theta_\mathrm{a}$ is employed, Eq. (7.5) provides the radius of the resulting aberration figure in the image plane. This disk reflects the astigmatic image point corresponding to a single point of the object. Hence, each object point is imaged as a disk of finite radius. The radius of the disk depends neither on the position of a particular point in the object plane nor on the position in the image plane. This is illustrated in Fig. 7.3. The quantity $C_3$ in Eq. (7.5) is the constant of spherical aberration, and $M$ is the lateral magnification. Because the impact of $C_3$ in the image plane scales with the third power of the ray parameter $|\omega|$, $C_3$ is the constant of spherical aberration of third order, i.e. it is the quantity $C_\mathrm{S}$ which is often used in the literature to refer to spherical aberration. Here, however, we would like to emphasize the order of the aberration which is denoted by the numerical subscript. Since the impact of $C_3$ is independent of the coordinate $\mathrm{w_o}$ in the object plane, the second numerical subscript, as introduced in the discussion of the chromatic aberrations, is equal zero and is thus omitted.

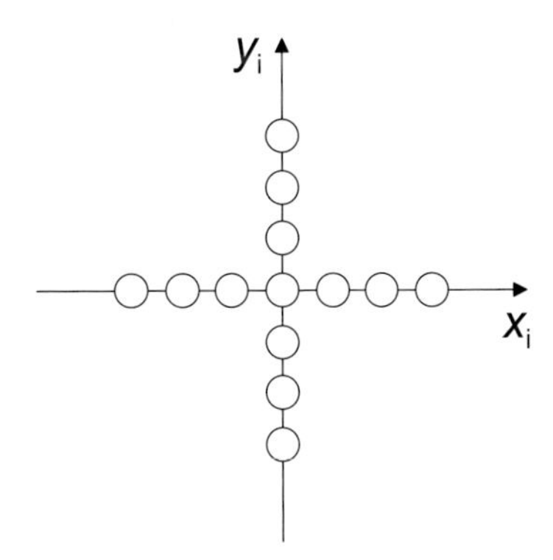


Fig. 7.3 The effect of spherical aberration $C_3$ on object points imaged onto the image plane $\mathrm{w_i} = x_\mathrm{i} + \mathrm{i}y_\mathrm{i}$. Each object point results in a disk. The radius of the disk depends neither on the position of the object point nor on the position of the image point in the image plane. The radius depends solely on the opening of the aperture, as described by Eq. (7.5).

The lateral magnification appearing in Eqs. (7.5) and (7.6) requires a short explanation. The aberration coefficients, such as $C_3$, are defined in terms of the scale in the plane where the specimen is located. Hence, in order to describe the impact of the aberration in the HRTEM image plane, one needs to consider the lateral magnification $M$ which is employed to record the micrograph. The magnification $M$ is determined by the objective lens and the (magnification) setting of the projector lens system. It is assumed that the coordinates in the object plane are related to the coordinates in the image plane by a linear magnification change. Under ideal imaging conditions, an image point $w_o$ in the object plane is imaged as a stigmatic image point $w_i$, with $w_i = -Mw_o$. However, if one wants to know the impact of the aberration $\Delta w_i$ in terms of the resolution of the micrograph, one needs to relate the impact of the aberration to the scale of the object. For this case, the magnification $M$ can simply be ignored ($M = 1$) in the above equations and one obtains the impact of the aberration on the scale of the specimen, i.e. the impact of the aberration referred back to the specimen (see, e.g. Hawkes, 2009b). Though this is the relevant measure for the resolution of the imaging process, it does not provide information on the extent of the aberration in the detector plane.

The following example shall illustrate this distinction. Let us assume that $C_3 = 1$ mm, i.e. a typical value of the third-order spherical aberration of a conventional objective lens. The rays that should be brought to focus by the lens shall pass the aperture in $|\omega| = |\omega_a| = 10$ mrad. Furthermore, we shall employ a microscope magnification in HRTEM mode of $M = 10^6$. According to Eq. (7.5), the radius of the corresponding aberration disk in the detector plane is 1 mm. However, from this we do not directly learn anything of its impact on the resolution of the imaging process. For this we need to know the impact of the aberration referred back to the specimen. Hence, ignoring $M$ in Eq. (7.5), or actually setting $M = 1$, yields the impact of $C_3$ in respect to the specimen. The radius of the aberration disk referred back to the object is then 1 nm. While the first value depends on the applied magnification, the second one is independent of the magnification and can be used to directly describe the impact of the aberration on specimen features that need to be resolved. Hence, one often uses this reduced effect, i.e. the effect of the aberration referred back to the specimen plane, in order to describe the impact of a given aberration, keeping in mind, however, that aberrations actually impact the image, which, in HRTEM, is a magnified image of the object.

In STEM, the situation is the same; the aberration coefficients are defined in the plane of the specimen. However, in STEM mode, the specimen plane is the plane where the electron beam is brought to focus and the electron probe is formed. Since this is the relevant plane where the impact of the aberrations needs to be known, the lateral magnification $M$ in Eqs. (7.5) and (7.6) can be set to $M = 1$.

Comparing Eq. (7.5) with the radius of the disk of least confusion given in Eq. (3.5) shows that the aberration figure in the Gaussian image plane is four times larger than the aberration figure in the plane where the disk of least confusion is located. The focal distance between the Gaussian image plane, for which Eq. (7.5) is valid, and the plane where the disk of least confusion is located, measures

$$C_{1\,\mathrm{DLC}} = -\frac{3}{4} M_{\mathrm{L}} C_3 |\omega|^2, \tag{7.7}$$

where $M_{\mathrm{L}}$ is the longitudinal magnification (see Eq. (6.97)).

The negative defocus offset $C_{1\,\mathrm{DLC}}$ indicates that the plane containing the disk of least confusion is in front of the Gaussian image plane. Comparing the defocus offset $C_{1\,\mathrm{DLC}}$ with the Scherzer focus given in Eq. (2.21) essentially shows that working at Scherzer focus implies adjusting the defocus in such a way that the disk of least confusion is moved towards the Gaussian image plane. In this way, the impact of $C_3$ on the imaging process can be minimized. This is indeed an alternative way of interpreting the Scherzer focus and the defocus associated with the Scherzer incoherent condition given in Eq. (3.9).

Hence, the fact that the disk of least confusion is located in front of the Gaussian image plane is the consequence that the spherical aberrations causes rays of finite $\omega$ to cross the optical axis in front of the Gaussian image plane.

#### 7.4.0.2 *Coma*

Coma is the second of the symmetry, permitted geometrical aberrations of a round lens, which we introduce here. As the name of the aberration suggests, the aberration figure caused by coma resembles that of a comet: a comet-like tail emerges from a sharp distinct point. This distinct point reflects the location of the Gaussian image point. Hence, in the presence of coma, each object point is imaged into a specific coma figure, as illustrated schematically in Fig. 7.4. A set of rays emerging from an object point $w_o$, which passes the aperture in a given angle $|\omega|$, leads to a circle in the image plane. The radius $r$ of the circle is given by $r = |MB_{3\,1} w_o| \omega\overline{\omega}$, where $B_{3\,1}$ is the coefficient of third-order coma and $M$ is the lateral magnification.

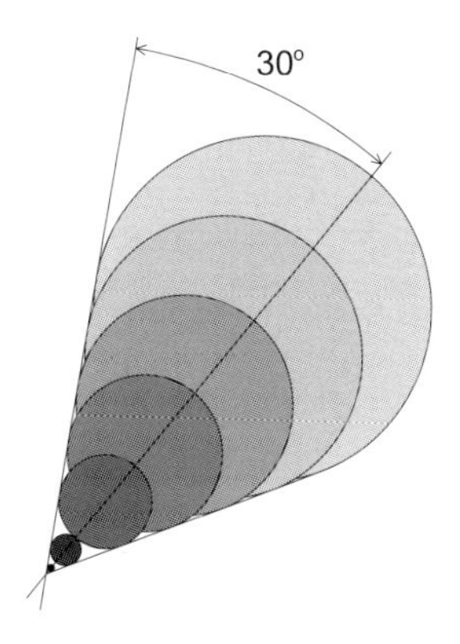


Fig. 7.4 Coma aberration figure.

The center of the circle formed by rays that pass the aperture in $|\omega|$ is displaced from the Gaussian image point by the length $l$, which is twice the radius of the circle, i.e. $l = 2 \cdot |MB_{31}\mathrm{w_o}|\omega\bar{\omega}$. For the whole range of $|\omega|$-values as determined by the opening of the aperture, a continuous series of circles is formed whose radii and distances to the Gaussian image point increase with increasing $|\omega|$. The individual circles lie on a line. The envelope of the aberration figure forms an angle of $60^\circ$ limited by the perimeter of the largest circle reflecting the largest $|\omega|$, which is determined by the diameter of the aperture opening.

While in light optics, coma can be described by one (real) aberration coefficient $B_{31}$, this is not generally possible in electron optics. In the presence of a magnetic field, which causes an electron to follow a helical trajectory defined by the Larmor rotation, coma in magnetic electron lenses comprises two contributions; *radial* (or *isotropic*) coma and *azimuthal* (or *anisotropic*) coma. Hence, in general two coefficients are necessary to describe coma of a round magnetic electron lens. However, here the complex expression of the geometrical ray parameters allows us to describe the aberration coefficients by one complex quantity. The radial and azimuthal coma is written as one complex coefficient $B_{31}$, whose real part $\Re(B_{31})$ reflects the radial coma and whose imaginary part $\Im(B_{31})$ describes the azimuthal coma.

Figure 7.5 schematically illustrates the effects of radial coma (a), azimuthal coma (b) and a combination of radial and azimuthal coma (b) of discrete object points imaged onto the Gaussian image plane $(x_\mathrm{i}, y_\mathrm{i})$. Each object point $\mathrm{w_o}$ is imaged into a coma figure. The size, or the length of the coma figure, increases with increasing $|\mathrm{w_o}|$. Radial coma leads to coma figures which are radially aligned, whereas azimuthal coma leads to coma figures which are aligned tangentially along rings whose center is the center of the Gaussian image plane. The combination of the two leads to coma figures which are in intermediate orientation. This is the

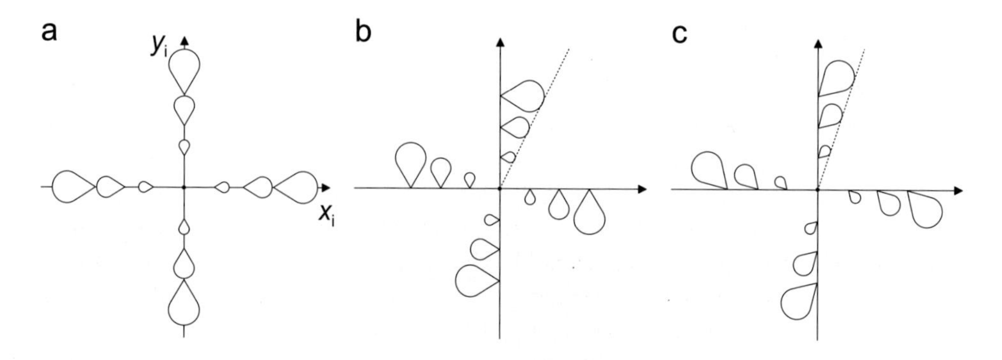


Fig. 7.5 Off-axial coma $B_{31}$. (**a**) Radial coma $\Re(B_{31})$, (**b**) anisotropic or azimuthal coma $\Im(B_{31})$, (**c**) combination of radial and anisotropic coma. As indicated by the dotted lines, the effect of $B_{31}$ vanishes for $\mathrm{w_i} = x_\mathrm{i} + \mathrm{i}y_\mathrm{i} \to 0$. Hence, in the center of the object plane, there is a stigmatic image point.

general case of a magnetic lens, which has a real and imaginary coma component in $B_{31}$. In the absence of magnetic fields, i.e. for pure electrostatic lenses, the azimuthal component vanishes such that $\Im(B_{31}) = 0$. Hence, a purely electrostatic lens behaves like a light optical lens which shows only radial coma.

In analogy to Eq. (7.5), the image aberration $\Delta w_{\mathrm{i}}$ associated with coma $B_{31}$ can be expressed as[6]

$$\Delta w_{\mathrm{i}} = -M\left[2B_{31}w_{\mathrm{o}}\omega\overline{\omega} + \overline{B}_{31}\overline{w}_{\mathrm{o}}\omega^2\right], \tag{7.8}$$

where $\overline{B}_{31}$ indicates the complex conjugate of $B_{31}$.

Comparing spherical aberration $C_3$ with coma $B_{31}$ reveals a fundamental difference (see, e.g. Figs. 7.3 and 7.5). While the impact of $C_3$ depends solely on the geometrical ray parameter $\omega$ in the aperture plane (see Eq. (7.5)), Eq. (7.8) shows that the effect of coma $B_{31}$ also depends on the location $w_{\mathrm{o}}$ of the object point. This simply means that coma $B_{31}$ causes each object point $w_{\mathrm{o}}$ to be imaged differently. Most importantly, the effect of $B_{31}$ vanishes if $|w_{\mathrm{o}}| \to 0$. Hence, while the effect of $C_3$ affects all image points equally, coma $B_{31}$ affects only off-axial object points. Coma $B_{31}$ is thus an off-axial aberration which impairs only object points that are not in the immediate neighborhood of the optical axis. Furthermore, the impact of coma increases linearly with $|w_{\mathrm{o}}|$. This is indicated by the dashed line in Fig. 7.5. The sum of the power of the ray parameters that describe the impact of coma in the image plane is three (see Eq. (7.8)). Therefore, coma $B_{31}$ is a third-order aberration which is expressed by the first numerical subscript. Moreover, since coma increases linearly with $w_{\mathrm{o}}$, the second numerical subscript is equal one, indicating that $B_{31}$ is an off-axial aberration that depends linearly on $w_{\mathrm{o}}$.

#### 7.4.0.3 *Field astigmatism and field curvature*

Astigmatism, or a bit more specifically, twofold astigmatism, is the most common correctable aberration in high-resolution electron microscopy. The kind of twofold astigmatism which is corrected to obtain and optimize atomic-resolution micrographs, however, is not the *field* astigmatism which is subject of this subsection. Nonetheless, if we consider the image of a single object point $w_{\mathrm{o}}$, the effect of field astigmatism, which as an off-axial aberration is denoted by $A_{32}$, is equivalent to the effect of the common twofold astigmatism which is an *axial* aberration denoted by $A_1$. Furthermore, while field astigmatism is a symmetry-permitted aberration of a round electromagnetic lens, twofold axial astigmatism is a parasitic aberration which is caused by deviations of the electromagnetic field from an ideal rotational symmetry. Still, if we consider a single object point ($w_{\mathrm{o}} \neq 0$) and not a finite field of view, the effect of field astigmatism and axial astigmatism is conceptually comparable.

[6]We employ the notation of the aberration coefficients according to Pöhner and Rose (1974), which was also adopted by Typke and Dierksen (1995). However, we include a factor 1/3 in $B_{31}$ to have the magnitude of $B_{31}$ in agreement with Rose (2009a).

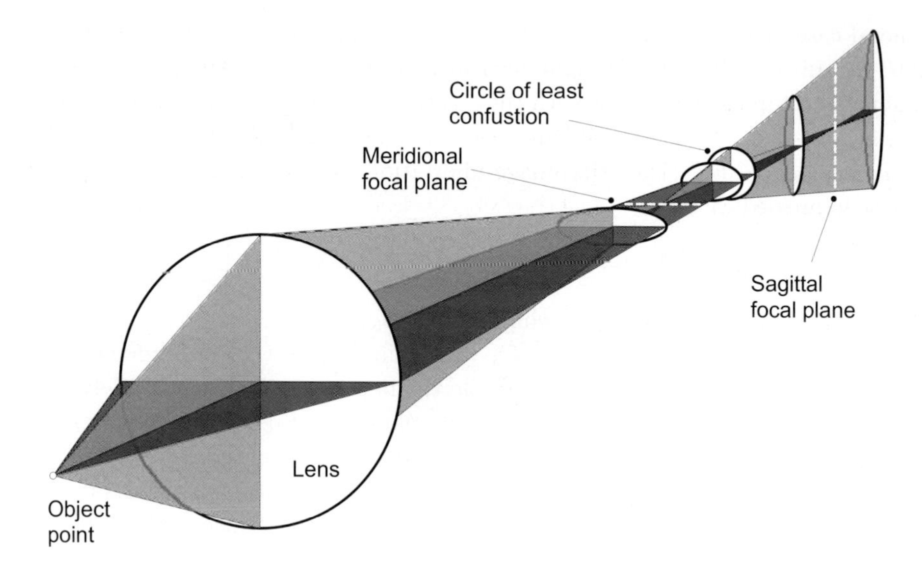


Fig. 7.6 Twofold astigmatism.

Figure 7.6 schematically illustrates the effect of twofold astigmatism on the image formation of a single object point. Electron trajectories emerge radially from the object point and enter the field of the lens, which is illustrated as a single plane corresponding to the aperture plane as well. The important point about astigmatism is that the focusing power of the lens depends on the azimuthal angle. For the case of twofold astigmatism, the focusing power is different in two orthogonal directions. This essentially means that rays described by the aperture coordinate $\omega = \theta_x + \mathrm{i}0$ are differently focused compared with rays which have the aperture coordinate $\omega = 0 + \mathrm{i}\theta_y$ (see Fig. 7.6). Hence, there is not a single focal point which would correspond to a stigmatic image. Instead, there are two orthogonal line foci, which lie in the *sagittal* focal plane and in the *tangential* or *meridional* focal plane, respectively. In Fig. 7.6, the meridional line focus is in front of the sagittal line focus. This, however, can be reversed if the focusing power of the lens is opposite to the illustration in Fig. 7.6. The Gaussian image plane lies between the meridional and sagittal focal planes. For an arbitrary coordinate along the optical path of the pencil, the image of the object point is an ellipse. For one particular position, the ellipse degenerates to a circle. In the sagittal and in the meridional plane, the ellipse degenerates to a line. Though the round beam between sagittal and meridional focal planes is not a stigmatic image point, in the presence of twofold astigmatism this is the best image that can be obtained from the object point. As in the presence of spherical aberration $C_3$, this circle is called the circle of least confusion. Hence, though a stigmatic image point is not feasible in the presence of twofold astigmatism, for a suitable focus setting it is still possible to have an object point imaged as a circular disk.

In the presence of *field astigmatism* $A_{32}$, Fig. 7.6 in principle still applies, provided, however, that only one single object point is considered, and that this object point is not identical to the point on the optical axis, i.e. $w_o \neq 0$. Yet, the situation is more complex. In the presence of field astigmatism, the twofold astigmatism increases quadratically with its off-axial distance. Therefore, in order to clearly differentiate twofold field astigmatism $A_{32}$, which is an off-axial aberration, from the twofold axial astigmatism $A_1$, $A_{32}$ is sometimes called Seidel astigmatism or Seidelian astigmatism, referring to the fact that it is one of the Seidel aberrations, whereas twofold axial astigmatism $A_1$ is a parasitic aberration of a round lens.

The image aberration $\Delta w_i$ associated with field astigmatism can be expressed as (see, e.g. Rose, 2009a)

$$\Delta w_i = -MA_{32}\overline{\omega}w_o^2. \tag{7.9}$$

Equation (7.9) reveals that the impact of $A_{32}$ increases quadratically with the off-axial distance $|w_o|$. Furthermore, because field astigmatism is a third-order aberration which quadratically increases with the off-axial distance of the object point, two numerical subscripts are employed in $A_{32}$.

In fact, the appearance of field astigmatism is intuitively understandable if we consider the lens field observed from a selected object point in front of the lens. For an object point on the optical axis, the field of the lens looks rotationally symmetric. No astigmatism is thus expected for an axial point. However, an off-axial object point sees the rotationally symmetric field from an angle, and what looks from the optical axis to be a circle appears as an ellipse for the off-axial position. The off-axial point thus experiences an 'elliptical' lens field, which, according to Fig. 7.6, explains the resulting astigmatic image formation.

Similar to the case of coma $B_{31}$, due to the Larmor rotation in the presence of a magnetic field, field astigmatism can have two components; a radial component and an azimuthal component. This circumstance can be accounted for by employing a complex aberration coefficient $A_{32}$, whose real part reflects the radial contribution and whose imaginary part corresponds to the azimuthal astigmatism caused by the magnetic field contribution.

In addition to field astigmatism, there is another aberration present whose impact increases quadratically with the off-axial distance of an object point. *Field curvature* (or image curvature) essentially causes the defocus to vary across the field of view. The corresponding image aberration can be described by

$$\Delta w_i = -MF_{32}\omega w_o \overline{w}_o. \tag{7.10}$$

Equation (7.10) reveals that field curvature is a third-order aberration similar to field astigmatism. This is expressed by the first numerical subscript in $F_{32}$. Furthermore, as the impact of field curvature increases quadratically with the off-axial distance of the object point, the second numerical subscript is 2. The aberration coefficient $F_{32}$ is real, i.e. it does not have an imaginary part.

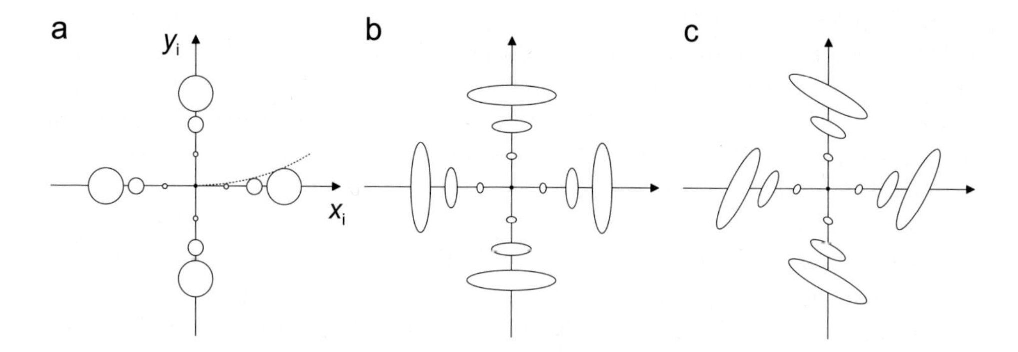


Fig. 7.7 Field curvature and field astigmatism. (**a**) Image aberration caused by field curvature, (**b**) the effect of radial field-astigmatism and image curvature, and (**c**) the combined effect of field curvature and radial and azimuthal field-astigmatism. The dotted line indicates that for $|\mathrm{w_i}| = |x_\mathrm{i} + \mathrm{i}y_\mathrm{i}| \rightarrow 0$, the effect of field curvature and field astigmatism vanishes.

By changing the defocus across the field of view, field curvature, as its name suggests, essentially causes the stigmatic image plane to bend. Hence, the effect of field curvature vanishes if the image is recorded on a suitably curved detector plane. However, for a plane detector placed in the Gaussian image plane, an object point $P_1$, which is located in a distance $|\mathrm{w_o}|$ from the optical axis, is imaged as a disk reflecting that it is out of focus. Furthermore, an object point $P_2$, which is located in a distance $2|\mathrm{w_o}|$ from the optical axis, i.e. twice the off-axial distance of $P_1$, is imaged as a disk which is four times larger than the disk corresponding to the object point $P_1$. This is illustrated in Fig. 7.7a. For vanishing off-axial distance, i.e. if $|\mathrm{w_o}| \rightarrow 0$, the corresponding object point is imaged unaffected from field curvature and field astigmatism. This is clearly different from the effect of spherical aberration, which affects all object points equally (see Fig. 7.3).

The co-presence of field astigmatism and field curvature causes the elliptical distortion of an image point to increase with increasing off-axial distance. This is illustrated for the case of radial field astigmatism in Fig. 7.7b, and for the case of a combination of radial and azimuthal field astigmatism in Fig. 7.7c.

#### 7.4.0.4 *Distortion*

All of the symmetry-permitted aberrations of a round lens that we have discussed so far, namely spherical aberration $C_3$, coma $B_{3\,1}$, field astigmatism $A_{3\,2}$ and field curvature $F_{3\,2}$, caused the image of an object point to appear as an astigmatic image point. The image points are either disks, coma figures, ellipses or combination of these figures. The effect of spherical aberration does not depend on the position of the object point, whereas for the cases of coma $B_{3\,1}$, field astigmatism $A_{3\,2}$ or field curvature $F_{3\,2}$, the image aberration increases with the off-axial distance. For these off-axial aberrations, the image aberration vanishes for an object point on the optical axis. An axial object point has a corresponding stigmatic image point.

*Image distortion*, as the fifth of the symmetry-permitted aberrations of a round lens, does not affect the formation of stigmatic image points. As the name suggests, image distortion causes a given field of view to appear as a distorted image of the object. Nevertheless, each object point finds a corresponding stigmatic image point in the distorted image. The image aberration associated with image distortion can be expressed as

$$\Delta \mathrm{w_i} = -M D_{33}\, \mathrm{w_o^2} \overline{\mathrm{w}}_\mathrm{o}. \tag{7.11}$$

This relation shows that image distortion is a third-order aberration whose impact increases with the third order of the geometrical ray parameter in the object plane $w_o$. On the other hand, the impact of image distortion does not depend on the ray parameter $\omega$ in the aperture plane. This explains why each object point finds a stigmatic image point as a counterpart, regardless of the resolution of the instrument.

Let us consider object points arranged on a square lattice, as illustrated by the dashed mesh in Fig. 7.8. *Radial* or *isotropic* image distortion cause the image of this mesh to appear either as a pincushion-type distorted mesh or as a barrel-type distorted mesh. These two types of distortions are illustrated in Figs. 7.8a and 7.8b, respectively. Pincushion and barrel distortions are the types of distortions that are feasible in round light optical lenses, as well as in purely electrostatic electron lenses. If, however, a magnetic field is present, which forces the electrons to follow helical trajectories, *spiral* or *azimuthal* distortion becomes feasible, too. This is illustrated in Fig. 7.8c. Hence, magnetic electron lenses generally lead to image distortions which are combinations of azimuthal distortion with either the barrel or pincushion distortion (see Fig. 7.8).

Similar to the complex aberration coefficients described above, the barrel or pincushion distortion is described by the real part of $D_{33}$, and the spiral or azimuthal distortion is reflected in the imaginary part of $D_{33}$.

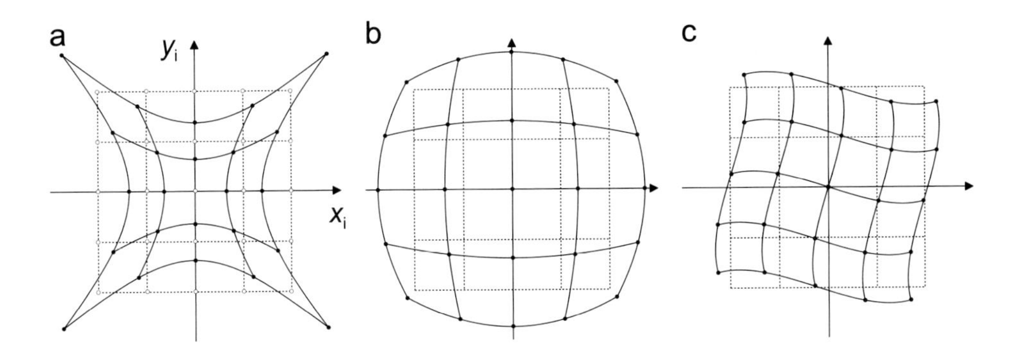


Fig. 7.8 Image distortions. The mesh indicated by the dashed lines represents the undistorted image. (**a**) Pincushion distortion, (**b**) barrel distortion and (**c**) spiral or azimuthal distortion.

It has to be emphasized that the contribution of the objective lens to the total image distortion in HRTEM imaging is small. The reason for this is simple: the object area is small. This, however, is not necessarily the case for the projector lenses behind the objective lens. The projector lenses stepwise increase the magnification of the transferred object. Hence, the size of the object that needs to be transferred and further magnified increases within the projector lens system. As a consequence, if image distortions are observable, they are caused mostly by the projector lenses.

#### 7.4.0.5 *Summary*

Spherical aberration $C_3$, coma $B_{31}$, field astigmatism $A_{32}$, field curvature $F_{32}$, and image distortion $D_{33}$ — these are the so-called Seidel aberrations, i.e. the symmetry-permitted geometrical third-order aberrations of a round lens. This is true for the case of light optical lenses (see, e.g. Born and Wolf, 2001) as well as for the round lenses used in electron optics (see, e.g. Hawkes and Kasper, 1989a). In light optics, and for the case of purely electrostatic electron lenses, all five aberrations are describable by real coefficients. Hence, five independent numerical values are needed to describe the aberrations of a round lens. However, in the case of magnetic electron lenses, coma $B_{31}$, field astigmatism $A_{32}$ and image distortion $D_{33}$ are described by complex values of non-vanishing imaginary parts. Hence, in electron optics one generally ends up with eight independent numerical aberration coefficients. Table 7.1 summarizes the Seidel aberrations and their respective image aberrations.

An important aspect of these aberrations concerns their off-axial dependence and their relevance for high-resolution imaging. While the impact of the off-axial aberrations, namely coma $B_{31}$, field astigmatism $A_{32}$, field curvature $A_{32}$ and image distortion $D_{33}$, vanishes for object points located on the optical axis (see Figs. 7.5, 7.7 and 7.8), the only geometrical aberration which impacts the image of object points on the optical axis is the spherical aberration $C_3$ (see Fig. 7.3). We recall the isoplanatic approximation, which simplifies the imaging process by assuming that the entire field of view (at high magnification) experiences the same phase contrast transfer function or, alternatively, the same electron probe, which

Table 7.1 Seidel aberrations; the symmetry-permitted geometrical third-order aberrations of a round lens.

| Aberration | Symbol | Value | Image aberration $\Delta w_i$ |
|---|---|---|---|
| Third-order spherical aberration | $C_3$ | real | $-MC_3\overline{\omega}\omega^2$ |
| Coma | $B_{31}$ | complex | $-M\left[2B_{31}\,w_o\omega\overline{\omega} + \bar{B}_{31}\overline{w}_o\omega^2\right]$ |
| Field astigmatism | $A_{32}$ | complex | $-MA_{32}\overline{\omega}w_o^2$ |
| Field curvature | $F_{32}$ | real | $-MF_{32}\omega w_o\overline{w}_o$ |
| Image distortion | $D_{33}$ | complex | $-MD_{33}\,w_o^2\overline{w}_o$ |

in fact is valid only for the axial point. Considering that all off-axial aberrations vanish on the optical axis, it is indeed solely the spherical aberration $C_3$ which remains effective for the imaging characteristics of near-axial object points. It is for this reason why the spherical aberration is said to be the limiting aberration in electron microscopy. Compared to the impact of the spherical aberration, the effects of coma, field astigmatism, field curvature and image distortion are negligible within the isoplanatic approximation.

Although it is valuable for high-resolution electron microscopy, the isoplanatic approximation is not necessarily suitable in other domains that employ round electron lenses. For electron optical devices other than transmission electron microscopes, it is not necessarily the spherical aberration which is the most troublesome aberration. In electron lithography, field curvature and field astigmatism are the most disturbing aberrations. These aberrations limit the field of view, i.e. the area of the lithographic mask, and thus complicate processing of large areas.

Besides the fact that the object area is small in transmission electron microscopy, the impact of off-axial aberrations can be minimized to some extent. For instance, it can be shown that by setting the aperture plane in HRTEM imaging to a plane, which is in between the object plane and the back focal plane, the radial coma, i.e. $\Re(B_{3\,1})$, can be annulled. The point on the optical axis in this coma-free plane is called the coma-free point (see, e.g. Rose, 2009a, Typke and Dierksen, 1995). The coma-free point is roughly in a distance $2/3f$, where $f$ is the focal distance of the lens. However, though this allows for minimizing the effect of radial coma, its azimuthal counterpart, which is unavoidable in the presence of a magnetic field, remains unaffected by this procedure. Hence, the azimuthal coma, i.e. $\Im(B_{3\,1})$, is invariant. Still, by placing the pivot point of the beam in a scanning transmission electron microscope in the coma-free point, the scanning beam does not produce any radial coma which would reduce the achievable resolution for object points that are in a finite distance from the optical axis.

A strategy to minimize the impact of the azimuthal components of coma $B_{3\,1}$, field astigmatism $A_{3\,2}$ and in particular of image distortion $D_{3\,3}$, is based on the fact that these azimuthal aberrations are caused by the Larmor rotation. As discussed in Chapter 6, changing the polarity of a magnetic lens does not affect the focusing property of the lens, however, it reverses the Larmor rotation and as a result it also reverses the azimuthal image aberrations (see, e.g. Glaser, 1952). Transferring an electron beam through a series of round electron lenses, as in the projector system of an electron microscope, it is advantageous to alternate the polarity of consecutive lenses. This minimizes the overall azimuthal image aberrations.

Though the impact of off-axial symmetry permitted aberrations is small or can be minimized to some extent, the spherical aberration, which has the order of magnitude of the focal length of a lens, remains, affecting the imaging characteristics of a (scanning) transmission electron microscope. Furthermore, the symmetry permitted aberrations are not the only aberrations which are of relevance in real electron optical systems. Due to mechanical imperfections, the ideal symmetry can never be fully established in a real electron optical unit. This leads to the appearance of geometrical (and chromatic) aberrations which one would not expect under ideal circumstances. Hence, even a round electron lens can lead to a larger variety of geometrical aberrations than the ones discussed above. Since the effect of off-axial aberrations vanishes for object points on the axis, the relevant parasitic aberrations are the ones which do not vanish for object points on the optical axis. These aberrations are the axial or aperture aberrations. They are also effective within the isoplanatic approximation. For the case of a round electron lens, the only symmetry-permitted third-order aberration, which is an axial aberration, is spherical aberration. Even in the simple case of rotationally symmetric devices, the non-ideality of a round electron lens leads to more axial aberrations, such as, for example, twofold axial astigmatism $A_1$ or axial coma $B_2$.

Moreover, due to the fact that aberration correctors employ multi-pole lenses (see Chapter 8), the variety of residual geometrical aberrations in such electron microscopes can be significantly larger than for conventional electron microscopes. Deciding whether a certain aberration is caused by a slightly misaligned multi-pole lens or by a mechanical imperfection of an optical element is in general not a trivial problem to solve. Hence, in the foregoing of this chapter, we do not strictly distinguish between parasitic aberrations and permitted geometrical aberrations. However, we would like to point out that this distinction is of fundamental importance in the discussion of the performance of an optical element of a given symmetry (see, e.g. Haider *et al.*, 1982), and also in the design of aberration correctors, where in fact permitted geometrical aberrations of multi-pole lenses are used to balance the intrinsic aberration of the round objective lens.

## 7.5 Wave Surface, Aberration Function and Image Aberrations

So far, we have been dealing with effects of the aberrations in the image plane, i.e. with image aberrations $\Delta w_{\mathrm{i}}$. For TEM imaging, this plane is the detector plane and for STEM, it is the specimen plane where an image of the demagnified electron source is formed. Describing the effects of the aberrations in the image plane is one way to deal with aberrations. Another approach, which is based on Fig. 5.13, is to explain the aberrations in terms of curved wave surfaces. In this section, we would like to make the connection between (axial) image aberrations and the concept of the geometrical wave front which we introduced in Chapter 5.

We have seen in Chapter 5 that the point eikonal $S$ is a function which describes the geometrical behavior of a pencil of electron trajectories (see Fig. 5.10). The point eikonal $S$ depends on two points, such that $S = S(P_0, P)$. These points can be any two points in space. According to the principle of Maupertius (or the principle of reduced action) (see Eq. (5.50)), rays are defined by $S \to$ extremum. The eikonal $S$ can be understood as a path length of a trajectory connecting point $P_0$ with $P$.

Though we can identify $S$ with a path *length*, we did not draw any attention to its physical dimension. In the definition of $S$ in Eq. (5.27) and similarly in the principle of Maupertius in Eq. (5.50), the unity of $S$ is given by the unity of the (canonical) momentum multiplied by the unity of a distance. Hence, in order to have the dimension of $S$ equal to a distance we must in principle normalize $S$ by the momentum $2\pi\hbar k_0$ of the incident electrons where $k_0$ is equal to the modulus of the wave vector of the incident electrons, i.e. $k_0 = 1/\lambda$. With this formal change, $S$ has the required dimension, i.e. $S$ shall describe the path length of an electron trajectory between two points. In the foregoing text we shall identify $S$ as being normalized such that its dimension corresponds to a distance.

Besides the interpretation of $S$ as a path length, we also explained that the set of points determined by

$$S(P_0, P) = \text{constant}$$

defines a surface which we called the *geometrical wave surface* (see Eq. (5.27)). Furthermore, Figs. 5.12 and 5.13, summarized in Fig. 7.9, illustrate that for an ideal imaging system which images a point of the object space into a stigmatic point in the image space, the geometrical wave surface has to be spherical. If deviations from a spherical wave surface occur, an object point is not imaged into a stigmatic image point (see Fig. 7.9b). An astigmatic image is thus the result of a deformed wave surface.

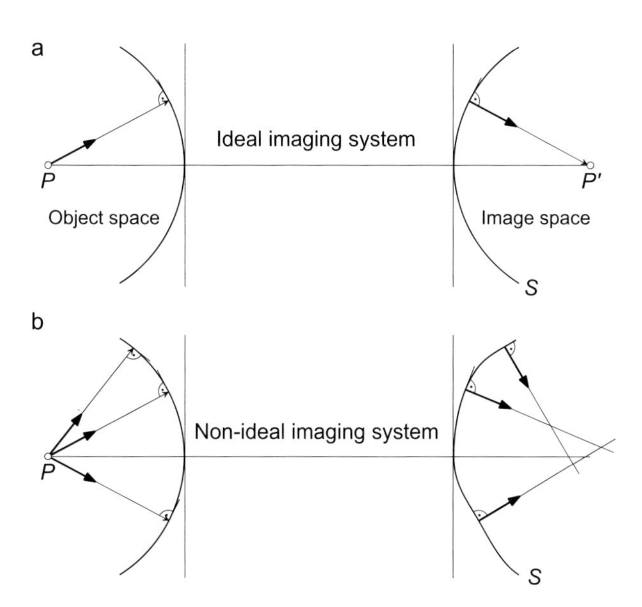


Fig. 7.9 Wave fronts. (**a**) An ideal imaging system translates a spherical wave surface into another spherical wave surface which forms a stigmatic image point $P'$ of the object point $P$. (**b**) For a non-ideal imaging system, the wave surface in the image space is not spherical. The formation of a stigmatic image point is not possible.

### 7.5.1 *HRTEM and STEM*

From this rather conceptional argument about the imaging properties of an ideal and a non-ideal imaging system, we have to identify where, i.e. in which plane of the electron microscope, the geometrical wave surfaces are of relevance for the discussion of the imaging properties of the microscope.

In HRTEM mode, we would like to form stigmatic images of all the points of the exit-plane wave, which is located in the object plane. The image shall be formed in the image plane where the electron detector is located. Hence, the relevant geometrical wave surfaces are behind the object plane.

A point $w_o$ in the object plane, where the specimen is located, can be considered to be the source point of a pencil of rays. In the field-free space, this pencil is described by a spherical wave surface (see Fig. 7.10a). Each ray of this pencil passes the aperture plane at a different location $\omega$. Depending on the source point $w_o$ and its coordinate $\omega$ in the aperture plane, i.e. depending on its geometrical ray parameters ($\omega$ and $w_o$), a ray can suffer from a specific phase shift due to the presence of geometrical aberrations. The phase shifts are induced by the non-ideality of the lens whose functionality it is to focus the pencil to a single point in the image plane. For a lens free of aberrations, the spherical wave surface in front of the lens is transferred into a spherical wave surface behind the lens. A stigmatic image point $w_i$ results. This is illustrated in Fig. 7.10a. However, on transferring the spherical wave surface from the object space into the image space, a non-ideal lens causes the spherical wave surface to warp. The quantity that describes the deviation from the spherical wave surface for an object point located in $w_o$ that emits a ray into the aperture plane at $\omega$ is the aberration function. Because of the aberration function, which will be introduced in the following section, a point in the object plane is not imaged as a stigmatic point in the image plane. Instead, the object point is imaged as a characteristic aberration figure described by the image aberration $\Delta w_i$.

In STEM mode, we would like to have a stigmatic image of the idealized point-like electron source. The image of the demagnified source shall be formed in the plane where the specimen is located. The focal point of the objective lens' pre-field defines the STEM probe. Blurring of this focal point due to geometrical or chromatic aberrations simply means that the STEM probe becomes larger. This, of course, impairs the achievable resolution. The relevant geometrical wave surface is thus in front of the specimen plane.

Although we have seen in Chapter 3 that the finite brightness prevents the formation of an infinitely small demagnified electron source, let us assume that the electron source is demagnified to a single point located in $w_o$ (see Fig. 7.10b). The source point in $w_o$ can be considered to emit a pencil of rays. This pencil is described by a spherical wave surface. Its angular range $\omega$ is defined by the aperture opening which provides the measure for the illumination semi-angle $\alpha$. By focusing the pencil onto the specimen plane, the pre-field of the objective lens images the

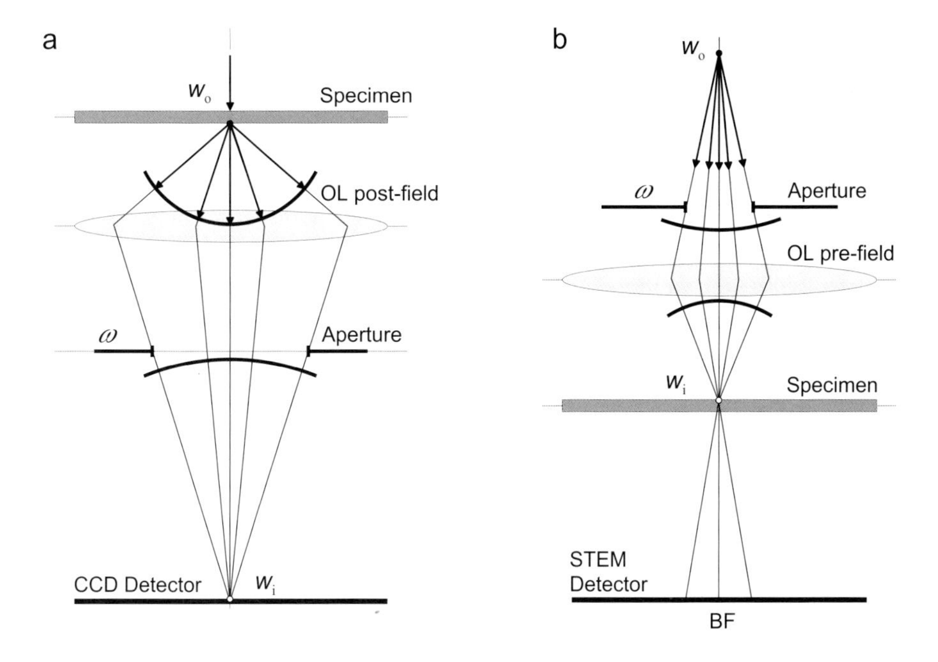


Fig. 7.10 The relevant geometrical wave surfaces in TEM and in STEM.

source point onto the specimen plane. The image of the source point in $w_o$ is the electron probe. For the ideal case of an aberration-free lens, the spherical wave surface in front of the objective lens' pre-field is transferred to a spherical wave surface behind the lens. As illustrated in Fig. 7.10b, under this idealized condition, a stigmatic image point $w_i$ results. However, geometrical aberrations of the lens cause the spherical wave surface behind the lens to curve, such that the pencil is not focused to a stigmatic image point.

According to the eikonal, the pencil of rays is fully defined by two geometrical ray parameters. Though we are essentially free in the choice of the position variables that shall describe the pencil of rays, we would like to know the wave surface in the relevant planes. In STEM mode, it is the wave surface in the front focal plane which defines the electron probe on the specimen plane. In HRTEM, it is the wave surface in the back focal plane which determines the imaging characteristics of a particular object point (see Fig. 7.10). Hence, it is common to choose the first position variable $P_0$ to be in the relevant aperture plane $\omega$ (front focal plane for STEM and back focal plane for HRTEM), and the other position $w_o$ to be in the object plane, which for STEM is the location of the (demagnified) source point and for HRTEM is the specimen plane (see Fig. 7.10). Hence, $S = S(w_o, \omega)$. The aberrations become apparent in the image plane $w_i$, which for STEM is the specimen plane and for HRTEM is the detector plane.

### 7.5.2 *Non-ideal geometrical wave surfaces*

As pointed out above, the curvature of a wave surface $S = S(\mathrm{w_o}, \omega)$ defines the aberration in the image space. While the geometrical wave surface of an ideal imaging system is a sphere (see Fig. 7.9a), a non-ideal imaging system does not have a spherical wave surface (see Fig. 7.9b). However, provided the deviations from the spherical surface are small, we can conveniently describe the warped wave surface of a non-ideal imaging system by the superposition of the ideal spherical wave surface $S^0$ and a function $W$, which takes into account the deviation from the spherical shape. Hence, the actual wave surface $S$ is then described by $S(\mathrm{w_o}, \omega) = S^0(\mathrm{w_o}, \omega) + W(\mathrm{w_o}, \omega)$. The function $W(\mathrm{w_o}, \omega)$ is called the *aberration function*. The relation between $S$, $S^0$ and $W$ is illustrated schematically in Fig. 7.11.

### 7.5.3 *The aberration function*

The aberration function $W(\mathrm{w_o}, \omega)$ is a function that depends on the position of the object point $\mathrm{w_o}$ and on the position $\omega$ where a particular ray passes the aperture plane. Each object point $\mathrm{w_o}$ thus has its characteristic wave surface $S$, as, for example, illustrated in Fig. 7.11, which describes how this particular object point is imaged onto the image plane. This general case implies that off-axial and axial aberrations are present, such that $W$ is a real function of $\mathrm{w_o}$ and $\omega$. For the Seidel aberrations, i.e. the symmetry-permitted geometrical aberrations of a round lens, the aberration function $W$ can be written as

$$W(\mathrm{w_o}, \omega) = \Re\left\{\frac{1}{2}C_1\omega\overline{\omega} + \frac{1}{4}C_3\left(\omega\overline{\omega}\right)^2 + B_{3\,1}\omega\overline{\omega}^2\mathrm{w_o} + \frac{1}{2}F_{3\,2}\omega\overline{\omega}\mathrm{w_o}\overline{\mathrm{w}}_\mathrm{o} + \frac{1}{2}A_{3\,2}\overline{\omega}^2\mathrm{w_o^2} + D_{3\,3}\overline{\omega}\mathrm{w_o^2}\overline{\mathrm{w}}_\mathrm{o}\right\}. \tag{7.12}$$

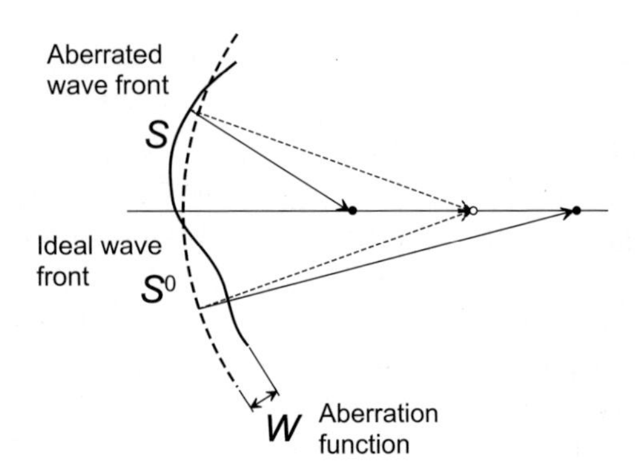


Fig. 7.11 The aberration function $W$ is the difference between the actual, i.e. the aberrated, wave front $S$ and the ideal spherical wave front $S^0$, $S = S^0 + W$.

The aberration coefficients, which were introduced in the previous section to describe image aberrations $\Delta w_{\rm i}$, are employed here to define the curvature of the wave front in respect to the spherical reference wave front. Besides the Seidel aberrations, namely spherical aberration $C_3$, coma $B_{3\,1}$, field astigmatism $A_{3\,2}$, field curvature $F_{3\,2}$ and image distortion $D_{3\,3}$, we also included a term with $C_1$ describing the adjustable defocus of the lens. Equation (7.12) thus enables us to describe the bending of the wave surface due to the Seidel aberrations of a round electromagnetic lens. For each point $w_{\rm o}$ in the object plane, which can be considered to be a source point of a pencil of rays, Eq. (7.12) describes the deviation from a spherical wave surface in the aperture plane $\omega$. Some of the aberrations in Eq. (7.12) are axial aberrations (i.e. depend only on $\omega$ but not on $w_{\rm o}$), and some, which depend on the location of the object point $w_{\rm o}$ (or $\overline{w}_{\rm o}$), are off-axial aberrations.

Since HRTEM imaging is carried out at high magnification, the object area is small. Similarly, if we work at high magnification in STEM mode, the scanned specimen area becomes small. This implies that the scan coils need to move the demagnified source image in front of the objective lens only in a very limited lateral range. Hence, regardless of HRTEM or STEM, for high-resolution imaging, the range of $w_{\rm o}$ that needs to be considered in the deviation of the imaging characteristics is small — small enough such that we can employ the isoplanatic approximation and thus neglect the dependence of $W$ on $w_{\rm o}$. Ignoring all the terms that contain $w_{\rm o}$ or $\overline{w}_{\rm o}$ in Eq. (7.12) simply means that all off-axial aberrations are neglected. Or better, under the isoplanatic approximation, $W$ is expressed for the specific case $w_{\rm o} = 0 + {\rm i}0$. It is assumed that $W(0 + {\rm i}0, \omega)$ is valid for the small range of $w_{\rm o}$ around $w_{\rm o} = 0 + {\rm i}0$, which defines the field of view in high-resolution imaging. In order to distinguish $W(0 + {\rm i}0, \omega)$ from the general geometrical aberration function $W(w_{\rm o}, \omega)$, we denote it by $\chi$, i.e. $\chi(\omega) = W(0 + {\rm i}0, \omega)$.

Under the isoplanatic approximation, i.e. neglecting coma $B_{3\,1}$, field astigmatism $A_{3\,2}$, field curvature $F_{3\,2}$ and image distortion $D_{3\,3}$, one ends up with the following (axial) aberration function of a round electromagnetic electron lens

$$W(w_{\rm o}, \omega) = W(0, \omega) = \chi(\omega) = \Re\left\{\frac{1}{2}C_1\omega\overline{\omega} + \frac{1}{4}C_3\left(\omega\overline{\omega}\right)^2\right\}. \tag{7.13}$$

Equation (7.13) is equivalent to the aberration function which we employed in Chapters 2 and 3 in order to describe the phase contrast transfer function in HRTEM and to include the lens effects in the derivation of the electron probe in STEM mode, respectively. However, while in the previous chapters $\chi$ was a rather abstract function missing a physical interpretation, in the context of geometrical wave surfaces, the aberration function $\chi$ (and $W$) possesses a clear physical interpretation; $\chi$ expresses the deviation of the aberrated wave surface from the spherical wave surface, as illustrated in Fig. 7.11.

Equation (7.13) describes the effect of the remaining geometrical aberrations of a rotationally symmetric imaging (or illumination) system. Provided that the field of view is sufficiently small and that the isoplanatic approximation can be employed, defocus $C_1$ and spherical aberration $C_3$ are the residual aberrations which need to be considered in the image formation process. Though $C_1$ can in principle be set equal zero, a finite (negative) value is of advantage in the presence of the positive $C_3$ of the objective lens. Of course, Eq. (7.13) is only valid if the effect of parasitic axial aberrations is small compared to the effect of $C_3$. This implies, for instance, that by employing a stigmator, twofold axial astigmatism, i.e. the dominant parasitic aberration of a round lens, is corrected to a precision where its residual impact can be neglected. Furthermore, the microscope is supposed to be properly aligned such that the influence of the symmetry permitted off-axial geometrical aberrations is minimized as well. If these conditions are met, the axial aberration function can be described by defocus $C_1$ and third-order spherical aberration $C_3$. Since both, $C_1$ and $C_3$, are isotropic aberrations, reducing the aberration function to the effects of $C_1$ and $C_3$ results in an isotropic information transfer in TEM mode and in a rotationally symmetric electron probe in STEM mode.

However, reducing the problem of the information transfer in HRTEM imaging as well as the formation of the electron probe to two isotropic geometrical aberrations is not necessarily correct. Prior to the successful implementation of aberration correctors into (scanning) transmission electron microscopes, it had already been recognized that Eq. (7.13) is not appropriate to describe the information transfer for *ultra*-high resolution imaging in conventional electron microscopes. Besides the correctable twofold axial astigmatism, additional anisotropic axial aberrations, such as threefold astigmatism and axial coma, would need to be considered in the image and probe formation process (see, e.g. Krivanek, 1994; Overwijk *et al.*, 1997; Stenkamp, 1998). More drastic is the situation for aberration-corrected microscopes. The aberration function given in Eq. (7.13) is simply no longer adequate. Firstly, the third-order spherical aberration $C_3$ of the objective lens is corrected by the aberration corrector. Hence, the 'dominant' geometrical aberration, i.e. $C_3$, can in principle be annulled. For this reason, it is likely that other geometrical aberrations become important. As will be shown in detail in the following chapter, aberration correctors employ multi-pole lenses. The amount of symmetry-permitted axial aberrations is larger for multi-pole lenses than it is for round lenses (see, e.g. Hawkes and Kasper, 1989a). Moreover, because of potential mechanical imperfections of multi-pole and round lenses, parasitic geometrical aberrations can arise. Of course, such aberrations can in principle also arise in conventional microscopes. However, they might be negligible for an instrument with a $C_3$ of about 1 mm. For a microscope whose $C_3$ can be set to a value which is nearly three orders of magnitude smaller than the initial value of the uncorrected lens, even small parasitic aberrations can become important. Hence, especially for the case of aberration-corrected microscopy, axial aberrations of higher order and higher symmetry need to be considered. The presence of anisotropic aberrations destroys the rotational symmetry of the transfer process in TEM imaging, as well as the rotational symmetry of the STEM probe.

In general, there are $(n+2)/2$ axial aberrations of $n$-th order if $n$ is an even number, i.e. $n = 2m$ with $m = 0, 1, 2, 3, ...$, and $(n+3)/2$ axial aberrations of $n$-th order if $n$ is an odd number, i.e. $n = 2m+1$. However, these numbers do not reflect whether the aberrations are real (one coefficient) or complex (two coefficients). Taking this distinction into account, there are $n+2$ independent axial aberration coefficients of order $n$. This implies that the total amount of independent axial aberration coefficients for a system where axial aberrations up to an order $n$ need to be considered, is $[(n+1)(n+4)]/2$.

Considering all feasible axial aberrations up to seventh order ($n = 7$), 44 independent aberration coefficients need to be taken into account. The corresponding (axial) aberration function $\chi(\omega)$ can be expressed as

$$\begin{aligned}\chi(\omega) = \Re\Big\{ & A_0\overline{\omega} + \frac{1}{2}C_1\omega\overline{\omega} + \frac{1}{2}A_1\overline{\omega}^2 + B_2\omega^2\overline{\omega} + \frac{1}{3}A_2\overline{\omega}^3 \\ & + \frac{1}{4}C_3\left(\omega\overline{\omega}\right)^2 + S_3\omega^3\overline{\omega} + \frac{1}{4}A_3\overline{\omega}^4 \\ & + B_4\omega^3\overline{\omega}^2 + D_4\omega^4\overline{\omega} + \frac{1}{5}A_4\overline{\omega}^5 \\ & + \frac{1}{6}C_5\left(\omega\overline{\omega}\right)^3 + S_5\omega^4\overline{\omega}^2 + R_5\omega^5\overline{\omega} + \frac{1}{6}A_5\overline{\omega}^6 \\ & + B_6\omega^4\overline{\omega}^3 + D_6\omega^5\overline{\omega}^2 + F_6\omega^6\overline{\omega} + \frac{1}{7}A_6\overline{\omega}^7 \\ & + \frac{1}{8}C_7\left(\omega\overline{\omega}\right)^4 + S_7\omega^5\overline{\omega}^3 + R_7\omega^6\overline{\omega}^2 + G_7\omega^7\overline{\omega} + \frac{1}{8}A_7\overline{\omega}^8 \Big\}, \end{aligned} \tag{7.14}$$

where the numerical subscript of each aberration indicates its order $n$. The absence of the second numerical subscript shows that the aberrations in Eq. (7.14) are all axial (or aperture) aberrations. The names and their respective symmetry are summarized in Table 7.2. An aberration $A_n$ is the $n$-th order astigmatism which has an $(n+1)$-fold symmetry, $C_n$ is the $n$-th order spherical aberration, which is an isotropic aberration (i.e. symmetry zero), $B_n$ is the $n$-th order axial coma of symmetry 1, $S_n$ is the $n$-th order star aberration with twofold symmetry, $D_n$ is the $n$-th order three-lobe aberration with threefold symmetry, $R_n$ is the $n$-th order rosette aberration of fourfold symmetry, $F_6$ is the sixth-order pentacle aberration with fivefold symmetry and $G_7$ is the chaplet aberration with sixfold symmetry. While the isotropic aberrations, i.e. defocus $C_1$ and the spherical aberrations $C_n$, are describable by a real aberration coefficient, the anisotropic aberrations are in general related to complex coefficients, which implies that their effect is described by two independent coefficients[7].

[7] Although Eq. (7.14) looks rather complex, especially the way each aberration is weighted by $\omega$ and $\overline{\omega}$, it has a systematic structure. For a given order $n$, the aberration of highest symmetry is the astigmatism $A_n$ of symmetry $N = n+1$. Its contribution to the aberration function scales with $\overline{\omega}^{(n+1)}$. An aberration of this particular symmetry appears again in the series of aberrations of order $n+2$, $n+4$ and so on (see Table 7.2). The impact of these higher-order aberrations of equal symmetry scales by $\omega^{(n+1)}(\omega\overline{\omega})^1$ for order $n+2$ and by $\overline{\omega}^{(n+1)}(\omega\overline{\omega})^2$ for order $n+4$, and

Table 7.2 Geometrical axial aberrations.

| Aberration | Symbol | Value | Symmetry $N$ | Wave aberration $\mathcal{R}(\cdots)$ |
|---|---|---|---|---|
| Beam/Image shift | $A_0$ | complex | 1 | $A_0\overline{\omega}$ |
| Defocus | $C_1$ | real | 0 | $\frac{1}{2}C_1\omega\overline{\omega}$ |
| Twofold astigmatism | $A_1$ | complex | 2 | $\frac{1}{2}A_1\overline{\omega}^2$ |
| Second-order axial coma | $B_2$ | complex | 1 | $B_2\omega^2\overline{\omega}$ |
| Threefold astigmatism | $A_2$ | complex | 3 | $\frac{1}{3}A_2\overline{\omega}^3$ |
| Third-order spherical aberration | $C_3$ | real | 0 | $\frac{1}{4}C_3\,(\omega\overline{\omega})^2$ |
| Third-order star-aberration | $S_3$ | complex | 2 | $S_3\omega^3\overline{\omega}$ |
| Fourfold astigmatism | $A_3$ | complex | 4 | $\frac{1}{4}A_3\overline{\omega}^4$ |
| Fourth-order axial coma | $B_4$ | complex | 1 | $B_4\omega^3\overline{\omega}^2$ |
| Fourth-order three-lobe aberration | $D_4$ | complex | 3 | $D_4\omega^4\overline{\omega}$ |
| Fivefold astigmatism | $A_4$ | complex | 5 | $\frac{1}{5}A_4\overline{\omega}^5$ |
| Fifth-order spherical aberration | $C_5$ | real | 0 | $\frac{1}{6}C_5\,(\omega\overline{\omega})^3$ |
| Fifth-order star-aberration | $S_5$ | complex | 2 | $S_5\omega^4\overline{\omega}^2$ |
| Fifth-order rosette aberration | $R_5$ | complex | 4 | $R_5\omega^5\overline{\omega}$ |
| Sixfold astigmatism | $A_5$ | complex | 6 | $\frac{1}{6}A_5\overline{\omega}^6$ |
| Sixth-order axial coma | $B_6$ | complex | 1 | $B_6\omega^4\overline{\omega}^3$ |
| Sixth-order three-lobe aberration | $D_6$ | complex | 3 | $D_6\omega^5\overline{\omega}^2$ |
| Sixth-order pentacle aberration | $F_6$ | complex | 5 | $F_6\omega^6\overline{\omega}$ |
| Sevenfold astigmatism | $A_6$ | complex | 7 | $\frac{1}{7}A_6\overline{\omega}^7$ |
| Seventh-order spherical aberration | $C_7$ | real | 0 | $\frac{1}{8}C_7\,(\omega\overline{\omega})^4$ |
| Seventh-order star-aberration | $S_7$ | complex | 2 | $S_7\omega^5\overline{\omega}^3$ |
| Seventh-order rosette aberration | $R_7$ | complex | 4 | $R_7\omega^6\overline{\omega}^2$ |
| Seventh-order chaplet aberration | $G_7$ | complex | 4 | $G_7\omega^7\overline{\omega}$ |
| Eightfold astigmatism | $A_7$ | complex | 8 | $\frac{1}{8}A_7\overline{\omega}^8$ |

so on. The numerical prefactors are chosen such that the prefactors of the corresponding image aberrations are either 1 or integers, but no fractions. As an example; the second-order aberration of highest symmetry is threefold astigmatism $A_2$ whose symmetry is $N = n + 1 = 3$. Aberrations of symmetry 3 appear again in the fourth- and sixth-order aberrations (see Fig. 7.12): fourth- and sixth-order three-lobe aberration $D_4$ and $D_6$. Since the impact of $A_2$ scales with $\overline{\omega}^3$, the impact of $D_4$ scales with $\omega^3(\omega\overline{\omega})$ and the impact of $D_6$ scales with $\omega^3(\omega\overline{\omega})^2$. The factor $(\omega\overline{\omega})$ is just a radial, parabolic angular factor. With these basic rules, the entire relation given in Eq. (7.14) can in principle be derived. Yet, the numerical prefactors need to be explained by the image aberrations, which we will relate to the wave aberrations in a subsequent part of this chapter.

The modulations of the aberrated wave surface caused by each of these aberrations are depicted in Fig. 7.12, which also reveals the symmetry of each aberration as listed in Table 7.2. Here we define the symmetry $N$ of an aberration as follows: an aberration $X_n$ has the symmetry $N$ if the distortion of the wave surface caused by the aberration is invariant by a rotation of $2\pi/N$ about the optical axis, with $N$ being the largest number for which this is fulfilled. Hence, if an aberration function $\chi$ associated by an aberration $X_n$ shows the characteristics that

$$\chi(\omega) = \chi\left(\omega \exp\left\{\frac{2\pi \mathrm{i}}{N}\right\}\right), \tag{7.15}$$

the aberration has the symmetry $N$. The symmetry $N$ associated with an aberration shall be the maximum number $N$ for which Eq. (7.15) is valid (see, e.g. Uno *et al.*, 2005). This formal definition is intuitively understandable by looking at Fig. 7.12. The symmetry $N$ is the number of times that the aberration function repeats on being rotated about the optical axis. For the case that $N \to \infty$, $X_n$ is an isotropic aberration and apparently not fully consistent, we denote the symmetry $N = 0$.

### 7.5.4 *Other notations for axial aperture aberrations*

The notation of the axial aberration coefficients we employ here is essentially based on the notation of Pöhner and Rose (1974), which is an expansion of the definition of the aberration coefficients introduced by Plies and Rose (1971). The numerical values, i.e. the prefactors of the coefficients, are however defined according to Haider *et al.* (2008b)[8]. This implies small numerical differences between the notation of Pöhner and Rose (1974) and the one employed here (Haider *et al.*, 2008b). Table 7.3 provides a comparison between the notations of Pöhner and Rose (1974) and Haider *et al.* (2008b). The notation of Haider *et al.* (2008b) is essentially used in the aberration diagnosis software of aberration-corrected STEM/TEM instruments which are equipped with hexapole-type aberration correctors.

There are other notations of the aberration coefficients in use. This, in fact, would not cause any problems as long as the aberration coefficients have the same absolute value[9]. But even this differs between different notations. For instance, depending on whether a factor of 1/3 is implicitly used in the definition of $B_2$

---

[8]The advantage of using the numerical prefactors according to Haider *et al.* (2008b) is that the corresponding image aberrations do not have fractions as prefactors, but solely integers.

[9]There are many notation schemes in use. We focus on the notations most frequently used in literature. The following references provide a brief overview of different notations employed in the literature and not covered in this text: Jiye and Crewe (1985); Thust *et al.* (1996); Krivanek (1994); Overwijk *et al.* (1997); Stenkamp (1998); Lupini (2001); Xiu and Gibson (2001); Sawada *et al.* (2009). The three most common notations in aberration-corrected electron microscopy are: the ones described in detail by Uhlemann and Haider (1998) and Haider *et al.* (2008b), which are employed in, e.g. Haider *et al.* (2000); Uno *et al.* (2005); Erni *et al.* (2009); Zach (2009). Secondly, the notation described in Krivanek *et al.* (2009a), which is based on Krivanek *et al.* (1999) and which is employed, for example, by Ramasse and Bleloch (2005); Krivanek *et al.* (2008a, 2009a). Thirdly, the notation used by Pöhner and Rose (1974), which is, for example, employed in Haider *et al.* (1995); Saxton (1995); Typke and Dierksen (1995); Haigh *et al.* (2009b).

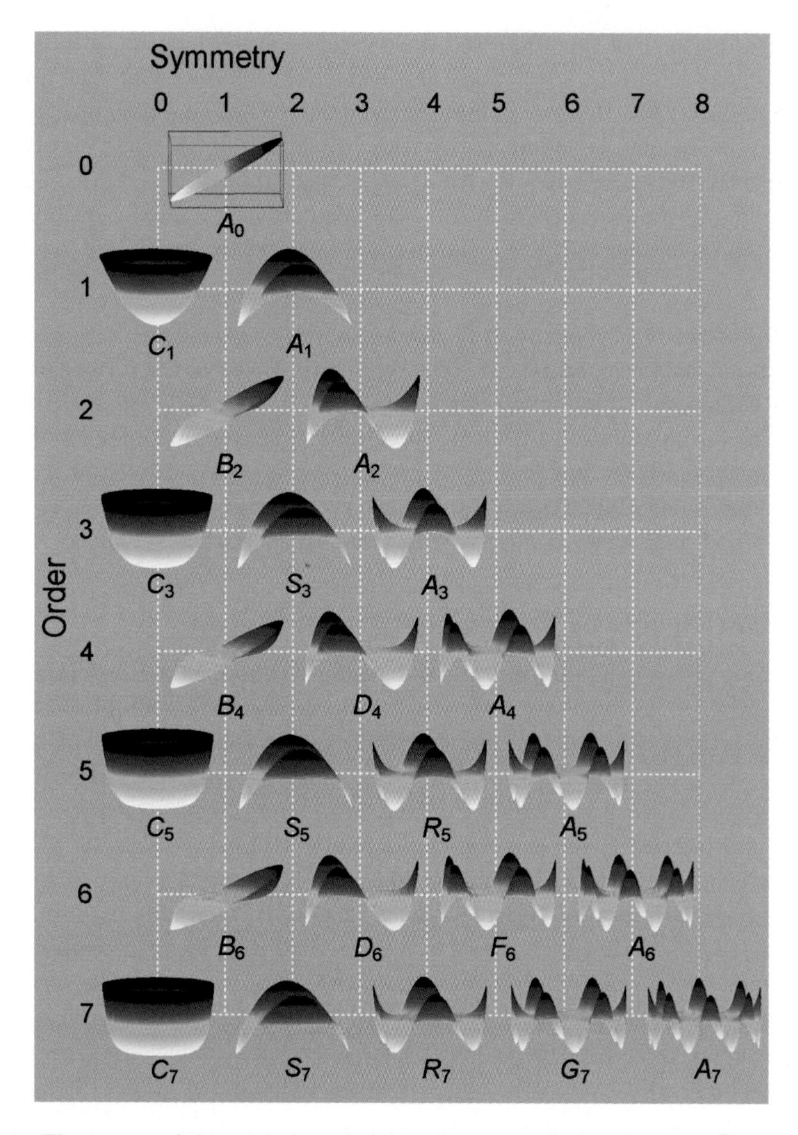


Fig. 7.12 The impact of the axial aberrations on the wave surface. Each part illustrates $\chi$ for the corresponding aberration coefficient.

(see, e.g. Eq. (7.14)) or whether the factor of 1/3 is written explicitly in the term of $B_2$, of course, changes the absolute value of the aberration coefficient and complicates the comparison between different instruments that measure the aberration coefficient based on different notations. However, the aberration function $\chi$ as given, for example, in Eq. (7.14), has to be invariant. The aberration function must be independent of the definition of the aberration coefficients and independent of the notation employed to describe a certain aberration coefficient. Furthermore, some notations do not employ complex coordinates or complex aberration coefficients but

Table 7.3 Comparison of different notations.

| Aberration | Haider *et al.* (2008) | Krivanek *et al.* (2009) | Pöhner and Rose* (1974) |
|---|---|---|---|
| Beam/image shift | $A_0$ | $C_{0,1}$ | $A_0$ |
| Defocus | $C_1$ | $C_{1,0}$ | $C_1$ |
| Twofold astigmatism | $A_1$ | $C_{1,2}$ | $A_1$ |
| Second-order axial coma | $B_2$ | $\frac{1}{3}C_{2,1}$ | $\frac{1}{3}B_2$ |
| Threefold astigmatism | $A_2$ | $C_{2,3}$ | $A_2$ |
| Third-order spherical aberration | $C_3$ | $C_{3,0}$ | $C_3$ |
| Third-order star-aberration | $S_3$ | $\frac{1}{4}C_{3,2}$ | $\frac{1}{4}B_3$ |
| Fourfold astigmatism | $A_3$ | $C_{3,4}$ | |
| Fourth-order axial coma | $B_4$ | $\frac{1}{4}C_{4,1}$ | |
| Fourth-order three-lobe aberration | $D_4$ | $\frac{1}{4}C_{4,3}$ | |
| Fivefold astigmatism | $A_4$ | $C_{4,5}$ | |
| Fifth-order spherical aberration | $C_5$ | $C_{5,0}$ | $C_5$ |
| Fifth-order star-aberration | $S_5$ | $\frac{1}{6}C_{5,2}$ | $\frac{1}{6}B_5$ |
| Fifth-order rosette aberration | $R_5$ | $\frac{1}{6}C_{5,4}$ | $\frac{1}{6}D_5$ |
| Sixfold astigmatism | $A_5$ | $C_{5,6}$ | $A_5$ |
| Sixth-order axial coma | $B_6$ | $\frac{1}{7}C_{6,1}$ | |
| Sixth-order three-lobe aberration | $D_6$ | $\frac{1}{7}C_{6,3}$ | |
| Sixth-order pentacle aberration | $F_6$ | $\frac{1}{7}C_{6,5}$ | |
| Sevenfold astigmatism | $A_6$ | $C_{6,7}$ | |
| Seventh-order spherical aberration | $C_7$ | $C_{7,0}$ | |
| Seventh-order star-aberration | $S_7$ | $\frac{1}{8}C_{7,2}$ | |
| Seventh-order rosette aberration | $R_7$ | $\frac{1}{8}C_{7,4}$ | |
| Seventh-order chaplet aberration | $G_7$ | $\frac{1}{8}C_{7,6}$ | |
| Eightfold astigmatism | $A_7$ | $C_{7,8}$ | |

(*)Though fourth-order aberrations are not considered, the notation of Pöhner and Rose (1974) is equivalent to the notation for the axial aberration coefficients employed by Typke and Dierksen (1995) and Saxton (1995).

are, for instance, related to a cylindrical coordinate system, where an aberration is then described by an absolute value and an azimuth angle (see, for example, the comparison in Typke and Dierksen, 1995). Still, even for such definitions, the aberration function $\chi$ needs to be invariant — invariant in respect to the coordinate system chosen and the notation of the aberration coefficients. Hence, comparing the wave aberration function $\chi$ in different notations provides a mean of comparing different notations.

Besides the notation of Haider *et al.* (2008b) which is employed here, another often-used notation for aberration coefficients is outlined by Krivanek *et al.* (1999). With a slight modification, this notation is employed and described by Krivanek *et al.* (2008b) and Krivanek *et al.* (2009a)[10]. In the notation of Krivanek *et al.* (2009a), each aberration coefficient is denoted by $C_{n,N}$. The first numerical subscript $n$ denotes the order $n$ of the aberration, while the second numerical subscript $N$ denotes its symmetry $N$. The order $n$ and the symmetry $N$ are in agreement with the definition introduced above. While aberration coefficients with $N > 0$ can be described by two independent coefficients $C_{n,N,a}$ and $C_{n,N,b}$ (Krivanek *et al.*, 2009a), they can be equally well be described as complex numbers such that $C_{n,N} = C_{n,N,a} + \mathrm{i}C_{n,N,b}$ (Hawkes, 2009b). Because the subscripts of the coefficients make it immediately apparent what the nature (order and symmetry) of the aberration is, the notation according to Krivanek *et al.* (2009a) is particularly useful in deriving combination aberrations which arise through the superposition of two or more geometrical aberrations. However, the notation according to Haider *et al.* (2008b) needs only one numerical subscript and describes the symmetry of the aberration by the (though randomly chosen) capital letter. Employing complex aberration coefficients ($C_{n,N} = C_{n,N,a} + \mathrm{i}C_{n,N,b}$) and complex coordinates, the aberration function $\chi(\omega)$ can be expressed in the notation of Krivanek *et al.* (2009a) as

$$\begin{aligned}\chi(\omega) = \Re\Big\{ & C_{0,1}\overline{\omega} + \frac{1}{2}C_1\omega\overline{\omega} + \frac{1}{2}C_{1,2}\overline{\omega}^2 + \frac{1}{3}\overline{C}_{2,1}\omega^2\overline{\omega} + \frac{1}{3}C_{2,3}\overline{\omega}^3 \\ & + \frac{1}{4}C_{3,0}\left(\omega\overline{\omega}\right)^2 + \frac{1}{4}\overline{C}_{3,2}\omega^3\overline{\omega} + \frac{1}{4}C_{3,4}\overline{\omega}^4 \\ & + \frac{1}{5}\overline{C}_{4,1}\omega^3\overline{\omega}^2 + \frac{1}{5}\overline{C}_{4,3}\omega^4\overline{\omega} + \frac{1}{5}C_{4,5}\overline{\omega}^5 \\ & + \frac{1}{6}C_{5,0}\left(\omega\overline{\omega}\right)^3 + \frac{1}{6}\overline{C}_{5,2}\omega^4\overline{\omega}^2 + \frac{1}{6}\overline{C}_{5,4}\omega^5\overline{\omega} + \frac{1}{6}C_{5,6}\overline{\omega}^6 \\ & + \frac{1}{7}\overline{C}_{6,1}\omega^4\overline{\omega}^3 + \frac{1}{7}\overline{C}_{6,3}\omega^5\overline{\omega}^2 + \frac{1}{7}\overline{C}_{6,5}\omega^5\overline{\omega} + \frac{1}{7}C_{6,7}\overline{\omega}^7 \\ & + \frac{1}{8}C_{7,0}\left(\omega\overline{\omega}\right)^4 + \frac{1}{8}\overline{C}_{7,2}\omega^5\overline{\omega}^3 + \frac{1}{8}\overline{C}_{7,4}\omega^6\overline{\omega}^2 + \frac{1}{8}\overline{C}_{7,6}\omega^7\overline{\omega} + \frac{1}{8}C_{7,8}\overline{\omega}^8\Big\}, \end{aligned} \quad (7.16)$$

where the bar in $\overline{C}_{n,N}$ denotes the complex conjugate of $C_{n,N}$.

In any case, whatever definition for the aberration coefficients is employed, the basic physical concept does not change. The aberration coefficients describe the curvature of an aberrated wave surface in respect to the ideal spherical one. Still, some boundary conditions have to be met; the number of independent aberration

[10]The notation introduced by Krivanek *et al.* (1999) has been further developed such that $C_{n,m}$ in Krivanek *et al.* (2008b) and Krivanek *et al.* (2009a) is equal to $C_{m,n}$ of Krivanek *et al.* (1999), i.e. the subscripts are interchanged.

coefficients is not arbitrary and their symmetry is related to the order of the aberration. For instance, there is no second-order aberration of sixfold symmetry. There kind of conditions are the invariant boundary conditions that have to be met, regardless of the notation and definition of the aberration coefficients. Which definition and which notation is chosen should depend upon its usability for a certain application. The important point is that the optics, which is described by the aberration coefficients, is independent of the definition of these coefficients. Table 7.3 compares the aberration coefficients of Haider *et al.* (2008b), Krivanek *et al.* (2009a) and the notation of Pöhner and Rose (1974).

### 7.5.5 *Aberration function and image aberrations*

While the impact of the aberrations on the geometrical wave surface can be described by the aberration function $W$, or alternatively by $\chi$, we are equally interested in the image aberrations $\Delta w_{\mathrm{i}}$ which describe the impact of the aberrations in the image plane for HRTEM imaging or in the specimen plane in the case of STEM imaging. In the previous sections, we introduced the image aberrations $\Delta w_{\mathrm{i}}$ and the aberration function $W$ independently. What remains to be addressed is the connection between image aberration and aberration function, i.e. the connection between the warping of the wave surface and its impact on the image or on the STEM probe. What is the relation between $\Delta w_{\mathrm{i}}$ and $W$?

The electron trajectories are perpendicular to the wave surface. For the case of a spherical wave surface, all trajectories stemming from the spherical wave surface intersect in one single point, which is the Gaussian image point. If the curvature of the wave surface deviates from a spherical surface, the trajectories do not intersect in a single point anymore — an astigmatic image point results. Hence, it must be the gradient of the wave surface which provides information on the position where a particular trajectory passing the aperture in $\omega$ intersects the Gaussian image plane. If the wave surface is spherical, the aberration function $W$ vanishes, i.e. $W = 0$, as well as its gradient. This is the reference state. For a warped aberration function, however, the gradient does not vanish. This means that a given ray does not intersect the Gaussian image plane in the Gaussian image point. This qualitative argument leads us to the quantitative formulation of the relation between the aberration function $W$ and the image aberration $\Delta w_{\mathrm{i}}$ (for a derivation see, e.g. Grivet, 1972; Hawkes and Kasper, 1989a), which can be expressed as

$$\boxed{\Delta w_{\mathrm{i}} = -M\left(\frac{2\partial W}{\partial \overline{\omega}}\right) = -M\left(\frac{\partial W}{\partial \theta_x} + \mathrm{i}\frac{\partial W}{\partial \theta_y}\right),} \tag{7.17}$$

where $\omega = \theta_x + \mathrm{i}\theta_y$ and $M$ denotes the lateral magnification. For the case that the effect of off-axial aberrations can be neglected, $W$ can be replaced by $\chi$.

Equation (7.17) is an important relation which directly relates the shape of the non-ideal geometrical wave surface expressed by $W$ to the effect of the aberrations in the relevant plane of observation. The image aberration is determined by the derivative of the aberration function in respect to the coordinates in the aperture plane. The image aberration of a particular ray is proportional to the gradient of $W$ or $\chi$. The greater the gradient of $W$ in a position $\omega$ of the aperture plane, the larger is the displacement from the Gaussian image point (see, e.g. Fig. 7.2). This is intuitively understandable by looking, for example, at Fig. 7.11. In Eq. (7.17), we essentially closed the missing connection between image aberrations and the shape of the corresponding wave surface, i.e. its wave aberration.

The derivative in Eq. (7.17) explains that for a given aberration, the sum of the exponents of the ray parameters in the aberration function is one unit larger than the order of the aberration, which is the sum of the exponents of the ray parameters describing the impact of the aberration in the image plane, i.e. its image aberration.

In the following chapter, it will be shown that the multi-pole optical elements employed in aberration correctors have characteristic geometrical (axial) aberrations, like round lenses have. As noticed by Otto Scherzer (1947), these aberrations or, alternatively, these imaging characteristics, can be employed to correct for the intrinsic spherical aberration of the objective lens. On the other hand, the violation of the rotational symmetry of the beam by making use of such multi-pole elements also means that there is a larger number of symmetry-permitted aberrations present. Furthermore, with the correction of the dominant geometrical aberration, i.e. the third-order spherical aberration $C_3$, the sensitivity towards parasitic and residual geometrical aberrations increases. Hence, in an aberration-corrected microscope, up to a certain order, all of the geometrical aberrations in Table 7.2 need to be considered and properly compensated, in order to approach the case of a rotationally symmetric aberration-corrected information transfer.

## 7.6 Summary

In this chapter, we discussed different types of aberrations: chromatic aberrations vs. geometrical aberrations; coherent vs. incoherent aberrations; and axial vs. off-axial aberrations. We also introduced the Seidel aberrations, i.e. the symmetry-permitted geometrical aberrations of a round lens. Based on the geometrical wave surface, we generalized the concept of the aberration function. We showed that aberrations can be described by either wave aberrations or image aberrations, and that these two descriptions can be linked to each other. Keywords for this chapter are: image aberration; wave aberration; aberration function; chromatic aberration; geometrical aberration; axial (or aperture) aberration; off-axial aberration; isoplanatic approximation; Seidel aberrations; order; rank and degree of an aberration; and aberration notations.

# Chapter 8

# Spherical Aberration Correctors

## 8.1 Overview

The impact of geometrical lens aberrations is of crucial importance for the imaging process in TEM and STEM. Nevertheless, as long as we restrict the imaging process to paraxial rays, rotationally symmetric electron lenses behave like ideal lenses which do not suffer from aberrations. This was essentially concluded in Chapter 6. The restriction to strictly paraxial rays means that the illumination semi-angle in STEM $\alpha \rightarrow 0$ is vanishingly small. For TEM, the restriction implies that only image frequencies $q \rightarrow 0$ are allowed to contribute to the image intensity. The limitation to strictly paraxial rays thus results in micrographs which have 'no' lateral resolution[1]. In order to enable a finite resolution, finite illumination angles and/or finite scattering angles need to be considered. As such, the limitation to paraxial rays is not suitable for describing the imaging process in electron microscopy. However, once we extend our considerations to rays that are not strictly paraxial, we need to account for the aberrations of electron lenses.

The symmetry-permitted third-order aberrations of a round electron lens are the Seidel aberrations (see Chapter 7). The only Seidel aberration which does not vanish on the optical axis is the third-order spherical aberration $C_3$. Therefore, within the isoplanatic approximation, where we express the imaging process in terms of the object point on the optical axis, the impact of coma $B_{3\,1}$, field astigmatism $A_{3\,2}$, field curvature $F_{3\,2}$ as well as image distortion $D_{3\,3}$ can be neglected.

The fact that $C_3$ is the only axial Seidel aberration explains why there are only two terms in the aberration function of conventional electron microscopes: the third-order spherical aberration $C_3$ (or $C_S$) and the defocus $C_1$, i.e. the adjustable focal length of the lens (Spence, 1981; Williams and Carter, 1996). In the first part of this book, we concluded that the third-order spherical aberration $C_3$ limits the point

[1]This is not generally correct. In STEM mode, one has to consider a parallel coherent electron probe of finite diameter with $\alpha \rightarrow 0$ which enables a finite resolution when being scanned across the specimen. This, for instance, is employed in coherent diffraction, which is applied to study electron diffraction of individual nanoparticles (see, e.g. Zuo *et al.*, 2003; Huang *et al.*, 2008). However, this is a different optical setup, which is not equivalent to the formation of the STEM probe as discussed in Chapter 3.

resolution in conventional atomic-resolution imaging. Regardless of the limitation of the point resolution due to $C_3$, the presence of the spherical aberration does not fundamentally impact the maximum achievable information transfer. If a STEM probe is formed with too large an illumination angle, or if image frequencies in HRTEM are allowed to contribute to the image intensity, which lie beyond the point resolution, the effect of the spherical aberration can impair a direct interpretation of the micrographs (see Chapter 4).

Nonetheless, as the resolution-limiting factor, the spherical aberration imposes a practical limitation in conventional atomic-resolution electron microscopy. Any stationary rotationally symmetric (i.e. round) electron lens, which is free of space charges, shows a positive third-order spherical aberration. This is the famous result of Otto Scherzer's (1936b), which is known as *Scherzer's theorem*. It simply means that the spherical aberration of practical round electron lenses always remains finite positive. While a positive $C_3$ is unavoidable, round electron lenses of minimal spherical aberration can be designed (see, e.g. Scherzer, 1936a, 1941).

Although the spherical aberration imposes an optical barrier on the achievable resolution, considering the early years in electron microscopy, it was not the optics that was limiting the resolution, but rather mechanical and electronic instabilities and the limited brightness of the available electron sources (Scherzer, 1939). Hence, in order to realize an electron microscope which was actually limited by the theoretical limit imposed by the spherical aberration, these secondary effects first needed to be tackled (see, e.g. Hillier, 1946). Only once these parasitic effects were sufficiently small the spherical aberration became of practical importance (Scherzer, 1939).

### 8.1.1 *Strategies to deal with the spherical aberration*

In the presence of spherical aberration as the resolution-limiting factor, the resolution power of an electron microscope can be increased either by decreasing the wavelength $\lambda$ of the electrons or by shortening the focal length $f$ of the relevant lens. These two strategies were already described by Scherzer (1939) as ways of enhancing the resolution limit of conventional electron microscopes. While the first approach reduces the apparent effect of the spherical aberration, the second approach reduces the aberration itself. By building high-voltage electron microscopes of acceleration voltages up to 3,000 kV and by designing dedicated high-resolution objective lenses of small focal length, both strategies have been followed successfully in the past.

The spherical aberration $C_3$ of an objective lens of focal length $f_o$ is typically of the order of $C_3 \approx f_o/4$. Hence, decreasing the focal length by increasing the strength of the electromagnetic field decreases the spherical aberration. This is realized in high-resolution lenses where the pole pieces are shaped such that the lens field and its decay are maximal (see, e.g. Fig. 6.15).

Decreasing the wavelength $\lambda$ as the second approach to deal with the spherical aberration essentially reduces the scattering angles. In order to transfer a certain spatial frequency $q$ in HRTEM mode, a scattering angle of $|\omega| = \lambda q$ needs to be allowed to pass the imaging part of the microscope. Hence, to achieve a certain resolution $\rho = 1/q$, the scattering angle becomes shorter with decreasing wavelength $\lambda$. Since the impact of an axial aberration solely scales with the scattering angle $|\omega|$, the smaller the scattering angle, i.e. the smaller the wavelength, the smaller the impact of the aberration on the imaging process. This means that it is 'easier' to achieve high resolution in high-voltage microscopes.

Although the deleterious impact of $C_3$ can be reduced, both strategies of dealing with the spherical aberration lead to severe limitations. Firstly, the knock-on radiation damage at high electron energies can be critical for many materials and can ultimately limit the application regime of the microscope. On the other hand, an objective lens of small focal length and low spherical aberration implies that the gap between the pole pieces is narrow. This geometrical restriction significantly reduces the tilt range of the specimen, as well as the usage of slightly larger specimen holders, as used for dedicated *in situ* experiments.

### 8.1.2 *Strategies to correct the spherical aberration*

Naturally, the question arises as to whether it is feasible to build an objective lens free of spherical aberration, or whether there are some kind of devices that can compensate for the dominant geometrical aberration of a round objective lens in conventional electron microscopy. Since the spherical aberration imposes a practical limitation, but not a limitation as a matter of principle (Scherzer, 1936b), there must be ways of overcoming the limitation imposed by the unavoidable spherical aberration of a round electron lens.

The above mentioned unavoidability of the positive spherical aberration is linked to the following three conditions, which in principle describe the characteristic of a typical round electron lens (Scherzer, 1949):

- the electromagnetic fields are free of space charges
- the fields do not vary in time, they are stationary
- the fields are rotationally symmetric.

Violating one of these conditions, it should be possible to overcome the limitation imposed by the spherical aberration (Scherzer, 1947). One could use electron lenses which contain space charges or which are so-called foil lenses (Scherzer, 1939). One could employ time-varying electromagnetic fields, or one could try to find suitable non-round electron optical elements, i.e. multi-pole lenses (Scherzer, 1949). In principle, all of these approaches do have a theoretical relevance, as discussed by Scherzer (1947) and Hawkes and Kasper (1989b). However, the only approach that turns out to be feasible for practical high-resolution electron microscopes is the usage of multi-pole lenses. Spherical aberration correctors employ non-round electron optical elements which add negative spherical aberration and thus can counterbalance the positive spherical aberration of the round objective lens.

In fact, after Scherzer's revolutionary publication in 1947, wherein he described and discussed possibilities for overcoming the resolution limit imposed by the spherical aberration, it took only four years till the first multi-pole lenses were manufactured and assembled as a complex optical unit, which was an aberration corrector (Seeliger, 1951). This first attempt has to be considered as a proof of principle. The primary goal of this first 'astigmatic attachment', as called by Robert Seeliger (1951), a student of Otto Scherzer, was not primarily to increase the resolution of the underlying electron optical instrument but to show that the spherical aberration can be controlled. The design of this unit was based on an idea of Otto Scherzer. It was clear that in order to get to a stage where the corrector would improve the resolution, the mechanical and electrical stabilities of the overall microscope needed to be enhanced. Two years later, Seeliger (1953) published electron micrographs recorded on the 'corrected' microscope operated at 25 kV, which showed that the image aberration due to the spherical aberration was significantly reduced. The reduction corresponded to a residual spherical aberration which was only 6% of its initial value. However, one has to consider that the objective lens employed by Seeliger (1953) had an intrinsic third-order spherical aberration of 92 mm. The round and cylinder lenses in the corrector unit added another 34 mm to the correctable spherical aberration. Although there was no actual resolution improvement, these were the first results of a microscope which was equipped with a unit that was supposed to correct the spherical aberration of the objective lens (Seeliger, 1953).

By employing a more stable high-tension supply in order to increase the high tension to 40 kV, Gottfried Möllenstedt (1956) could improve the performance of the first aberration-corrected electron microscope built by Seeliger (1951). Möllenstedt (1956) reported that with an objective aperture of 20 mrad, which at that time was about an order of magnitude larger than a normal objective aperture, an improvement of the resolution by a factor of seven was possible compared to the uncorrected objective lens, which had a spherical aberration of 92(+34) mm. Möllenstedt (1956) employed an extra-large aperture just to increase the impact of the spherical aberration and thus to simplify the assessment of the degree of correction.

The aberration corrector, which Seeliger (1951, 1953) and Möllenstedt (1956) employed in their studies, consisted of two round lenses, a stigmator, two cylindrical lenses and, as the core units, three octupole lenses, which are the optical elements that would compensate for the spherical aberration.

Though this first attempt was actually quite successful in a sense that, within a period of ten years, a first unit was built that was capable of partially correcting the (large) spherical aberration of a round lens, the scale bars on the micrographs were still given in units of 0.1 $\mu$m. The performance of this microscope was thus far from realizing atomic-resolution imaging. The partially spherical aberration-corrected microscope of Seeliger (1951) was suffering from a significantly larger residual spherical aberration than present-day conventional high-resolution electron microscopes, which are equipped with objective lenses that typically have a spherical aberration $C_3$ of 0.5–2.0 mm. Hence, one could argue that an improved objective lens would have been of greater benefit at the time Seeliger (1951) was working on the multi-pole aberration corrector. Still, the important aspect about these studies was that even by employing a series of multi-pole lenses in a correction unit, micrographs could be formed which did not give any negative indication about the complex optics between objective lens and image plane.

The first attempt to correct for the spherical aberration (Seeliger, 1951) marked the beginning of a process that lasted almost half a century. Driven by new theoretical insights, various correctors were developed and to some extend also brought to application. Indeed the development of aberration correctors in transmission electron microscopes, and a bit later in the electron probe-forming scanning transmission electron microscopes as well, has a rather long and complex history. Peter W. Hawkes entitled this period *Forty Disappointing Years* (Hawkes, 2009a). What Peter Hawkes' expression suggests actually summarizes the period quite well: none of the approaches was successful throughout. Without knowing that the end of this gray period was close, the ambivalence of the results led to quite sceptical opinions about the feasibility of aberration correction, not even mentioning its implementation in commercial microscopes. However, independent of their (partial) success, all the approaches which aimed at solving the problem of spherical aberration in electron microscopy must be considered as extremely valuable for present-day aberration correctors. It is not the purpose of this text to present this history in detail, which from various points of views is described in literature (see, e.g. Hawkes, 2004, 2009a; Rose, 2008, 2009b; Smith, 2009; Müller *et al.*, 2008).

From a very practical and oversimplifying point of view, one could say that the pioneering work of Seeliger (1951) marked the beginning of one type of aberration corrector. This type of corrector is essentially based on the functionality of octupole lenses. Another type of a spherical aberration corrector is based on hexapole elements[2]. This type of corrector was first proposed by Beck (1979) and then further investigated and advanced by Crewe and Kopf (1980), Crewe and Salzman (1982) and Rose (1981). While the first hexapole-type corrector was considered for a probe-forming instrument (Beck, 1979), the first octupole corrector was considered for the broad-beam imaging mode in a transmission electron microscope (Seeliger, 1951). Interestingly, the first *workable* aberration correctors, i.e. correctors which would improve the resolution, were brought to application opposite to their original purposes. The first workable spherical aberration corrector, which was employed for HRTEM imaging, was a hexapole corrector (Haider *et al.*, 1998), whereas the first workable aberration corrector employed in a probe-forming instrument was a quadrupole–octupole corrector. This first workable quadrupole–ocutpole corrector,

[2]Though we realize that for consistency with the naming of quadrupole and octupole lenses, one would need to call a lens with six poles a sextupole lens (Krivanek *et al.*, 2008b), we decided to employ the naming as used in the original work about such units (Beck, 1979). This is also the naming employed in the description of the first workable *hexapole* corrector (Haider *et al.*, 1998).

which was based on a design proposed by Rose (1971a), would correct for the spherical as well as for the chromatic aberration in a low-voltage scanning transmission electron microscope (Zach and Haider, 1995). The first workable spherical aberration corrector in a dedicated scanning transmission electron microscope was a quadrupole–octupole corrector as well (Krivanek *et al.*, 1997, 1999).

## 8.2 Multi-Pole Lenses

The functionality of present-day spherical aberration correctors is based on magnetic multi-pole lenses. There are essentially two types of correctors in use: correctors which employ the effect of two extended magnetic hexapoles and correctors which are based on combining quadrupole and octupole elements.

A magnetic multi-pole element of multiplicity $2m$ (with $m = 2, 3, 4, ...$) is a lens which has $2m$ magnetic poles arranged in equidistant azimuthal angles about the optical axis. Neighboring poles have opposite polarity, either S (south) or N (north). For a quadrupole, $m = 2$; for a hexapole, $m = 3$; for a octupole, $m = 4$; and for a dodecapole, $m = 6$. Since there are two types of alternating poles, $m$ is the number of times that the magnetic field repeats on being rotated about the optical axis. Hence, the number $m$ reflects the symmetry of the magnetic field. Figure 8.1 shows a schematic of a magnetic hexapole element. All of the magnetic poles are identical. The polarity of the poles is controlled by the electric current running through the coils. For neighboring magnetic poles, the current runs in opposite direction.

The primary effect of a multi-pole lens of multiplicity $2m$ is to induce an aberration of order $n = m - 1$ and of symmetry $N = m$. A multi-pole of multiplicity $2m$ essentially introduces $m$-fold astigmatism $A_{m-1}$ (see Table 7.2). A hexapole element ($m = 3$) as depicted in Fig. 8.1 leads to an aberration of order $n = m - 1 = 2$ and symmetry $N = m = 3$, which according to Fig. 7.12 is threefold astigmatism $A_2$. Similarly, a quadrupole causes twofold astigmatism $A_1$. Therefore, quadrupoles and hexapole elements are used to correct twofold and threefold astigmatism, respectively, which can both severely affect the image contrast in conventional transmission electron microscopes (Krivanek *et al.*, 1995; O'Keefe *et al.*, 2001b). However, since by proper control of a unit containing more than six magnetic poles a hexapole field can be generated, a corrector for threefold astigmatism does not necessarily require a hexapole element (Overwijk *et al.*, 1997). In any case, an optical element which is used to correct for a certain type of astigmatism is called a *stigmator*. In fact, a stigmator can be considered as a simple aberration corrector. Yet, when we talk about aberration correctors we usually mean a corrector which corrects for the limiting aberration of a round electron lens, i.e. which corrects the spherical aberration $C_3$. As can be shown (see, e.g. Krivanek *et al.*, 2009a), the $m$-fold astigmatism induced by a $2m$-pole element increases with the length of the element, the strength of the magnetic field, the diameter of the beam within the multi-pole element and with decreasing electron

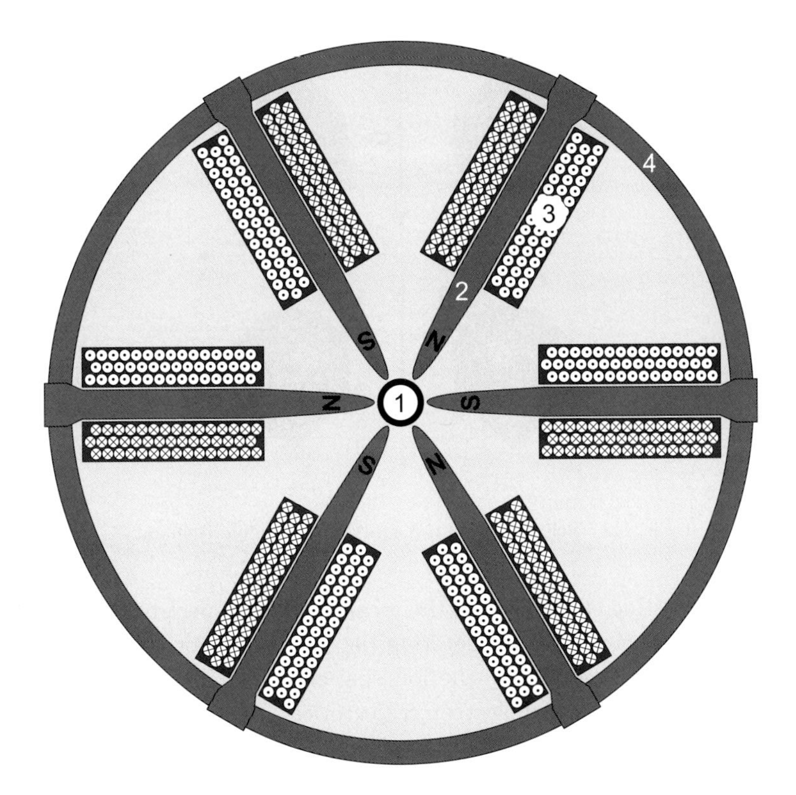


Fig. 8.1 Schematics of a magnetic hexapole: 1; liner tube, 2; magnetic pole, 3; coil, and 4; yoke. Six (identical) magnetic poles made of soft magnetic material are arranged such that neighboring magnetic poles have opposite polarity. This is achieved by inverting the current through the coils. The magnetic flux lines run from the N- to the S-poles and create a magnetic field of threefold symmetry within the liner tube.

energy. Besides the primary effect, multi-pole elements can show secondary effects.

### 8.2.1 *Quadrupoles*

Although a weak quadrupole can be used as a corrector for twofold astigmatism, if we consider a magnetic quadrupole as a lens, we have to consider its focusing properties. A magnetic quadrupole consists of four poles, as depicted in Fig. 8.2. The magnetic field lines depicted using the magnetic induction $\boldsymbol{B}$ run from the N- to the S-poles. The force the field exerts on an electron entering the quandrupole element is determined by the Lorentz force (see Eq. (5.2)). What essentially results for the setup depicted in Fig. 8.2 are a force $\boldsymbol{F}_y$ in $y$-direction, which drives the electrons towards the optical axis, and a force $\boldsymbol{F}_x$, which deflects the electron away from the optical axis. Hence, a magnetic quadrupole can be considered as a convex (or convergent) lens in one direction and as a concave (or divergent) lens in the

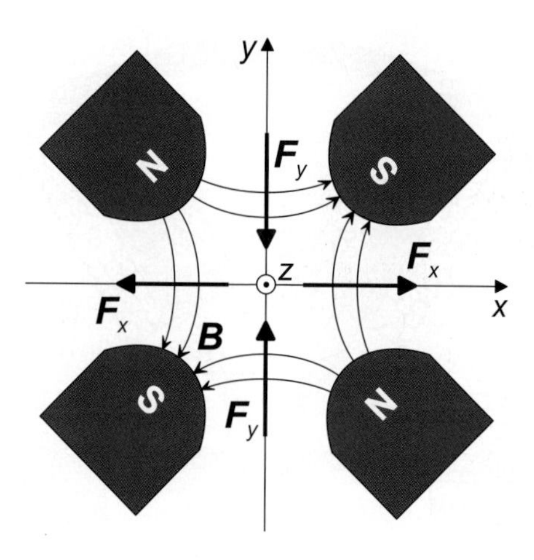


Fig. 8.2 Schematics of a magnetic quadrupole element.

perpendicular direction. For Fig. 8.2, the focal point in respect to the $y$-axis is real, while the focal point for electrons entering the quadrupole on the $x$-axis is virtual (see, e.g. Rose, 2009a). Similar to the focusing effect of a round electron lens, the quadrupole affects the paraxial electron trajectories. Therefore, the focusing action of a magnetic quadrupole is a first-order effect.

The focusing power of a quadrupole lens exceeds that of a round electromagnetic lens. It is for this reason that quadrupoles are used as focusing lenses in particle accelerators, where the focusing power of round lenses would be insufficient. However, one quadrupole forms a line focus (see Fig. 8.3), and not a focal point. Therefore, in order to be able to transform a round electron beam into a single focal point, one needs to employ more than just one single quadrupole. In principle, two quadrupoles with perpendicular focusing actions are sufficient to form a stigmatic focal point (Rose, 2009a). However, in order to maintain a certain degree of flexibility in forming the focal point, typically four, i.e. a quadruplet of quadrupoles, is used (Hawkes and Kasper, 1989b).

The main purpose of quadrupole lenses in aberration correctors is to form strongly elliptical beams or line foci. Such line foci, or *anamorphotic* images[3] as they are also called, are needed to take advantage of the imaging properties of octupole elements which are used to induce negative spherical aberration.

### 8.2.2 *Hexapoles*

Unlike round electron lenses and quadrupoles, hexapole fields do not alter the paraxial trajectories. As such, they neither induce aberrations nor act as focusing lenses

[3] An anamorphotic image reflects an image contained in a round beam which collapses to a line (see, e.g. Rose, 2010).

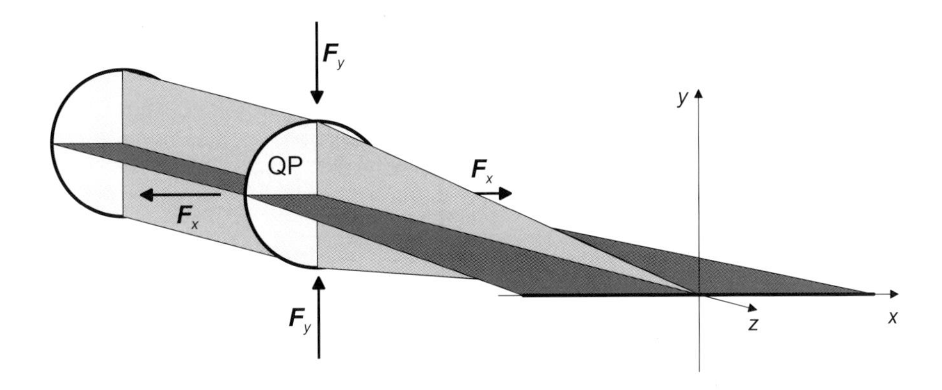


Fig. 8.3 A quadrupole (QP) element focuses a round beam into a line.

for paraxial trajectories. Though hexapoles do not have such a first-order effect, hexapoles induce threefold astigmatism $A_2$, a second-order aberration. This second-order aberration is the primary effect of hexapole fields. As such, they find application as stigmators for threefold astigmatism.

However, in the context of hexapole elements, it is important to mention that *combination* aberrations can occur. Combination aberrations are not limited to hexapole elements. Combining any two optical elements of arbitrary symmetry, aberrations can occur which are not characteristic for either of the optical units and which are not just given by the sum of the aberrations of the individual elements (see, e.g. Krivanek *et al.*, 2009a).

In order to elucidate the combination aberration of two hexapole elements, let us assume an electron is travelling parallel to the optical axis and enters a short hexapole field in a position $y_{\mathrm{HP}}$ on a meridional plane $(y, z)$. If the electron follows a paraxial trajectory ($y_{\mathrm{HP}} \to 0$), the electron does not experience the hexapole field at all and keeps on continuing its journey unaffected by the field. However, for (significantly) finite values of $y_{\mathrm{HP}}$, the electron is deflected such that its trajectory after passing the hexapole field is inclined by an angle $\Delta\gamma$ in respect to its original direction. If we consider only second-order effects, the deflection $\Delta\gamma^{(2)}$ of an electron entering the field in $y_{\mathrm{HP}}$ is identical to the deflection $\Delta\gamma^{(2)}$ of an electron that enters the field in $-y_{\mathrm{HP}}$ (see Fig. 8.4). The deflection $\Delta\gamma^{(2)}$, though equal for the electrons in $y_{\mathrm{HP}}$ and $-y_{\mathrm{HP}}$, is, however, a function of the azimuth angle. Indeed, the azimuthal and radial dependence of $\Delta\gamma^{(2)}$ causes a round beam entering the field to end up as a distorted beam of threefold symmetry. This characteristic distortion reflects the primary aberration of the hexapole field. Hence, a thin hexapole lens essentially causes threefold astigmatism $A_2$.

However, arranging two thin hexapole fields in series, a combination aberration occurs which is of third order. As illustrated in Fig. 8.4, in addition to the deflection $\Delta\gamma^{(2)}$, the electrons leaving the second hexapole field have a total deflection angle

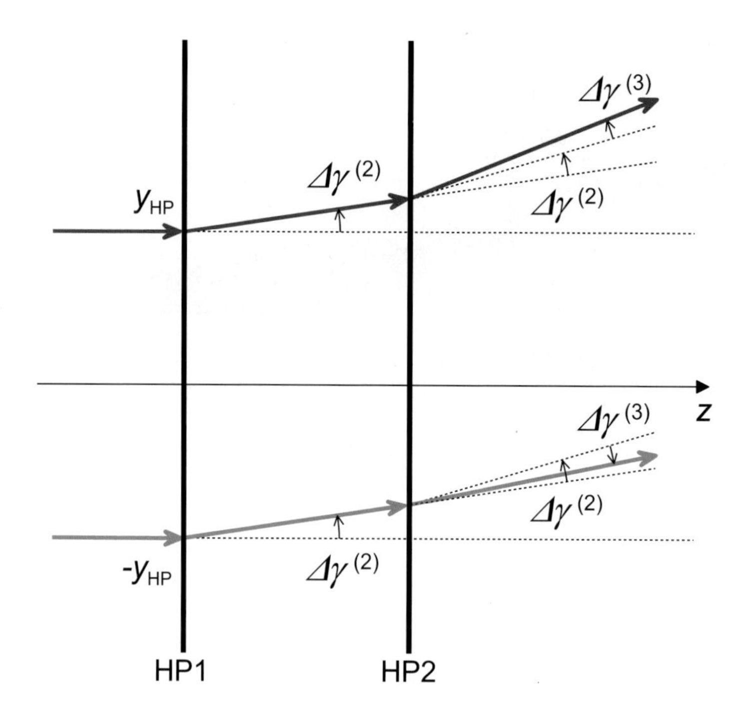


Fig. 8.4 The effect of two thin hexapoles fields (HP1 and HP2) on two non-paraxial electrons entering the first hexapole field in $y_{\mathrm{HP}}$ and $-y_{\mathrm{HP}}$, respectively. Because the off-axial distances of both electrons in the first hexapole field are identical, they experience the same deflection $\Delta\gamma^{(2)}$. However, because of the deflection induced by the first hexapole, the second hexapole affects the two electrons differently. This leads to a divergent beam reflecting the third-order focusing characteristics which corresponds to negative spherical aberration.

$\Delta\gamma = 2\Delta\gamma^{(2)} + \Delta\gamma^{(3)}$. While the deflection $2\Delta\gamma^{(2)}$ is equal for both electrons, the deflection $\Delta\gamma^{(3)}$ is always directed away from the optical axis. Hence, the deflection angle $\Delta\gamma^{(3)}$ is of equal magnitude for both electrons but points in opposite directions. This third-order deflection is independent of the azimuthal position of an electron in the hexapole field. It is an isotropic effect. Hence, the third-order deflection $\Delta\gamma^{(3)}$ causes the electron beam to diverge (see Fig. 8.4). This third-order effect is the *negative* spherical aberration which reflects the combination aberration of two thin hexapole fields. It is this third-order effect which is employed in spherical aberration correctors that are based on hexapole elements. However, instead of arranging a series of thin hexapoles in series to amplify the combination aberration, one can employ a long (or extended) hexapole field, i.e. a hexapole field of a defined extension along the optical axis. Such an extended hexapole equivalently shows the third-order (negative) focusing effect (Rose, 1994; Haider *et al.*, 2008a).

The third-order effect of an extended hexapole field can be illustrated on the basis of Fig. 8.5. First of all, Fig. 8.5 shows that paraxial electrons travelling on the optical axis do not experience the Lorentz force. This explains the absence of any

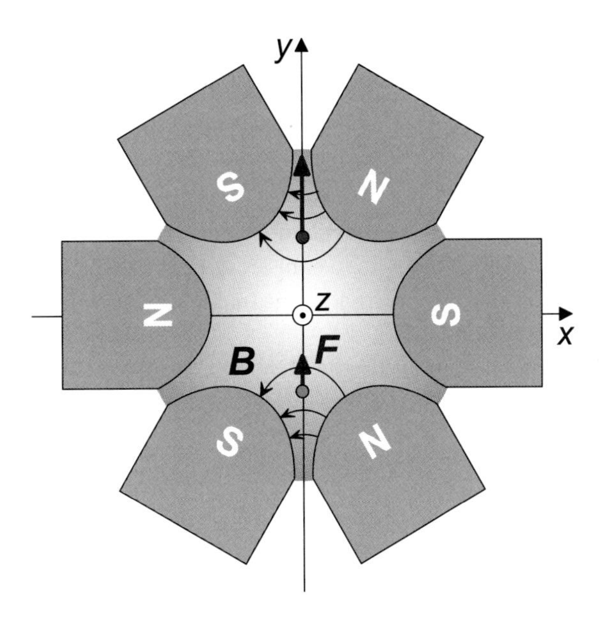


Fig. 8.5 The effect of the extended hexapole field. While the red electron entering the hexapole field in $y > 0$ is deflected into an area of the hexapole field where the deflection force increases, the green electron entering the hexapole field in $y < 0$ is deflected into an area of the hexapole field where the deflection force is weaker. Hence, for an extended hexapole field, the beam, though experiencing a threefold distortion, becomes divergent, which reflects the third-order focusing effect of the extended hexapole field (see also Fig. 8.4). The blue background shall qualitatively indicate the increase of the field strength with increasing distance to the optical axis.

first-order effect in a hexapole field. Secondly, in analogy to Fig. 8.4, two electrons travelling parallel to the optical axis enter the hexapole field in a finite distance to the optical axis. The electrons travel on a meridional plane $(y, z)$. While one electron enters the hexapole at $y_{\mathrm{HP}}$ (red), the other electron enters the hexapole at $-y_{\mathrm{HP}}$ (green) (see Fig. 8.5). Hence, both electrons have the same distance to the optical axis when they enter the hexapole field. Once in the hexapole, the electrons experience the Lorentz force which for the arrangement in Fig. 8.5 deflects them in positive $y$-direction. As long as their radial distances are equal $(y_{\mathrm{HP}})$, the instantaneous deflection of the first electron coincides with the deflection that the second electron experiences. Hence, if the hexapole is short, nothing more than a small deflection occurs. However, in an extended hexapole, the electron that enters in $y_{\mathrm{HP}}$ (red) is driven into an area of the hexapole field where the magnetic force is stronger, while the other electron entering the hexapole in $-y_{\mathrm{HP}}$ (green) is deflected into an area of the hexapole field where the magnetic deflection force is weaker. Hence, the Lorentz force the first electron experiences becomes stronger, whereas the Lorentz force of the second electron becomes weaker. The longer the electrons are exposed do the hexapole field, i.e. the longer the hexapole, the larger is the difference of the deflections that the two electrons experience. The trajectories of the two electrons is divergent. This effect is isotropic, i.e. it does not matter in which direction we set the $y-$axis. The (isotropic) divergence of the beam reflects the third-order focusing effect of the extended hexapole field. This is negative spherical aberration.

Still, the negative spherical aberration is essentially a side product; the primary and dominant effect of a hexapole, independent of whether this is a short or a long hexapole, is threefold astigmatism. Hence, in order to take advantage of the negative spherical aberration, one needs to find a suitable setup in order to neutralize this second-order effect.

The important point about hexapole elements in aberration correctors is that short hexapoles only induce threefold astigmatism $A_2$, whereas long hexapoles in addition produce negative spherical aberration. The induced negative spherical aberration can be described by a combination aberration.

### 8.2.3 *Octupoles*

The primary effect of an octupole field ($m = 4$) is to induce an aberration of order $n = m - 1 = 3$ and of symmetry $N = m = 4$, which, according to Fig. 7.12 and Table 7.2, is fourfold astigmatism $A_3$.

The variety of effects of the octupole field can be enlarged by placing a quadrupole in front of it. The combination aberrations of the quadrupole–octupole unit can be employed to correct for all third-order aberrations. The quadrupole, i.e. the first element, forms an elliptical distortion of the beam. If an elliptical beam is exposed to an octupole field, it leads to a combination aberration of the primary aberrations of the individual units. While the primary aberration of a quadrupole is twofold astigmatism $A_1$ (a first-order aberration), the primary aberration of an octupole is $A_3$, i.e. fourfold astigmatism (a third-order aberration). The combination of these two elements can be employed to generate combination aberrations which cover all three third-order aberrations: spherical aberration $C_3$, star aberration $S_3$ and fourfold astigmatism $A_3$. The ratio between these combination aberrations depends on the ellipticity of the beam that enters the octupole field.

The important point about the combination of a magnetic quadrupole with a magnetic octupole is that (negative) spherical aberration can be induced. Of course, depending on the orientation of the axes of the elliptical beam with respect to the octupole field, either positive or negative spherical aberration is induced as the resulting combination aberration. Still, an octupole induces fourfold astigmatism and a quadrupole induces twofold astigmatism. A corrector which employs the third-order focusing effect of a quadrupole–octupole system thus needs to correct for these primary aberrations.

## 8.3 The Basic Principles of Spherical Aberration Correctors

Geometrical aberrations as discussed in Chapter 7 have one very useful characteristic: they add up. Apart from combination aberrations, which can occur, one has to consider that if a first optical element that causes an aberration $X_n$ is put in series with a second element that shows an aberration $Y_n$, the resulting aberration is the sum of the two. This simply means that the warping of the wave surface is additive. Though this seems to be a rather obvious circumstance, which indeed is known as the *addition theorem* (Born and Wolf, 2001), it has a very practical consequence for aberration correction. It is not necessary to design lenses which are free of aberrations. Alternatively, we can add optical elements such that the sum of all the individual aberrations annuls. Hence, to correct for the spherical aberration of the objective lens, one needs to add a unit to the existing system, which has negative spherical aberration and thus compensates for the positive spherical aberration of the round lens. Indeed, what needs to be corrected is the sum of all the spherical aberrations of all the relevant round lenses. This sum, however, is dominated by the spherical aberration of the objective lens. The preceding brief discussion about the effects of multi-pole lenses and the possibility of combination aberrations provides evident strategies on how to tackle the problem of correcting the spherical aberration of the objective lens. Before we lay out these strategies, we need to discuss at which position in the microscope column an aberration corrector needs to be positioned.

The spherical aberration of a twin-type objective lens, consisting of two pole pieces, affects the imaging process in two ways. Though we are aware that the specimen is actually immersed in the magnetic field of the objective lens, we can model the effect of the objective lens by an objective lens pre-field, which defines the formation of the STEM probe, and an objective lens post-field, which essentially defines the imaging performance in TEM mode. These two situations are depicted in Fig. 2.1 and Fig. 3.1, respectively. In order to correct the spherical aberration in STEM mode, one needs to correct the spherical aberration of the pre-field of the objective lens. While the spherical aberration of the pre-field of the objective lens is largely irrelevant for TEM imaging, it is the spherical aberration of the post-field of the objective lens which needs to be corrected for HRTEM.

The specimen is located in the pole-piece gap of the objective lens. Hence, in order to add a unit which compensates the positive spherical aberration of the pre-field, we need to place the corrector unit just in front of the objective lens. Analogously, in HRTEM mode the corrector unit is located just behind the objective lens. Indeed, one can consider a corrector in combination with the post- or the pre-field of the objective lens as an extended objective lens. We call a microscope which is equipped with an aberration corrector for STEM imaging a probe- or an illumination-corrected microscope, and an image-corrected instrument shall be an instrument which has an aberration corrector placed behind the objective lens, i.e. an imaging aberration corrector. Moreover, a microscope, which is equipped with two aberration correctors, one for STEM and one for HRTEM imaging, i.e. one in front and one behind the objective lens, is called a double-corrected microscope (see, e.g. Hutchison *et al.*, 2005).

From the discussion of the effect of multi-pole lenses, it seems to be obvious that there are (at least) two types of spherical aberration correctors; spherical aberration correctors which are based on hexapole lenses, and spherical aberration correctors which are based on systems employing quadrupole and octupole lenses. The first workable spherical aberration corrector for STEM imaging was a quadrupole–octupole corrector (Krivanek *et al.*, 1997), and the first workable spherical aberration corrector for TEM imaging was a hexapole corrector (Haider *et al.*, 1998). While the hexapole corrector, which is used for HRTEM imaging, can also be employed for STEM imaging (Haider *et al.*, 2000), only very recently, the first quadrupole–octupole aberration correctors was employed for HRTEM imaging. Indeed, the first quadrupole–octupole aberration corrector for HRTEM imaging, which employs magnetic and electrostatic elements, corrects for geometrical aberrations as well as for the chromatic aberration (Haider *et al.*, 2008b; Kabius *et al.*, 2009).

### 8.3.1 *Hexapole spherical aberration corrector*

An extended hexapole field induces negative spherical aberration. This effect is equivalent to the combination aberration of a series of independent thin hexapole fields. If we consider a corrector element, which shall compensate for the positive spherical aberration of the objective, the presence of negative spherical aberration in such extended hexapole fields can be of use. However, the negative spherical aberration $C_3$ as a third-order effect comes at the expense of a large amount of threefold astigmatism $A_2$, which indeed is the primary aberration of a hexapole field. Feeding a round beam through a single (extended) hexapole field, the amount of $A_2$ that is induced deteriorates the imaging performance such that the advantage of having the spherical aberration annulled would be insignificant. Furthermore, besides the third-order spherical aberration $C_3$, an extended hexpole leads to additional secondary aberrations; one of them is the fourth-order three-lobe aberration $D_4$ and sixfold astigmatism $A_5$. Hence, in order to make use of hexapole fields to correct the spherical aberration, a setup needs to be found which does not induce second-order aberrations or any other aberration which could potentially impair the imaging performance (Rose, 1990). The fact that the primary effect of the hexapole field, namely the threefold astigmatism $A_2$, needs to be cancelled out in order to be able to benefit from its secondary effect, i.e. the negative spherical aberration $C_3$, has led to various corrector designs (Beck, 1979; Crewe and Kopf, 1980; Crewe, 1982, 1984; Jiye and Crewe, 1985; Shao, 1988; Rose, 1990; Mitsuishi *et al.*, 2006).

Let us first consider a simple corrector design which makes use of a single extended hexapole field. A rotationally symmetric beam is propagated into an extended hexapole field. At the end of the hexpole field, the beam would show negative spherical aberration as well as a large amount of threefold astigmatism. However, if we manage to form a cross-over of the beam at the center of the extended hexapole field, the situation is different. The cross-over can be thought as an inversion of magnification $M = -1$. Regardless of whether there is an inversion within the hexapole field or not, the effect of the negative spherical aberration of the hexapole field is essentially the same. The spherical aberration is an isotropic aberration of even azimuthal symmetry, which is thus invariant to an inversion. However, this is not the case for the threefold astigmatism, which has odd azimuthal symmetry. Inverting a triangular beam with the aid of the cross-over inverts the direction of the triangular distortion. Hence, in front of the cross-over, the hexapole field induces threefold astigmatism leading to a particular threefold distortion. Behind the cross-over, the threefold distortion induced by the hexapole field remains unchanged, however, it does now alter the inverted beam and thus could in principle fully compensate the triangular distortion which the beam experienced in the first half of the hexapole field. Therefore, at the exit plane of the hexapole field, the initially round beam would only show negative spherical aberration, and the second-order distortion would be annulled. For a probe-forming system, the only step that remains to be done is to feed this beam with negative spherical into the objective lens, and to adjust the strength and the length of the hexapole field such that the induced negative spherical aberration equilibrates the positive spherical aberration of the objective lens. One would end up with an electron probe of zero spherical aberration. Indeed, this principle was described as one possible solution of employing extended hexapole fields to cancel out the positive spherical aberration in a probe-forming instrument (Crewe and Kopf, 1980). Though this concept seems to be a viable approach, because the optical setup does not provide any flexibility in dealing with side effects, its practicability turns out to be a major obstacle.

The only workable hexapole-type spherical aberration corrector makes use of two extended hexapole fields and four round transfer lenses (Rose, 1990). These are the main, i.e. strong, elements of the hexapole corrector. Starting with the objective lens (OL), there is a first telescopic transfer doublet, made of two round magnetic lenses TL11 and TL12, which images the coma-free point of the objective lens into the first magnetic hexapole field (HP1) (see Fig. 8.6a). The coma-free point is approximately at a distance $2/3f_o$ from the object plane, where $f_o$ is the focal length of the objective lens. In analogy to the first transfer doublet, the second pair of round transfer lenses (TL21 and TL22) telescopically images the first hexapole into the second hexapole field (HP2) with a magnification of $M = -1$. These six elements form the aberration corrector; a first transfer doublet (TL11 and TL12), the first hexapole field (HP1), the second transfer doublet (TL21 and TL22) and the second hexapole (HP2). The setup of the hexapole spherical aberration corrector for HRTEM is illustrated in Fig. 8.6a (Rose, 1990). The corresponding schematics for the STEM mode is shown in Fig. 8.6b (Müller *et al.*, 2006).

The hexapole corrector essentially consists of two telescopic transfer doublets as illustrated in Fig. 6.33, which transfer the coma-free point of the objective lens into the first and the second hexapole field. As such, the hexapole spherical aberration corrector as depicted in Fig. 8.6 is an $8f$-system, where $f$ is the focal length of the (identical) transfer lenses.

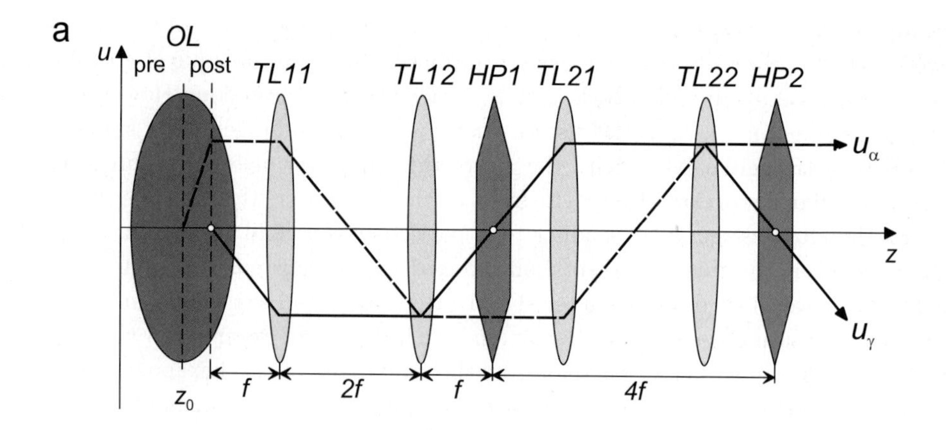


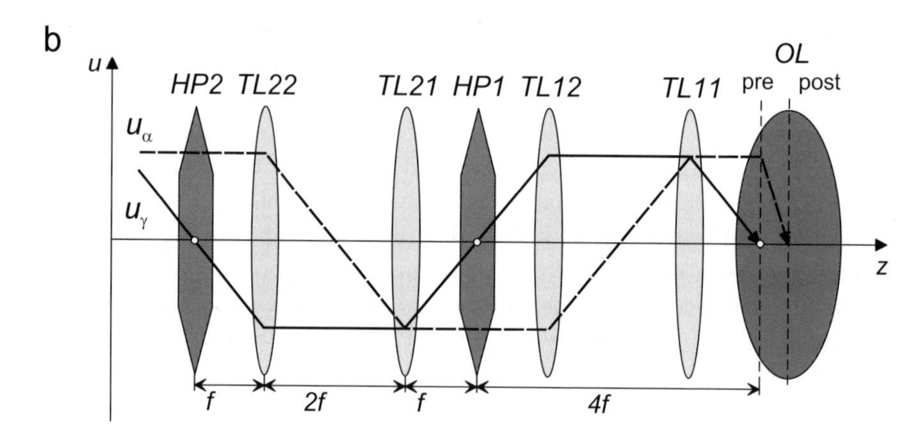


Fig. 8.6 Schematics of the hexapole spherical aberration corrector for (**a**) HRTEM and for (**b**) STEM. The ray diagrams are reproduced from Rose (1990) (HRTEM) and Müller *et al.* (2006) (STEM). Indicated are the field ray $u_\gamma$ and the axial ray $u_\alpha$.

Both extended hexapole fields induce threefold astigmatism $A_2$ and negative spherical aberration $C_3$. The crucial point of this corrector design is that the second-order effect, i.e. the threefold astigmatism $A_2$, is cancelled out. The second transfer doublet images the beam of the first hexapole field with a magnification of $M = -1$ into the second hexapole field. Since both hexapole fields have the same azimuthal orientation, the transfer from the first to the second hexapole field by a magnification of $M = -1$ ensures that the second-order effect vanishes. This is, in principle, equivalent to a situation where a cross-over inverts the beam inside of an extended hexapole field. However, the transfer doublets warrant the necessary flexibility such that side effects and mechanical imperfections, including mechanical misalignments, can be compensated by properly adjusting the transfer lenses and the strengths of the hexapole fields, which only under idealized conditions can be considered to be

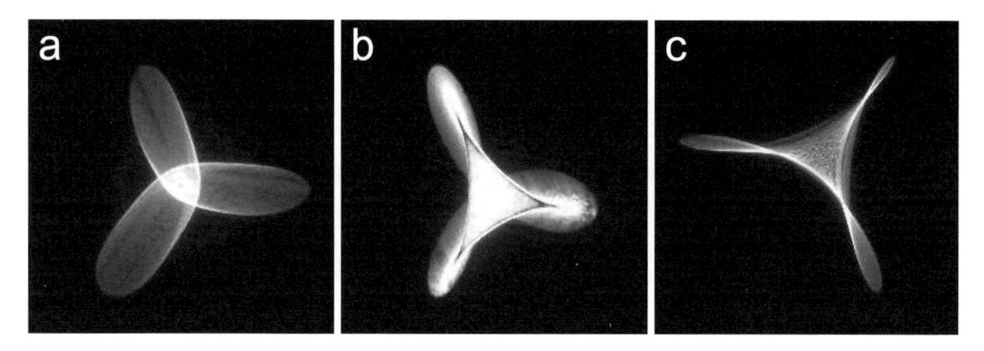


Fig. 8.7 The effect of the two hexapoles on the caustic image in a probe aberration-corrected microscope (see Fig. 8.6b). (**a**) HP1 is excited to its nominal value. The threefold distortion of the beam, i.e. the primary aberration $A_2$, is apparent (underfocused imaged). (**b**) HP2 is excited without having the transfer doublet TL21, TL22 activated (overfocused image). HP1 is not excited. Both hexapoles, HP1 and HP2, cause a characteristic threefold distortion of equal azimuthal orientation. (**c**) HP2 is excited and the transfer doublet TL21, TL22 is activated (less underfocused than in (**a**)). HP1 is off. The transfer doublet ensures that the threefold distortion caused by HP2 is inverted (see (**c**)) such that it can be counterbalanced by HP1 when both hexapoles are excited. The main resulting effect is the negative spherical aberration. Since threefold astigmatism (and the applied focus) strongly dominates the caustic images, the negative spherical aberration induced by the hexapoles is not directly visible.

identical. The effect of the hexapoles on the caustic image and the effect of the inversion controlled by the second transfer doublet of a probe aberration corrector are illustrated in Fig. 8.7. The inversion causes the primary effect of the two hexapole fields to cancel out. What remains is the negative spherical aberration given by both extended hexapole fields. Furthermore, similar to the cancellation of $A_2$, the second transfer doublet also warrants that the three-lobe aberration $D_4$ vanishes. Yet, the sixfold astigmatism $A_5$, which can be considered as a ternary aberration of an extended hexapole field (Müller *et al.*, 2008), cannot be annulled, i.e. the sixfold astigmatism $A_5$, induced separately by the first and the second hexapole, adds up.

This type of hexapole spherical-aberration corrector has originally been employed in aberration-corrected transmission electron microscopes used for HRTEM (Haider *et al.*, 1998) and later also in probe-forming instruments used for STEM imaging (Erni *et al.*, 2006), as well as in double-corrected microscopes (Hutchison *et al.*, 2005).

### 8.3.2 *Quadrupole–octupole spherical aberration corrector*

Although the hexapole spherical aberration corrector was the first aberration corrector employed for HRTEM imaging, it was originally an electrostatic octupole-type aberration corrector that was suggested (Scherzer, 1947) and tested (Seeliger, 1949, 1951, 1953) as a possible unit for correcting the spherical aberration in TEM imaging. In contrast to its original purpose, quadrupole–octupole aberration correctors were first brought to application in probe-forming instruments (Zach and Haider,

1995; Krivanek *et al.*, 1997). Yet, there is a distinct difference between the octupole spherical aberration corrector which was originally proposed by Scherzer (1947) and the first working quadrupole–octupole aberration corrector employed in a dedicated scanning transmission electron microscope (Krivanek *et al.*, 1997). While the original design in 1947 was based on cylindrical lenses and electrostatic octupole fields, the first workable aberration corrector in a dedicated scanning transmission electron microscope was based upon pure magnetic quadrupole and octupole lenses.

However, apart from providing one of the basic concepts for correcting the spherical aberration, Scherzer (1947) also laid out a way to correct the chromatic aberration of a round lens. This led to corrector designs based upon combined electrostatic-magnetic quadrupole fields (Rose, 1967) which would correct both the spherical and the chromatic aberration (Rose, 1971a, 1971b). The idea of correcting simultaneously the spherical and the chromatic aberration was realized in the first aberration corrector for a scanning electron microscope (Zach and Haider, 1995) and in the first quadrupole–octupole imaging aberration corrector (Haider *et al.*, 2008b; Kabius *et al.*, 2009). In the present context, however, we draw our attention solely to spherical aberration correctors which employ magnetic octupole fields.

Why octupoles? The dominant aberration of the objective lens, which does not vanish on the optical axis, i.e. the spherical aberration $C_3$, is of third order. The primary aberration of an octupole lens is fourfold astigmatism $A_3$, which is a third-order aberration, too. The bending of the wave aberration function due to the presence of spherical aberration as well as due to the fourfold astigmatism depends on the forth power of the angular coordinate in the aperture plane. Though the azimuthal dependence of fourfold astigmatism differs from spherical aberration, it shows an identical radial dependence in certain directions (see Fig. 7.12).

While fourfold astigmatism $A_3$ reflects a wave surface which has positive and negative modulations, the spherical aberration $C_3$ corresponds to an isotropic wave surface which is homogeneously bent in positive direction only (see Fig. 7.12). The wave aberration function $\chi$ due to $A_3$ can be smaller or larger than zero, whereas $\chi$ for $C_3$ is always positive. Furthermore, wave surfaces and aberration functions are additive. Hence, balancing the absolute values of the two aberrations $C_3$ and $A_3$ and adding up the respective aberration functions yields an aberration function which shows the double positive value in two orthogonal directions where $A_3$ causes a positive bending of the wave surface. Along these two directions, the aberrations add up in a 'destructive' way. On the other hand, along the directions $A_3$ has its minimal values, i.e. where the aberration function $\chi$ of $A_3$ is negative, the positive value of the spherical aberration can compensate for the negative value of the aberration function of $A_3$. Whether the aberration functions fully compensate each other in these directions depends on the ratio between $|A_3|$ and $C_3$. The important point is that by properly adjusting $|A_3|$ against $C_3$ in two orthogonal directions, the sum of the aberration functions can be annulled, while in the directions which are rotated by 45° in respect to the 'flat' directions, the aberration function becomes

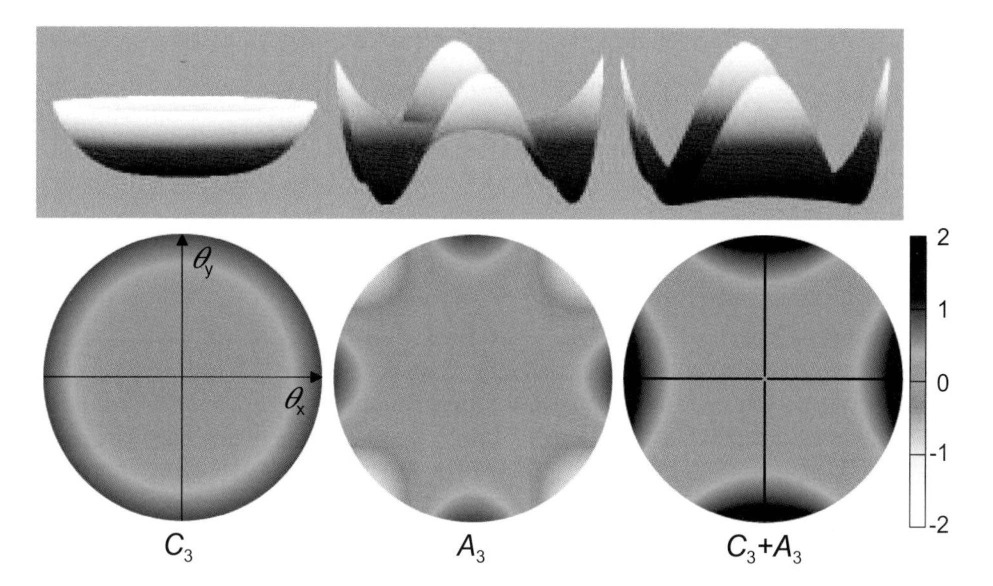


Fig. 8.8 Adding spherical aberration $C_3$ and fourfold astigmatism $A_3$. The maximum value of the wave aberration functions of $C_3$ and $A_3$ is $|\chi| = 1$. Adding up these two aberrations yields the wave aberration function labelled $C_3 + A_3$. This summed wave aberration function shows the double value of the individual aberration functions in the directions indicated by the black lines. In the directions indicated by the white lines, the sum of $C_3$ and $A_3$ leads to a flat aberration function, indicating that in these two directions the positive spherical aberration of a round electron lens can be fully corrected by an octupole field with a primary aberration $A_3$.

twice the value of the individual surfaces of $C_3$ and $A_3$ (see Fig. 8.8). This simply means that along two directions, indicated by the white lines in Fig. 8.8, spherical aberration can be fully compensated by fourfold astigmatism. Yet, along two other directions, indicated by the black lines in Fig. 8.8, the deleterious effects of $C_3$ and $A_3$ add up.

As impracticable as it seems on first sight, such a *partial*-spherical aberration corrector which consists of a single octupole attached to a round objective lens has been theoretically analyzed by Scherzer and Typke (1967/1968). Though a single octupole cannot isotropically correct the spherical aberration, Scherzer and Typke (1967/1968) concluded that a resolution improvement by about a factor of 1.3 is feasible by employing a single octupole. The resolution gain is solely due to the circumstance that the spherical aberration is correctable in two perpendicular directions. On the basis of the corresponding aberration functions, the situation of combining fourfold astigmatism $A_3$ with spherical aberration $C_3$ is illustrated in Fig. 8.8.

In Fig. 8.8 it is assumed that $|A_3| = A_3 = C_3$, i.e. the imaginary part of $A_3$ is arbitrarily set equal zero. The individual aberration functions due to $C_3$ and $A_3$ are given by (see Table 7.2)

$$\chi_{C3}(\omega) = \frac{1}{4} C_3 \left(\omega\overline{\omega}\right)^2 \quad \text{and} \quad \chi_{A3}(\omega) = \mathcal{R}\left(\frac{1}{4} A_3 \overline{\omega}^4\right),$$

with $\omega = \theta_x + i\theta_y$. For the direction $\theta_y = 0$, we obtain from the above equations that

$$\chi_{C3}^{\oplus}(\omega) = \frac{1}{4}C_3\theta_x^4 \quad \text{and} \quad \chi_{A3}^{\oplus}(\omega) = \frac{1}{4}A_3\theta_x^4.$$

Along this direction, as well as along the direction $\theta_x = 0$, the sum of the aberration functions reflects twice the value of the individual aberrations. On the other hand, for the directions $\theta_x = \pm\theta_y$, the aberration functions yield

$$\chi_{C3}^{\otimes}(\omega) = C_3\theta_x^4 \quad \text{and} \quad \chi_{A3}^{\otimes}(\omega) = -A_3\theta_x^4.$$

Hence, provided $|A_3| = A_3 = C_3$, along these two directions, indicated by the superscript $^{\otimes}$, the aberration functions annihilate each other.

From this we can conclude that the combination of an octupole lens with a rotationally symmetric electron lens has a double-edged consequence. While in two directions the effect of the spherical aberration of the round lens can be fully annulled, in two other directions, the deleterious effects of $C_3$ and $A_3$ add up. From the perspective of correcting aberrations, we do not gain much by combining an octupole with a round electron beam. However, if we consider a strongly (twofold) astigmatic beam, ideally a line focus, which enters the octupole field such that the main axis of the elliptical beam coincides with one of the azimuthal directions the octupole field annihilates the spherical aberration, one can compensate for the spherical aberration in this particular azimuthal direction. Having corrected the spherical aberration in one direction, we can employ another stigmator to invert the elliptical shape of the astigmatic electron beam. This reshaped elliptical beam can now be transferred into another octupole element whose azimuthal orientation coincides with the first octupole. With this procedure, we can correct the spherical aberration in the direction orthogonal to the one corrected before. Using another stigmator, the elliptical beam, which is now free of spherical aberration in two orthogonal directions, can be transferred into a round beam again. However, what remains is a beam which still has a fourfold modulation, i.e. the azimuthal dependence of the octupole fields is reflected in the residual warping of the wave surface. This fourfold modulation is the residual fourfold astigmatism which in principle is correctable by a third octupole element.

This straightforward idea pictures the main principle of spherical aberration correctors that make use of magnetic octupole fields. As will be shown below, in reality the situation is a bit more complex. An octupole field influences all three third-order aberrations, i.e. $C_3$, $S_3$, $A_3$. Hence, all of them need to be considered in the setup of the corrector. At least three octupoles are needed to have all three third-order aberrations under control. Besides the octupoles, elements are needed which form the necessary astigmatic ray path through the octupoles (Rose, 1967; Krivanek *et al.*, 2009a).

The importance of the optical elements, which induce the elliptical distortion of the beam, is reflected in the chronological sequence of building the first aberration corrector. Of course, it was not clear whether an optical path through octupole fields would be feasible or even at all helpful. But first of all, the question need to be addressed regarding whether it is possible to form a suitable stigmatic image from an astigmatic ray path, even without having octupole fields involved. Therefore, besides testing the imaging properties of an octupole element, the second critical test of Scherzer's (1947) corrector design that Seeliger (1949) carried out was simple in its idea: one had to show that after distorting the beam in one and then in the other direction, and finally back to a supposedly broad beam, an image quality can be achieved which is equivalent to the quality of an image formed by a ray path that does not involve strongly anamorphotic image planes. This test was carried out (Seeliger, 1949) and provided motivation to implement the octupole lenses in the previously tested astigmatic ray path (Seeliger, 1951, 1953; Möllenstedt, 1956).

Yet there is a distinct difference in the way the astigmatic ray path was formed by Seeliger (1949) and the way such a ray path was realized in subsequent corrector designs (see, e.g. Deltrap, 1964). As suggested by Scherzer (1947), Seeliger (1949, 1951) employed electrostatic cylindrical lenses in his test of the astigmatic optical path as well as in the first corrector. Cylindrical lenses are optical units in which the field is constant in a direction perpendicular to the optical axis (Rose, 1972; Grivet, 1972; Hawkes and Kasper, 1989a, 1989b). Such cylindrical lenses can be used to form line foci. However, from the above discussion of the effect of a magnetic quadrupole, it is clear that instead of cylindrical lenses, quadrupoles can be employed to form an astigmatic ray path as well. A corrector, which employs quadrupoles instead of cylindrical lenses, was first suggested by Archard (1955). Adjustable quadrupole fields enhance the flexibility of precisely controlling the ellipticity of the beam. This flexibility is of particular importance for adjusting all three axial third-order aberrations: spherical aberration $C_3$, star aberration $S_3$ and fourfold astigmatism $A_3$. Indeed, being able to control the ellipticity of the beam makes it possible to control the ratio between the three third-order aberrations which are induced by an octupole element. In the limiting case of a line focus in a short octupole, the ratio between $C_3$:$S_3$:$A_3$ induced by the octupole is 3:4:1, and in the limiting case of a round beam interacting with an octupole, it is solely fourfold astigmatism $A_3$ which is induced by the octupole (Lupini, 2001).

Designing a quadrupole–octupole spherical aberration corrector means answering two questions: how many quadrupoles are needed for an astigmatic ray path which results in a stigmatic image, and secondly: how many octupoles are needed in order to have all third-order aberrations under control such that the net effect of the corrector is to solely induce negative spherical aberration $C_3$, which can compensate the positive spherical aberration of the objective lens. As mentioned above, three quadrupoles and three octupoles are needed to simultaneously satisfy the condition of a stigmatic image formation from an astigmatic ray path and the condition of having all third-order aberrations adjustable (Deltrap, 1964). The first octupole acting on a line focus, formed by the first quadrupole, induces third-order aberrations in the ratio $C_3$:$S_3$:$A_3$= 3:4:1. A line focus perpendicular to the first one, formed by the second quadrupole, is now transferred into the second octupole, which induces third-order aberrations as $C_3$:$S_3$:$A_3$= 3:-4:1. Hence, while the values of $C_3$ and $A_3$ add up, the star aberration $S_3$ induced by the first octupole is cancelled out by the second octupole. The engineered (negative) $C_3$ shall compensate the spherical aberration of the objective lens, while the resulting $A_3$ unavoidably stemming from the first two octupoles is annulled by adding a third octupole. For this, a broad beam is necessary, which is formed by a third quadrupole.

This principle seems to be quite straightforward. However, in order to enable a maximum degree of symmetry with a minimum amount of multi-poles, Deltrap's quadrupole–octupole corrector, i.e. one of the first designs of a probe aberration corrector, made use of four multi-pole elements of eight poles where each pole had two independent windings. The quadrupole windings causing the poles to be N and S in pairs, while the octupole windings enabled alternating poles to be either N or S. Hence, what results are four identical elements which can produce a quadrupole field, an octupole field or a superposition of both. The idea of such combined quadrupole–octupole elements goes back to Archard (1955). Deltrap's corrector can be considered as a partial success, since it enabled to annul the spherical aberration. However, it was only a partial success since the corrector did not allow for improving the resolution of the microscope.

The first spherical aberration corrector in a scanning transmission electron microscope which enabled a smaller electron probe and thus increased the resolution is based on a magnetic quadrupole–octupole design (Krivanek *et al.*, 1997). This first probe corrector employs six identical multi-pole lenses, which, similar to Deltrap's multi-pole elements, allow for independently forming quadrupole and octupole fields. These six units are arranged in two groups, each containing three multi-poles. This prototype quadrupole–octupole spherical aberration corrector was soon replaced by another design which employs four quadrupoles and three octupoles in alternating order (see, e.g. Batson *et al.*, 2002). This second generation probe corrector is depicted in Fig. 8.9. The electron beam is transferred by the last condenser lens into the corrector's quadrupole Q4. This quadrupole forms an elliptical beam with its major axis along the $y$-direction and transfers this ellipse into octupole O3. This octupole essentially induces negative spherical aberration $C_3$ in $y$-direction. The next quadrupole field the beam experience is formed by Q3. It undoes the elliptical distortion of Q4 such that a round beam enters octupole O2. The primary purpose of this octupole is to induce fourfold astigmatism $A_3$. This fourfold astigmatism $A_3$ is needed to counterbalance the residual fourfold astigmatism which results from the two-step spherical aberration correction in the octupoles O3 and O1. The following quadrupole Q2 again forms an elliptical beam with its major axis now along the $x$-axis. This ellipse enters octupole O1, whose purpose it is to induce negative spherical aberration in $x$-direction. Finally, there is

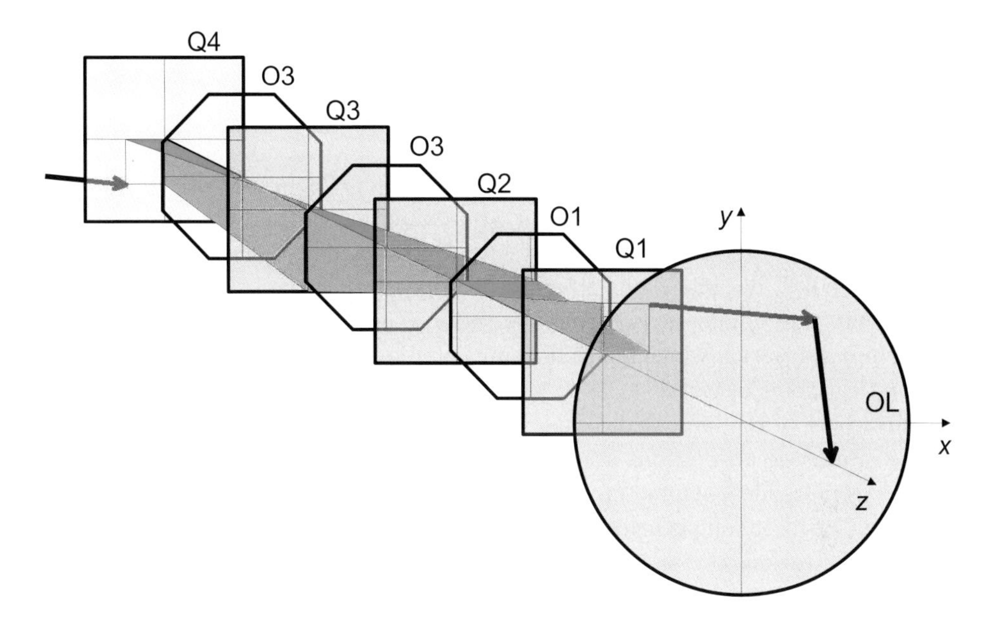


Fig. 8.9 Quadrupole–octupole aberration corrector according to Krivanek *et al.* (1999). A field ray coming from the last condenser lens enters the corrector, which consists of four quadrupoles (Q4, Q3, Q2, Q1) and three octupoles (O3, O2, O1). Within the corrector, i.e. within the ray path connecting Q4 with Q1, the projection of the ray onto the $x, z$ plane (red) and onto the $y, z$ plane (green) are shown. The objective lens OL focuses the ray onto the specimen plane.

quadrupole Q1 forming again a broad electron beam, which, after running through the series of multi-pole fields, suffers from negative spherical aberration. Due to the symmetrical setup, the star aberration $S_3$ induced by the elliptical beam interacting with octupole O3 is counterbalanced by the star aberration $-S_3$ induced by O1. The negative spherical aberration finally counterbalances the positive spherical aberration of the objective lens OL such that the focused electron probe is free of third-order aberrations (Krivanek *et al.*, 2009a).

This type of spherical aberration corrector has successfully been employed in dedicated scanning transmission electron microscopes (Krivanek *et al.*, 1997, 1999). Installed on a 300 kV scanning transmission electron microscope, this corrector enabled for the first time resolving of a crystal spacing of less than 0.1 nm in scanning transmission mode (Nellist *et al.*, 2004) and thus opened the way for direct sub-Ångström imaging in electron microscopy. Because an octupole-type aberration corrector makes use of the primary aberration of the octupole field, i.e. the fourfold astigmatism, it is called a *direct-action* corrector, while a hexapole-type corrector, whose correction is based upon a secondary effect, is thus called an *indirect-action* corrector (Krivanek *et al.*, 2009a).

## 8.4 Beyond the Basic Principles — Parasitic Aberrations and Other Complications

### 8.4.1 *The multitude of axial aberrations*

The primary purpose of a spherical aberration corrector is to correct the spherical aberration which within the isoplanatic approximation is the only non-vanishing geometrical axial aberration of a round lens. However, employing multi-pole fields like quadrupoles, hexapoles or octupoles, it appears that the original situation of dealing with one dominant isotropic aberration, i.e. the spherical aberration, has turned into a situation which might be more complex from a practical but also from a theoretical point of view.

One aspect of the increased complexity concerns the presence of parasitic aberrations. Parasitic aberrations arise when the real symmetry of a multi-pole field deviates from its ideal symmetry. Such deviations can be caused by mechanical imperfections or by inhomogeneities of the material employed for the magnetic poles. Parasitic aberrations can also occur if the real optical path of the beam through the system deviates from its ideal trajectory. Misalignment essentially leads to the appearance of parasitic lower-order and fourth-order aberrations (Batson, 2009). For these two reasons, the amount of observable (parasitic) aberrations in a real instrument always exceeds the amount of symmetry allowed aberrations (Müller *et al.*, 2008). Indeed, the proper alignment of a real electron optical instrument is simply given by the path of the beam which minimizes parasitic effects. This makes it necessary to employ deflectors and stigmators. Only such secondary optical elements enable the proper setup of complex electron optical instruments.

However, in order to efficiently correct the dominant parasitic aberrations, it is essential to correct them at the location where they actually occur. It is thus of crucial importance to understand the effect of potential misalignments and to be able to identify the source of significant parasitic aberrations. For instance, if the two hexapoles in a hexapole spherical aberration corrector are not properly imaged onto each other, but with some lateral shift instead, the beam suffers from strong twofold astigmatism. Of course, it is in principle possible to correct twofold astigmatism using a (weak) quadrupole-type stigmator. Yet, it is much more efficient to have deflectors available which allow matching of the axis of the first hexapole with the one of the second. Another example of a misalignment aberration is second-order axial coma $B_2$. If two optical units which both have spherical aberration $C_3$ are misaligned in respect to each other, second-order axial coma $B_2$ is produced. The presence of residual axial coma is indeed typical in aberration-corrected microscopes. The round objective lens, as well as the corrector unit, both have spherical aberration. Regardless of whether it is negative or positive spherical aberration, what results if these two units are not properly imaged onto each other is axial coma $B_2$. Of course, there are means of matching the axis of the corrector with the one of the objective lens, or the axis of one hexapole with the one of the other.

Such adjustments make it possible to reduce the effect of parasitic aberrations at the source where they occur. Still, an aberration-corrected electron microscope, which employs several multi-pole lenses, is a highly complex instrument. This complexity demands highly sophisticated alignment strategies in order to tackle, for instance, fourth-order aberrations, which essentially arise because of slight misalignments (Batson, 2009).

The crucial point is that the correction of the third-order spherical aberration does not warrant that the instrument is free of geometrical axial aberrations. It is of fundamental importance to be able to control and also to adjust, lower-order axial aberrations in particular (see Table 7.2). In daily work with spherical aberration-corrected microscopes, some of these aberrations can be tuned by auto-alignment routines and some of them need to be adjusted manually. Some of these adjustments need to be performed regularly, whereas other adjustments are stable over a long time over which the corrector is in use. As a matter of fact, the tolerable limits of all of these aberrations decrease drastically with the increased resolution, which in principle comes with the correction of the spherical aberration (Uhlemann and Haider, 1998; Haider *et al.*, 2000). As a consequence, it is often the case that the actual resolution is limited by lower-order aberrations and not by the residual intrinsic aberrations of a corrector (Müller *et al.*, 2008; Erni *et al.*, 2010).

### 8.4.2 *Residual intrinsic axial aberrations*

Though parasitic aberrations occur, they can largely be controlled by various alignment routines and procedures. Provided that the quality of the optical components is sufficiently high, all of these parasitic aberrations can in principle be tuned such that they are small enough and do not alter the point resolution of the instrument. But even if all the residual parasitic aberrations are insignificant, there are still geometrical aberrations present which limit the ultimate optical resolution, i.e. the geometrical resolution limit. Though the aberration corrector corrects the dominant intrinsic geometrical aberration of the uncorrected microscope, the aberration-corrected microscope equivalently has its characteristic *residual intrinsic aberrations*. For a hexapole third-order aberration-corrected instrument, the residual intrinsic axial aberrations are of fifth order: the fifth-order spherical aberration $C_5$ and the sixfold astigmatism $A_5$ (Müller *et al.*, 2008). For an equivalent quadrupole–octupole aberration corrector (Batson *et al.*, 2002), it is the fifth-order spherical aberration $C_5$ as well as the fifth-order star aberration $S_5$ and the fifth-order rosette aberration $R_5$ which ultimately can define the geometrical resolution limit (Dellby *et al.*, 2001). Hence, once the third-order aberrations are under control and the impact of parasitic aberrations is minimized, the resolution-limiting axial aberrations are then of fifth order.

### 8.4.3 *Balancing the effect of higher-order aberrations*

Being able to control parasitic aberrations does not necessarily mean that they have to be zeroed. For instance, in a third-order aberration-corrected instrument, one assumes that all first-, second- and third-order aberrations can be corrected and annulled. However, having the whole set of these aberrations zeroed does not mean that the instrument is in its optimal state. Once all third-order aberrations are correctable, parasitic fourth-order aberrations or the intrinsic fifth-order aberrations might become important. Provided that these higher-order aberrations are known and that they cannot directly be controlled, it can be beneficial to (partially) counterbalance higher-order aberrations by adjusting lower-order aberrations of equal symmetry to suitable values. This strategy of partial compensation is indeed very similar to the (negative) Scherzer focus which counterbalances the effect of the (positive) third-order spherical aberration. The method works equally well for anisotropic aberrations. For instance, second-order threefold astigmatism $A_2$ can partially balance the effect of fourth-order three-lobe aberrations $D_4$ (see Fig. 7.12). Although their radial dependence is different, they have the same symmetry. Hence, it is possible to balance the effect of $D_4$ within a certain $\omega$-range by a suitable choice of $A_2$. Of course, threefold astigmatism $A_2$ never fully compensates for a finite value of $D_4$, but still within a given $\omega$ or spatial-frequency range, $A_2$ can largely minimize the impact of the fourth-order aberration, which might not be adjustable at all.

Furthermore, it has already been pointed out by Scherzer (1949) that on a third-order aberration-corrected instrument, where $C_3$ can be annulled, the best setting is not given by $C_3 = 0$ but by a $C_3$ setting that considers the intrinsic fifth-order spherical aberration $C_5$. This has been realized in one of the first working (third-order) spherical aberration correctors for a dedicated scanning transmission electron microscope, where the spherical aberration is set to $-26$ $\mu$m in order to compensate the effect of the fifth-order spherical aberration $C_5$ of 63 mm (Dellby *et al.*, 2001).

### 8.4.4 *Off-axial aberrations*

Controlling all relevant axial aberrations is already a rather difficult task. However, besides the axial aberrations, another aspect that requires our attention is the validity of the isoplanatic approximation for the case of spherical aberration-corrected instruments. There are two cases where the isoplanatic approximation might fail: in case of ultra-high resolution imaging and for the case that a large field of view is needed. For these two cases, the impact of potential off-axial aberrations needs to be considered. Though there are indeed higher-order off-axial aberrations feasible (see, e.g. Gaj, 1971), the relevant off-axial aberration is the third-order coma $B_{3\,1}$, which, similar to the spherical aberration, is inherent for round lenses. Though its impact vanishes on the optical axis, it does affect image points which are not on the optical axis. The importance of off-axial aberrations increases with the required resolution for a given, but finite, field of view and with the size of the field of view.

The effect of third-order coma $B_{3\,1}$ can be minimized by matching the coma-free point of the objective lens with the coma-free point of the corrector (Rose, 1990). For a hexapole aberration corrector (Fig. 8.6), the coma-free point is located in the center of the hexapoles, while the coma-free point of the objective lens is at a distance of about $2/3\ f_o$ of the objective lens' focal length, which is of the order of a few millimeters. Obviously, it is not possible to have the center of an extended hexapole lens in such close proximity to the objective lens. Therefore, for an image aberration-corrected microscope, the coma-free point of the objective lens needs to be imaged into the coma-free point of the corrector. Analogously, for a probe aberration-corrected instrument, the coma-free point of the corrector needs to be imaged into the coma-free point of the objective lens. This condition reasons the presence of the first telescopic transfer doublet (TL11 and TL12) of the aberration corrector (see Fig. 8.6). While for the probe corrector the coma-free point in front of the objective lens is of relevance, for the image-corrected microscope it is the coma-free point behind the objective lens. Similar to the hexapole corrector (see Fig. 8.6), where a telescopic transfer doublet of round lenses is used to match the coma-free points of the objective lens and the corrector, alternatively, an astigmatic ray path through four quadrupole lenses can be used for this purpose. This is realized in an advanced quadrupole–octupole probe aberration corrector (see Krivanek *et al.*, 2009a).

However, what exactly can be minimized by matching the coma-free point of the objective lens with the coma-free point of the corrector? The third-order off-axial coma $B_{3\,1}$ has a radial (or isotropic) and an azimuthal (or anisotropic) component. The latter is caused by the Lorentz rotation the electrons undergo while being exposed to the magnetic fields of the lenses. What is actually minimized by matching the coma-free point of the objective lens with the coma-free point of the corrector is the radial component of $B_{3\,1}$ only (Rose, 1999). Indeed, while the radial component can be considered to be annulled, the azimuthal component of $B_{3\,1}$ is not fundamentally influenced by this setup (Rose, 1990; Krivanek *et al.*, 2009a).

The important point about reducing the effect of the dominant off-axial aberration, i.e. $B_{3\,1}$, is that such an optical instrument, which is partially free of coma, allows for increasing the maximum number of equally well resolved image points, and thus for enhancing the field of view. Though the impact of the radial coma vanishes, the contribution of the azimuthal coma is still present. A microscope, whose third-order off-axial coma is completely corrected, is called an *aplanatic* instrument or an *aplanator*, while a microscope whose radial coma contribution is corrected only is called a *semi-aplanatic* microscope or a *semi-aplanator* (Haider *et al.*, 2008b). The first workable hexapole aberration corrector (Haider *et al.*, 1998), as illustrated in Fig. 8.6, is a semi-aplanatic instrument.

### 8.4.5 *The chromatic aberration*

So far, we essentially assumed that a spherical aberration corrector enables control of geometrical aberrations to a certain degree, i.e. a third-order spherical aberration corrector allows for correcting axial geometrical aberrations up to third order, and for partially minimizing the effect of off-axial geometrical aberrations. Since the (temporal) information limit of a microscope does not depend on geometrical aberrations, this assumption basically implies that the information limit is unaffected by the aberration corrector. This would mean for HRTEM imaging that the modulations of the phase contrast transfer function due to the spherical aberration can be considered to be removed, such that up to the information limit (of the uncorrected microscope), direct object information can be imaged. Hence, under this assumption, the point resolution essentially equalizes the temporal information limit of the uncorrected microscope, which is defined by the chromatic aberration and the energy spread of the electron beam. This is not completely true.

Implementing a spherical aberration corrector into a transmission electron microscope means that the amount of round electron lenses, multi-poles, deflectors and stigmators increases. The result of adding such a complex electron-optical unit to a conventional electron microscope simply means that the overall chromatic aberration of the instrument increases. Regardless of probe or image aberration corrector, each additional optical unit contributes with its intrinsic chromatic aberration to the instrument's total chromatic aberration. Furthermore, because of the implementation of multi-pole units, anisotropic chromatic aberration contributions become feasible. Hence, while a spherical aberration corrector allows for reducing the effect of geometrical aberrations, it increases the incoherent aberration of the instrument, and thus decreases the information limit. In workable state-of-the-art spherical aberration correctors, the chromatic aberration of the corrected instrument is a factor of about 1.5–2.0 larger than the chromatic aberration of the uncorrected instrument. For HRTEM, this simply means that the information limit is a factor of about 1.2 larger than the information limit of the uncorrected microscope (see Eq. (2.33)). For STEM, the enhanced chromatic aberration $C_C$ essentially reduces the achievable image contrast. However, due to the increased probe current and the improved point resolution feasible with an aberration-corrected probe-forming instrument, this effect is not readily visible. Still, the increased chromatic aberration in HRTEM and STEM can be (over-)counterbalanced by working with an electron source of lower energy spread, or by employing an electron monochromator which can reduce the energy spread of the beam down to less than 100 meV.

### 8.4.6 *Dealing with the fifth-order spherical aberration*

The coherent information transfer of third-order aberration correctors is intrinsically limited by fifth-order aberrations. However, there is a way to correct the fifth-order spherical aberration $C_5$ in a third-order aberration-corrected electron microscope.

Combination aberrations can occur if two *aberrated* optical units are interacting with each other. In general, if an optical unit with an axial geometrical aberration of order $n$ and symmetry $N$ interacts with an optical unit of order $m$ and symmetry $M$, combination aberrations can occur which either have order $n+m-1$ and symmetry

$|N - M|$ or which have order $n + m - 1$ and symmetry $N + M$ (Krivanek *et al.*, 2009a). If there are two units which both have third-order spherical aberration $C_3$, i.e. $n = m = 3$ and $N = M = 0$, what results as their combination aberration is an aberration of fifth order and symmetry 0. This is fifth-order spherical aberration $C_5$.

If the coma-free point of the object lens coincides with the coma-free point of the corrector unit (see, e.g. Fig. 8.6), no fifth-order spherical aberration is produced, i.e. no combination aberration occurs. However, if the transfer lenses, which image the coma-free point of the objective lens into the coma-free point of the corrector are changed, some type of defocus occurs, i.e. the coma-free points do not match. If such a deviation occurs, the combination aberration $C_5$ between the two $C_3$ producing units, i.e. the objective lens and the corrector, appears. Hence, using the transfer optics between the objective lens and the corrector and the excitation of the objective lens, the fifth-order spherical aberration, which indeed is a residual intrinsic aberration, can be tuned such that it zeroes out. This applies to the hexapole aberration corrector as well as to quadrupole–octupole-type correctors, provided, of course, suitable transfer optics is available, which enables the flexibility of varying the transfer between objective lens and aberration corrector.

However, there is one fundamental drawback: if the coma-free plane of the objective lens is not imaged onto the coma-free plane of the corrector, the radial component of the third-order off-axial coma $B_{3\,1}$ cannot be annulled. Hence, the correction of the fifth-order spherical aberration comes at the expense of off-axial coma. The presence of off-axial coma due to the adjustment of $C_5$ reduces the number of equally well resolved image points in TEM mode. The usable field of view becomes smaller. For STEM, the scan deflectors can still be adjusted such that the larger off-axial coma is not apparent (Müller *et al.*, 2006, 2008). For this reason, the reduction of the fifth-order spherical aberration by means of employing the combination aberration of objective lens and corrector is of practical advantage only for STEM imaging (Müller *et al.*, 2008).

Image aberration correctors are usually setup such that the number of equally well resolved image points is maximal and that the third-order off-axial coma is minimized, whereas third-order spherical aberration-corrected probe-forming instruments preferably are setup such that $C_5$ is annulled. This alignment strategy, laid out by Rose (1971a), implies that the residual fifth-order spherical aberration in imaging aberration-corrected instruments remains finite.

## 8.5 Improved Correctors

The successful realization of third-order spherical aberration correctors in scanning electron microscopy (Zach and Haider, 1995), TEM (Haider *et al.*, 1998) and in STEM (Krivanek *et al.*, 1997) did not constitute the happy end of a 60-year continuous progress. It rather paved the way into a new area of aberration-corrected electron optics.

The hexapole-type aberration corrector and the quadrupole-octupole aberration corrector, as described above and illustrated in Figs. 8.6 and 8.9, respectively, essentially allow for controlling first-, second- and third-order geometrical axial aberrations. Fourth-order (parasitic) aberrations can be minimized. In the absence of information loss due to incoherent aberrations caused by instabilities, partial temporal and partial spatial coherence, fifth-order aberrations are limiting the resolution. Fifth-order aberrations are the residual intrinsic aberrations of third-order correctors. The uncorrected fifth-order aberrations are of the orders of millimeters (Uhlemann and Haider, 1998; Haider *et al.*, 2000; Dellby *et al.*, 2001) and fundamentally limit the point resolution of a 200–300 kV transmission electron microscope around 0.08-0.10 nm (see, e.g. Nellist *et al.*, 2004; Erni *et al.*, 2006). In order to overcome the limitation imposed by fifth-order aberrations, the geometrical axial aberrations of fifth order need to be corrected or sufficiently minimized such that either higher-order geometrical aberrations or incoherent aberrations become the limiting factors.

*Direct* correction of fifth-order aberrations, i.e. not an indirect correction via a secondary effect like the way $C_3$ is corrected in the hexapole-type corrector, requires optical units whose primary aberration is of fifth order. As explained above, a multi-pole with $2m$ poles has a primary aberration which is $m$-fold astigmatism. Astigmatism of symmetry $m$ is an aberration of order $n = m - 1$. Therefore, an optical unit whose primary aberration is of fifth-order is a dodecapole, i.e. a twelve-pole field (Haider *et al.*, 1982). Dodecapoles would enable correcting for geometrical axial aberrations inclusively fifth-order aberrations, similarly to the way the third-order quadrupole–octupole corrector corrects for $C_3$ by employing an astigmatic ray path through octupole fields.

### 8.5.1 *Fifth-order correction with hexapole fields*

A probe aberration corrector based on three dodecapole lenses has been proposed, and preliminary results showed that this corrector minimizes the intrinsic fifth-order sixfold astigmatism which is characteristic for hexapole-type aberration correctors (Sawada *et al.*, 2009). However, it is not the primary effect of the dodecapoles which is employed in this dodecapole corrector to minimize sixfold astigmatism $A_5$. A dodecapole can be used to generate a hexapole field which is rotatable about the optical axis. It is this flexibility which is employed in the dodecapole corrector to minimize $A_5$ (Sawada *et al.*, 2009). The actual hexapole fields in the dodecapoles have the same functionality as the hexapole fields in the hexapole corrector. Still, the strategy of employing dodecapoles allows for minimizing the residual intrinsic fifth-order aberrations.

Another way of reducing the effect of the intrinsic sixfold astigmatism was laid out by Müller *et al.* (2006), who showed that the induced sixfold astigmatism $A_5$

linearly scales with the length $\ell_{\text{HP}}$ of the hexapole

$$A_5 \propto \ell_{\text{HP}}. \tag{8.1}$$

Reducing the length of the hexapoles also affects the induced negative spherical aberration $C_3$. However, increasing the current $I_{\text{HP}}$ through the coils in the hexapole lens, i.e. increasing the strength of the hexapole, can counterbalance this effect. Expectably, the necessary current $I_{\text{HP}}$ to keep $C_3$ corrected increases with decreasing length of the hexapole

$$I_{\text{HP}} \propto \frac{1}{\ell_{\text{HP}}^{3/2}}. \tag{8.2}$$

These two relations show that without affecting the primary functionality of the spherical aberration corrector it is possible to reduce the length of the hexapole fields of a hexapole spherical aberration corrector in order to significantly reduce the residual intrinsic sixfold astigmatism $A_5$. The improved hexapole probe aberration corrector thus contains shorter hexapole lenses as well as weak quadrupole fields superimposed on both hexapole fields, which, in particular, are used to reduce fourfold astigmatism $A_3$ and fourth-order axial coma $B_4$ (Müller *et al.*, 2006, 2008). These two rather small modifications of the original design make it possible to further minimize the fourth-order aberrations as well as the residual intrinsic fifth-order aberration, i.e. sixfold astigmatism $A_5$.

Using this improved hexapole spherical aberration corrector on a 300 kV scanning transmission electron microscope, Lupini *et al.* (2009) show that the residual intrinsic sixfold astigmatism $A_5$ can be reduced to a value of less than 100 $\mu$m, whereas on the same microscope equipped with a standard hexapole corrector, i.e. with longer hexapole lenses, a value of about 1.7 mm for $A_5$ was measured. In addition, all three (parasitic) fourth-order aberrations, i.e. axial coma $B_4$, three-lobe aberration $D_4$ and fivefold astigmatism $A_4$, can be reduced by roughly one order of magnitude. Their residual values are between 1 and 2 $\mu$m (Lupini *et al.*, 2009; Erni *et al.*, 2009) and are thus not critical. Since for a probe aberration-corrected instrument the residual intrinsic fifth-order spherical aberration $C_5$ can be annulled by changing the first transfer doublet and by making use of the resulting combination aberration of the objective lens and the corrector, both of the dominant fifth-order aberrations can be minimized without having to entirely redesign the hexapole corrector. Though reducing the fifth-order spherical aberration comes at the expense of off-axial aberrations, a suitable setup of the scan deflectors between corrector and objective lens can prevent the deleterious effect of off-axial coma (Müller *et al.*, 2006).

### 8.5.2 *Fifth-order correction with quadrupole–octupole fields*

Not only the hexapole spherical aberration corrector has been advanced to minimize geometrical axial aberrations of fifth order; the quadrupole–octupole probe aberration corrector has also been brought to a level where seventh-order axial aberrations define the geometrical resolution limit (Dellby *et al.*, 2009). The improved quadrupole–octupole aberration corrector consists of 16 quadrupole lenses which are used to form the astigmatic ray path and to transfer the beam into the objective lens. The quadrupoles are arranged in four quadruplets: one quadruplet behind the condenser lens forms an elliptical beam for the first of the three octupoles (O). Thereafter follows another quadruplet of quadrupoles (QQ), which forms an elliptical beam for the second octupole. What follows behind the second octupole is another quadruplet of quadrupoles, which feeds the beam into the third octupole. Finally, the quadruplet of quadrupoles behind the third octupole transfers the beam into the objective lens (OL). Hence, coming from the condenser lens (CL), the setup of the corrector can be described as CL-QQ-O-QQ-O-QQ-O-QQ-OL. This corrector design allows for minimizing all geometrical axial aberrations up to sixth-order, such that the seventh-order axial aberrations define the geometrical resolution limit, which is about 40 pm at 100 kV and 30 pm at 200 kV, respectively.

Dellby *et al.* (2009) show that this improved quadrupole–octupole $C_3/C_5$ aberration corrector can be employed as an imaging aberration corrector, which should provide equivalent results than the hexapole spherical aberration corrector commonly used in TEM imaging.

### 8.5.3 *Correction of geometrical and chromatic aberrations*

Overcoming the limitation imposed by the fifth-order aberrations does not necessarily imply that the observable resolution of the instrument is improved. What can be corrected with the correctors described above are geometrical aberrations. As such, what is improved is the geometrical resolution limit. This is not necessarily the resolution limit of the instrument. Incoherent aberrations, like the limited coherence of the electron beam as well as mechanical and electric disturbances, can limit the resolution of an instrument. Indeed, the better the geometrical resolution, the greater the chance that small incoherent disturbances become limiting. Hence, in order to take advantage of the improved geometrical resolution provided by spherical aberration correctors, one has to tackle incoherent aberrations.

#### 8.5.3.1 *Stability*

One crucial point concerns the overall instrument stability. If the mechanical and electric disturbances exceed the optical resolution limit, the resolution is simply defined by the disturbances. In this case, improving the optics does not help to gain resolution. Hence, dedicated aberration-corrected high-resolution instruments need to be built on dedicated microscope platforms which warrant the increased stability necessary to enable an information transfer of high spatial frequencies. Of course, the ideal noise-free microscope does not exist. But the disturbances need to be small enough such that they do not (significantly) affect the optical capabilities of the instrument.

The increased number of (complex) optical elements requires a larger number of (stable) power supplies. Equivalently, the increased physical size of the instruments demands an enhanced mechanical stability. Therefore, the stability requirements of each of the sub-units increases. Furthermore, since an aberration-corrected microscope enables an enhanced optical resolution, the overall stability needs to exceed the stability of the uncorrected microscope. Having all potential sources of disturbances under control is a challenging, but mandatory task, in order to be able to benefit from the improved optical capabilities of an aberration-corrected microscope.

#### 8.5.3.2 *Spatial coherence*

Although the partial spatial coherence can be considered as a factor which can limit the information limit of an electron microscope, it is not usually the limiting factor in aberration-corrected TEM imaging. The reason for this lies in the nature of the damping envelope function due to partial spatial coherence, which is given in Eq. (2.26). This damping envelope function essentially scales with the gradient of the aberration function. Therefore, if the geometrical aberrations are corrected or can be controlled such that the aberration function $\chi$ (see, e.g. Eq. (7.14)) ideally becomes a flat function, the damping of the information transfer due to partial spatial coherence is strongly reduced. Because of this circumstance, partial spatial coherence is generally not the factor that defines the information limit in aberration-corrected microscopy. This is, of course, valid only if the brightness of the source allows for a sufficiently small beam divergence angle $\theta_{\mathrm{S}}$. This is, however, fulfilled for common field-emission electron sources.

On the other hand, as will be shown in Chapter 9, the spatial coherence and the brightness of the electron source are crucial parameters for aberration-corrected STEM imaging. Only an electron source of high brightness enables a sufficiently small effective source size, which still carries sufficient current such that the probe broadening due to the source size is not the probe-size limiting factor.

#### 8.5.3.3 *Monochromatic electron beams*

While the impact of partial spatial coherence can be reduced by working with an electron source of sufficiently high brightness, there are two ways to deal with the impact of partial temporal coherence. One way is to correct the chromatic aberration of the objective lens. The other way is to significantly reduce the energy spread of the electron beam such that the focus spread due to the chromatic aberration is small enough and does not impair the information transfer.

The energy spread of the electron beam can be reduced by employing an electron monochromator in the gun area of an electron microscope. The monochromator acts prior to the main acceleration of the electrons to their nominal energy. By making use of an energy slit in an energy dispersive plane formed by the monochromator, an electron monochromator filters out electrons of quasi-equal energy and thus reduces the characteristic energy spread of the electron source. The energy spread of Schottky field-emission electron source is of the order of 1 eV. This essentially defines the energy spread of the beam. Employing an electron monochromator, the energy spread of the beam can be reduced to values below 100 meV. This is, of course, beneficial to reduce the effect of the chromatic aberration. Yet, filtering out electrons simply means that the total beam current is reduced, often by a factor of 60 to 80%. Hence, with decreasing energy spread, the current of the beam decreases. Therefore, one has to find a compromise between the tolerable energy spread of the beam and the required beam current to obtain a certain signal-to-noise ratio in TEM and STEM images.

There are basically two electron monochromators in use. One is a double-focusing Wien filter (Tiemeijer, 1999a, 1999b) and the other is an electrostatic $\Omega$-type filter (see, e.g. Benner *et al.*, 2004). These monochromators allow for overcoming the limitation imposed by the chromatic aberration of the microscope and thus to enhance the information limit in TEM mode (Freitag *et al.*, 2004, 2005; Walther *et al.*, 2006).

In addition to these two proven electron monochromators, Krivanek *et al.* (2009b) proposed a design for a monochromator dedicated for a scanning transmission electron microscope. The monochromator is based on a curved optical setup wherein electrons follow $\alpha$-type electron trajectories. This proposed monochromator is conceptually different from the Wien-filter type and the $\Omega$-filter type monochromators mentioned above. The monochromator is located between the condenser lens and the probe-forming optics, i.e. it acts on electrons which already have their nominal energy. Furthermore, by making use of the chromatic aberration of hexapole lenses, the monochromator can also be employed to counterbalance the chromatic aberration of the objective lens. As such, the monochromator can reduce the energy width of the beam and correct the chromatic aberration in the probe-forming optics. Its realization and functionality in a scanning transmission electron microscope have to be shown. Yet, the proposed design is promising as it is planned to enable small electron probes of narrow energy spread. Such electron probes are particularly useful for analytical measurements involving electron energy-loss spectroscopy of high energy resolution (see, e.g. Erni and Browning, 2005). Furthermore, the reduction of the chromatic effect reduces the chromatic broadening of the electron probe due to the focal spread which is caused by the chromatic aberration of the objective lens.

#### 8.5.3.4 *Spherical and chromatic aberration corrector*

Besides employing a monochromatic beam to reduce the effect of the chromatic aberration on the image performance, one could aim at actually correcting the chromatic aberration of the objective lens. A strategy to correct for the chromatic aberration has already been laid out by Scherzer (1947). The idea of correcting for the chromatic aberration was further developed by Rose (1971a), who emphasized the necessity of overcoming the limitation imposed by the chromatic aberration in order to achieve atomic resolution in TEM imaging for microscopes operated between 50 and 100 kV. A first design of a chromatic and spherical aberration corrector was proposed which would enable about 100 equally well resolved image points (Rose, 1971a). Shortly after the first corrector design, an improved design was suggested (Rose, 1971b), which, apart from the chromatic and the spherical aberration, also corrects third-order off-axial coma and thus enhances the number of equally well resolved image points. Because a microscope which is equipped with such a corrector is, in terms of geometrical aberrations, an aplanatic instrument and, in terms of the chromatic aberration, an achromatic instrument, it is called an *achroplanatic* microscope (Haider *et al.*, 2008b).

An aberration corrector based on the principles of Rose (1971a, 1971b), which enables achroplanatic imaging in TEM mode, has been developed and brought to application (Kabius *et al.*, 2009). This advanced aberration corrector is based on quadrupole and octupole fields. The chromatic aberration is corrected by means of electric-magnetic quadrupole elements. The spherical aberration is corrected by octupole fields and the off-axial coma can in principle be corrected by means of four skew octupoles.

Starting with the objective lens, a telescopic transfer doublet images the coma-free point of the objective lens into the coma-free point of the corrector, analogously to the hexapole aberration corrector depicted in Fig. 8.6a. The main part of the corrector consists of two multi-pole quintuplets which are connected with each other by a telescopic transfer doublet of round lenses (Rose, 2009a). Considering schematically to have each hexapole replaced with a quintuplet of quadrupoles and octupole fields, the symmetrical setup is similar to the hexapole-type corrector, as depicted in Fig. 8.6a. Yet, the function of each quintuplet is not comparable with the function of an extended hexapole field. At the midplane of each quintuplet there is an electric-magnetic quadrupole field as well as an octupole field. These two fields are used to correct for the chromatic and the spherical aberration. In front and behind the midplane there are two magnetic quadrupoles which are essentially used to form an astigmatic ray path, or indeed an anamorphotic image in the electric-magnetic quadrupole and octupole field. Behind the second quintuplet there is a third telescopic transfer doublet which contains in its center an octupole. As there are at least three octupole fields needed to correct for all third-order geometrical aberrations, this octupole is used to correct for the resulting fourfold astigmatism of the two-step, octupole-based spherical aberration correction.

Though this corrector contains many more elements, the underlying principles, which are employed to correct for the geometrical aberrations, are identical to the ones described above for the quadrupole–octupole type corrector. Yet, the stability requirements for the individual units and the electronics are very demanding (Haider *et al.*, 2008b). One important aspect concerns the fact that the large amount of optical elements in the corrector adds a large amount of chromatic aberration to the total system. This is similar to the case of a spherical aberration corrector which adds a certain amount of chromatic aberration to the overall system. Hence, the chromatic aberration corrector must correct the chromatic aberration of the objective lens *plus* its intrinsic chromatic aberration determined by the elements of the corrector. The corrector's intrinsic chromatic aberration should at most be similar to the chromatic aberration of the objective lens. Therefore, the corrector must approximately compensate for a chromatic aberration which is twice the chromatic aberration of the objective lens.

The realization of this corrector and the successful correction of the chromatic and the spherical aberration has been shown by Kabius *et al.* (2009). Correction of the chromatic aberration and the dominant geometrical aberrations is certainly advantageous for HRTEM imaging, and particularly for atomic-resolution imaging at reduced microscope high tensions. Achromatic imaging overcomes the limitation imposed by the partial temporal coherence and, in addition, disturbing geometrical aberrations are absent. Yet, there are many aspects that can benefit from such a corrector. This concerns Lorentz microcopy, energy-filtered imaging, applications which require thick specimens, or ultra-fast microscopy (Kabius and Rose, 2008).

## 8.6 Summary

On the basis of the effect of quadrupoles, (extended) hexapole fields and octupole fields, this chapter discussed the main principles used in present-day spherical aberration correctors. Apart from the basic principles of these corrector types, aspects were addressed which are of importance for the practical handling of aberration correctors. Some of these aspects are: how to deal with off-axial aberrations; the limiting geometrical aberrations; how to deal with higher-order aberrations. Keywords of this chapter are: hexapole aberration corrector; quadrupole–octupole aberration corrector; combination aberrations; intrinsic aberrations; parasitic aberrations; misalignment aberrations; and improved aberration correctors.

# Chapter 9

# Aberration-Corrected Imaging

## 9.1 Aberration Diagnosis

The spherical aberration of a dedicated high-resolution objective lens is in the range of 0.5 to 1.5 mm. If not measured by the user (see, e.g. Spence, 1981), the spherical aberration of conventional electron microscope is often known from the manufacturer of the microscope and can be considered as a characteristic instrumental constant. The value of the twofold astigmatism that is correctable with a stigmator does not need to be known, since it can be manually corrected to a precision where its impact compared to the spherical aberration is negligible. However, on a spherical aberration-corrected microscope this approach is no longer justifiable. A spherical aberration corrector corrects the spherical aberration of the objective lens as well as residual parasitic aberrations that arise mainly because of the complex optics. The spherical aberration is an adjustable parameter and its value essentially depends on the setup of the corrector, for instance on the strength of the hexapole fields and on the diameter of the beam within the hexapoles. Moreover, though many of the parasitic aberrations can be controlled, the necessary precision to which they need to be adjusted often exceeds the feasibility of a manual alignment. Hence, considering the fact that the precision and accuracy of experimental values are always finite, what does it actually mean to have the aberrations corrected, i.e. what are the achievable values and how can these values be controlled? These type of questions are addressed by a branch of aberration-corrected electron microscopy that is called *aberration diagnosis*.

Aberration diagnosis deals with the measurement and also indirectly with the controlling of the aberration coefficients. Of course, one can set up an aberration corrector according to electron optical calculations which essentially provide for each element a nominal value of its excitation. This theoretical setup should then be adequate to nullify the correctable aberrations. However, in practice, where the symmetry and the mechanical alignment of the optical units are never ideal, the theoretical setup might not correspond fully with the real setup that actually minimizes all of the controllable aberrations. Hence, it is of crucial importance to be able to measure and quantify the aberrations of an advanced electron optical instrument.

Let us assume that the intrinsic third-order spherical aberration $C_3$ of the objective lens is of the order of 1 mm. Given this starting value for an aberration-corrected microscope, is it sufficient to have $C_3$ corrected to a value of 0.1 mm or is even a reduction to 1 pm $C_3$ necessary to consider a real microscope to be spherical aberration corrected? Indeed, the residual geometrical aberrations essentially define the geometrical resolution limit of the microscope, but there are also residual intrinsic aberrations which cannot be controlled. Hence, considering the residual intrinsic aberrations it might not be necessary to correct the spherical aberration to a level of 1 pm in order to consider the instrument 'free' of spherical aberration. Indeed, to achieve a certain resolution limit, each aberration has a tolerable range. If it exceeds this range it becomes limiting, but for any arbitrary value the aberration measures within the tolerable range, its effect can be considered to be insignificant. In order to judge on theoretical grounds whether a residual aberration is significant or whether it is within the tolerable range, we need to be able to measure the aberration coefficients accurately. The higher the required resolution, the smaller the tolerable limits and the more difficult it is to actually measure the aberration coefficients with sufficient precision. First of all, however, we need to address the question: what does resolution mean in the case of aberration-corrected microscopy? Since the resolution criteria introduced in Chapters 2 and 3 and expressed in Eqs. (2.32) and (3.11) are based on the spherical aberration $C_3$, they are not *per se* applicable to spherical aberration-corrected electron microscopy.

### 9.1.1 *Geometrical resolution limit*

The wave aberration function $\chi$ describes the deviation from a spherical wave surface. The warping of the non-spherical wave surface is due to the presence of geometrical aberrations. The unit of the aberration function $\chi$ is a length. Let us consider two electrons emerging from the same object point. Under ideal coherent imaging conditions, i.e. $\chi = 0$, they are supposed to end up in one single point which is the stigmatic image point representing the object point. However, in the presence of geometrical axial aberrations, $\chi \neq 0$. Hence, the first electron, which passes the aperture in $\omega_1$, follows an optical path which is $\Delta\chi = \chi(\omega_2) - \chi(\omega_1)$ longer than the optical path the second electron follows, which passes the aperture in $\omega_2$. If $\omega_1 \to 0$, $\Delta\chi \to \chi$. Though $\chi$ can be expressed as a length, considering the fact that electrons can be described by waves of a given wavelength, the aberration function can also be expressed in terms of the phase of the electrons. The aberration phase shift $\gamma(\omega)$, which can be written as

$$\gamma(\omega) = \frac{2\pi}{\lambda}\chi(\omega), \tag{9.1}$$

enables us to compare optical systems of different wavelengths $\lambda$. In analogy to light optics (see, e.g. Born and Wolf, 2001), the resolving power of an optical instrument limited by geometrical aberrations can be defined as the smallest value

$|\omega|$ for which the aberration function is equal to $\lambda/4$. In terms of the phase $\gamma$, the resolution requirement demands a maximum tolerable aberration phase shift of $\pi/2$ across the diameter of the aperture opening. This analysis is essentially based on Lord Rayleigh's analysis of a light optical system which is limited by spherical aberration. However, this rule is only a rough guide for what is tolerable concerning the maximum deviation from the ideal spherical wave front (Born and Wolf, 2001). In electron optics, the criterion is commonly applied to the radius of the aperture opening. Hence, the geometrical resolution limit can then be defined as the smallest value of $|\omega|$ for which the aberration phase shift is $|\gamma| \leq \pi/4$. This $\pi/4$-criterion is used to define upper limits of the individual aberration coefficients (Uhlemann and Haider, 1998; Haider *et al.*, 2000). The target resolution defines the necessary aperture angle $|\omega|$ for which the aberration phase shift should be smaller than $\pi/4$. Once this angle is known, for each aberration a maximum tolerable value can be calculated using, for example, Table 7.2.

Considering incoherent aberrations such as instabilities, partial temporal and partial spatial coherence, the geometrical resolution limit does not necessarily have to be the actual resolution limit of the microscope which can be measured. The observable resolution can be determined by incoherent aberrations. Furthermore, even in the case that the impact of incoherent aberrations is sufficiently reduced, it is not necessarily the case that the geometrical resolution limit becomes the measurable resolution limit. Experimental data are always affected by noise. Hence, apart from measuring a certain image feature, which might indicate a certain resolution, it must be warranted that this feature of interest is of statistical significance in regard to the experimental noise present (Van Aert and Van Dyck, 2006). Still, in order to optimize the optical alignment of a microscope as well as to assess the geometrical resolution of a given optical system, it is necessary to be able to measure the residual geometrical aberrations.

### 9.1.2 *Zemlin-tableau method*

The most common method employed to measure geometrical axial aberrations in transmission electron microscopy is the so-called Zemlin-tableau method. The method was developed in order to minimize the axial coma in conventional high-resolution transmission electron microscopy (Zemlin *et al.*, 1978). Though originally applied to conventional electron microscopy, the method has been expanded in order to assess all of the relevant geometrical axial aberrations which need to be considered in aberration-corrected microscopy (Uhlemann and Haider, 1998).

The impact of axial aberrations does not depend on the position of the object point. However, the impact of axial aberrations increases with the tilt of the beam in respect to the optical axis of the image forming lens. Hence, the main idea of the Zemlin-tableau method is to make use of the characteristic change of the coherent contrast transfer under tilted illumination.

A measure for the contrast transfer is a diffractogram or a power spectrum of an image of an amorphous specimen area. Such a diffractogram is simply derived via a fast Fourier transform (FFT) of the underlying real space image (Kirkland, 1998). A diffractogram can be considered to represent a two-dimensional plot of the modulus of the phase contrast transfer function[1]. In general, a diffractogram shows concentrical ellipses of high and low intensities. The spacing and sequence between these ellipses depend on the applied defocus $C_1$ and on the constant of spherical aberration $C_3$. Ellipses of vanishing intensity represent areas where the phase contrast transfer function approaches zero, while ellipses of high intensity reflect passbands of the phase contrast transfer function which are extrema, either negative or positive. Furthermore, the ellipticity reflects the twofold astigmatism $A_1$. For $A_1 \rightarrow 0$, the ellipses of constant intensity degenerate to circles. This circumstance is used to precisely correct and monitor twofold astigmatism in HRTEM imaging. Hence, from a single diffractogram, we can measure defocus $C_1$, spherical aberration $C_3$ and twofold astigmatism $A_1$ (see, e.g. Spence, 1981). The impact of other aberrations is not directly measurable from a single diffractogram. Yet, the effect of other axial aberrations becomes apparent by recording a tilt tableau, i.e. a series of diffractograms obtained from images of an amorphous object recorded under well defined tilted illuminations. A series of diffractograms recorded under constant illumination tilt but varying azimuth tilt is called a Zemlin tableau. Figure 9.1 shows a Zemlin tableau recorded on a conventional transmission electron microscope operated at 200 kV with $C_3 \approx 1.2$ mm. Though the central diffractogram of the supposedly untilted illumination shows concentrical rings, the diffractograms of the tilted condition show characteristic changes which indicate that the contrast transfer severely depends on the illumination direction. The beam tilt applied to record the tableau shown in Fig. 9.1 is roughly 4 mrad. Hence, on a conventional electron microscope of finite spherical aberration, the transfer function strongly depends on the direction of the illumination. Small deviations from the optical axis substantially impair the information transfer. This fact underlines the importance of working with a small beam divergence angle in conventional HRTEM.

What can we actually learn from such a Zemlin tableau? Recording an image under tilted illumination means that the aberration function $\chi(\omega)$ valid for the untilted illumination becomes a function $\chi(\omega + \tau)$, where $\tau$ expresses the applied illumination tilt written in the complex coordinates of $\omega$. Hence, a tilt of the illumination is simply described by an offset $\tau$ in the aperture plane behind the object. Though $\chi(\omega)$ is in general a function which has an even and an odd part, what a diffractogram or an FFT of an image plots is the even part of the function only (see footnote 1). Hence, in respect of a diffractogram analysis, the crucial component of $\chi$ is its even part.

[1]Yet, there is a distinct difference between a diffractogram and the contrast transfer function: because of the Fourier transform, a diffractogram is centrosymmetric while the actual contrast transfer function does not have this symmetry restriction.

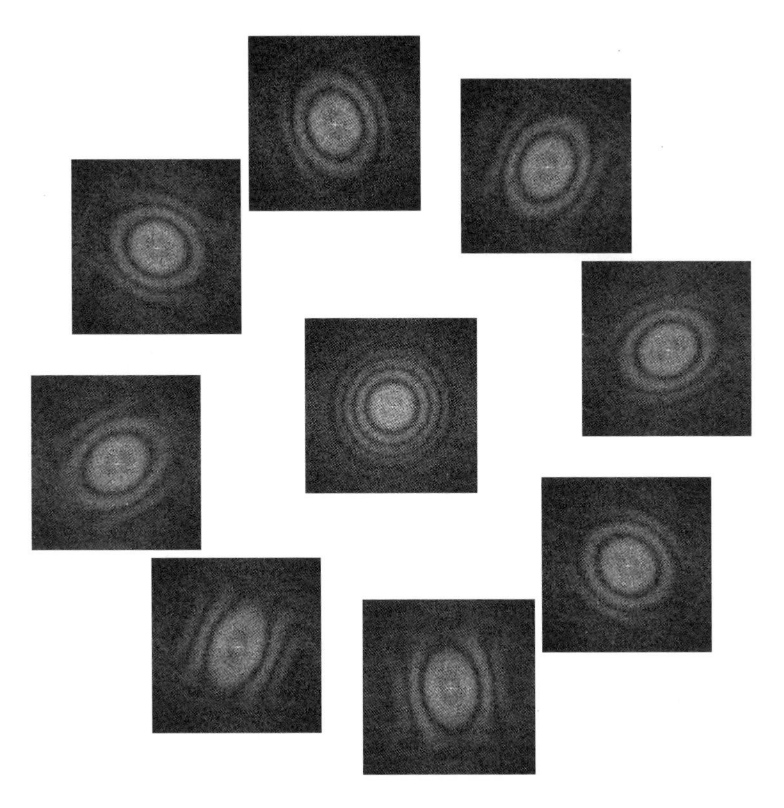

Fig. 9.1 Tilt tableau recorded on a conventional transmission electron microscope operated at 200 kV. The third-order spherical aberration of the instrument is $C_3 \approx 1.2$ mm. The tableau was recorded in underfocus. The tilt angle of the beam was $|\tau| = 4$ mrad; the azimuth angle was changed in steps of $\pi/4$, i.e. 45°.

The even part of $\chi(\omega, \tau)$, i.e. the even part of the aberration function $\chi$ under tilted illumination, denoted by a subscript $\chi_+$, can be written as (Uhlemann and Haider, 1998)

$$\chi_+(\omega, \tau) = \frac{1}{2}\chi(\omega + \tau) + \frac{1}{2}\chi(-\omega + \tau) - \chi(\tau). \tag{9.2}$$

This modified aberration function $\chi_+(\omega, \tau)$ can be written explicitly in terms of the aberration coefficients using, for example, Eq. (7.14). However, the crucial point about Eq. (9.2) is that $\chi_+(\omega, \tau)$ contains an *effective* defocus $\tilde{C}_1$ and an *effective* twofold astigmatism $\tilde{A}_1$. The tilde $\tilde{}$ shall indicate that the corresponding aberration is not the actual axial aberration but the aberration of equal symmetry measured under tilted illumination. The effective defocus and the effective twofold astigmatism reflect the tilt-induced defocus and the tilt-induced twofold astigmatism. These are the measurable quantities which reflect the axial aberrations. Hence, by measuring the effective first-order aberrations, higher-order axial aberrations can be deduced.

Considering axial aberrations up to fifth order, the effective defocus $\tilde{C}_1$ can be written as (Uhlemann and Haider, 1998)

$$\begin{aligned}\tilde{C}_1(\tau) &= C_1 + 4\,\Re\left(\tau B_2\right) + 2\tau\bar{\tau}C_3 + 6\,\Re\left(\tau^2 S_3\right) \\ &+ 12\,\Re\left(\tau^2\bar{\tau}B_4\right) + 8\,\Re\left(\tau^3 D_4\right) + 3\left(\tau\bar{\tau}\right)^2 C_5,\end{aligned} \tag{9.3}$$

and similarly, the effective twofold astigmatism $\tilde{A}_1$ is given by

$$\begin{aligned}\tilde{A}_1(\tau) &= A_1 + 2\tau\bar{B}_2 + 2\bar{\tau}A_2 + \tau^2 C_3 + 6\tau\bar{\tau}\bar{S}_3 + 3\bar{\tau}^2 A_3 \\ &+ 2\tau^3 B_4 + 6\tau^2\bar{\tau}\bar{B}_4 + 12\tau\bar{\tau}^2 D_4 + 4\bar{\tau}^3 A_4 \\ &+ 2\tau^3\bar{\tau}C_5 + 5\bar{\tau}^4 A_5,\end{aligned} \tag{9.4}$$

where all the aberration coefficients $X_n$ are the axial aberrations for the untilted illumination. The equations for $\tilde{C}_1$ and $\tilde{A}_1$ provide the necessary information to deduce all the axial aberration coefficients contained in the equations, provided of course that a suitable large set of diffractograms is available to determine the effective defocus and the effective twofold astigmatism. Each diffractogram of a given tilt $\tau$ provides measures for $\tilde{C}_1(\tau)$ and $\tilde{A}_1(\tau)$. Having a list of $\tilde{C}_1(\tau)$ and $\tilde{A}_1(\tau)$ for various tilt values $\tau$, the axial aberration coefficients contained in Eqs. (9.4) and (9.5) can be used as fitting parameters. The best fit provides the set of experimentally determined axial aberration coefficients. According to Eqs. (9.4) and (9.5), a microscope free of geometrical aberrations should show a tilt tableau whose diffractograms are invariant and do not dependent on the applied beam tilt.

Although the actual quantification of the individual aberration coefficients is essential for the proper alignment of a microscope, the equations above can also be employed for a qualitative interpretation of a Zemlin tableau. Provided that only a few aberrations dominate the information transfer, the impact of certain axial aberrations can be read directly from the characteristics of the diffractograms in the tableau. The effect of twofold astigmatism $A_1$, axial coma $B_2$, threefold astigmatism $A_2$, spherical aberration $C_3$, star aberration $S_3$ and fourfold astigmatism $A_3$ are illustrated in Fig. 9.2. Of course, depending on the ratio between the real and imaginary part of the complex aberration coefficients, the characteristics of the corresponding tableau can rotate in azimuthal direction.

Without aiming at quantifying the aberrations apparent in the tableau shown in Fig. 9.1, on the grounds of Eqs. (9.4) and (9.5) the tableau can be analyzed qualitatively in order to estimate which of the aberrations are present. The diffractogram of the untilted beam in Fig. 9.1 shows concentrical rings of high and low intensity. This indicates that the twofold astigmatism $A_1$ is properly minimized. If the information transfer is affected by $A_1$, it would show up in the untilted as well as with equal strength in all the tilted diffractograms (see Fig. 9.2). Hence, $A_1$ is not dominant. On tilting the beam, the diffractograms show ellipses instead of circles. The ellipses are directed mainly towards the center of the tilt axis. This corresponds to the tableau characteristic in the presence of spherical aberration $C_3$, which contributes substantially to the effective defocus and the effective twofold

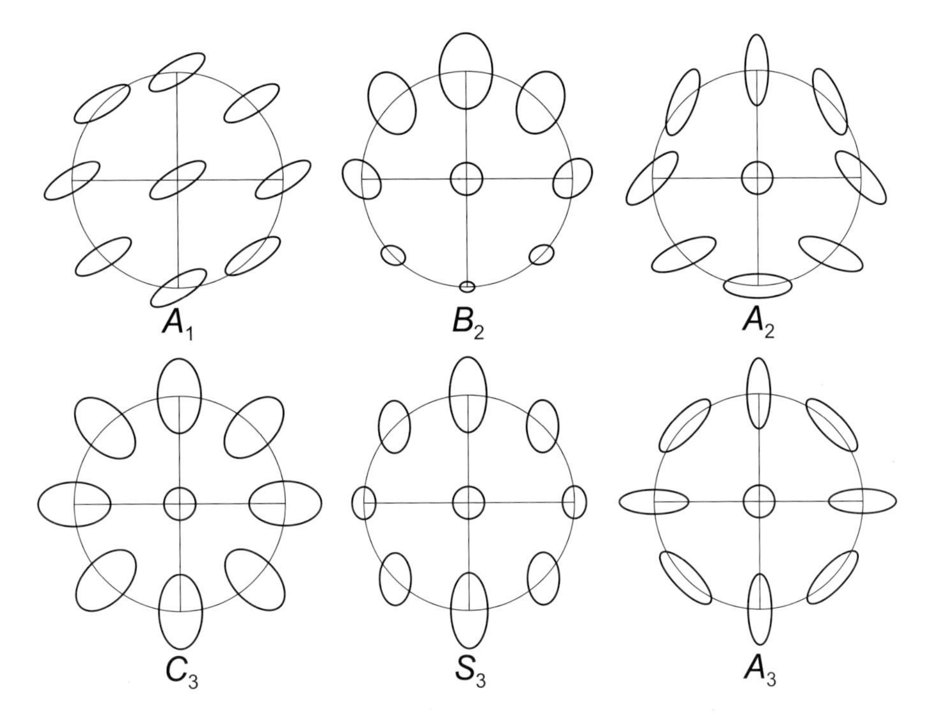


Fig. 9.2 Schematic representation of Zemlin-tableau characteristics for the axial aberrations up to third order. The impact of first-order aberrations, such as defocus and twofold astigmatism, does not depend on the tilt angle. For aberrations of order $n \geq 2$, such as axial coma $B_2$, threefold astigmatism $A_2$, spherical aberration $C_3(> 0)$, star aberration $S_3$ and fourfold astigmatism $A_3$, there is no elliptical distortion observable for the untilted case. Only by applying an illumination tilt is the characteristic distortion of each of the aberrations observable. Though the tableaux shown represent $B_2$, $A_2$, $C_3$, $S_3$ and $A_3$, they are indeed equally valid for higher-order aberrations of equal symmetry, like $B_4$ for the tableau indicated with $B_2$, $D_4$ for the $A_2$ tableau, $C_5$ for the $C_3$ tableau, $S_5$ for the $S_3$ tableau and, for example, $R_5$ for the $A_3$ tableau.

astigmatism. Yet, the diffractograms at 4 through 7 o'clock position seem to be closer to the Gaussian focus than the diffractograms on the opposite side of the tableau. This clearly indicates the presence of axial coma $B_2$. Hence, from a simple visual inspection of the tilt tableau shown in Fig. 9.1, we can conclude that the coherent information transfer of the microscope underlying this particular Zemlin tableau is affected by spherical aberration $C_3$ and by axial coma $B_2$. This behavior can be expected for a conventional electron microscope. Threefold astigmatism, as another potential aberration of a conventional electron microscope, cannot readily be identified from the tableau shown in Fig. 9.2. This does not mean that it is absent. Threefold astigmatism might be present but the imaging characteristics are clearly dominated by spherical aberration and axial coma.

Making the change from the characteristic tilt tableau of a conventional electron microscope, as shown in Fig. 9.1, to one recorded on an aberration-corrected microscope operated at 80 kV, Fig. 9.3 reveals that the diffractograms of an

Fig. 9.3 Zemlin tableau recorded with an aberration-corrected electron microscope operated at 80 kV. The tilt tableau angle is 32 mrad. Small elliptical distortions can still be observed, yet the information transfer hardly depends on the tilt angle. None of the features characteristic for having a dominant third-order spherical aberration can be found (for comparison see Fig. 9.2).

aberration-corrected microscope hardly depend on the applied illumination tilt. Although minor elliptical distortions are observable in the diffractograms recorded under tilted illumination, considering the fact that the tableau was recorded with a beam tilt of 32 mrad (compared to 4 mrad in Fig. 9.1), it is clear that the optical behavior of an aberration-corrected microscope is less dependent on the applied illumination tilt. This reflects that the aberration corrector effectively corrects the third-order spherical aberration $C_3$ of the objective lens, and that parasitic aberrations are also largely absent.

In conclusion, from the analysis of the effective defocus and the effective twofold astigmatism in a series of diffractograms of a Zemlin tableau, axial aberration coefficients can be quantitatively determined. Provided one or two axial aberrations are dominant (regardless of whether this is an aberration-corrected or a conventional electron microscope), the impact of such aberrations is directly reflected in the main characteristics of the tableau. This enables a qualitative evaluation of a tilt tableau.

### 9.1.3 *Tilt tableaux for probe-forming microscopes*

Consider a STEM imaging microscope which suffers from twofold astigmatism. Provided that a certain defocus is applied, twofold astigmatism leads to a twofold distortion of the electron probe. Being, however, close to the Gaussian focus, the electron probe is round, but not as small as it could get if the twofold astigmatism was absent (see Fig. 7.6). Recording a STEM micrograph close to Gaussian focus would thus yield a micrograph of somewhat lower resolution compared to what would be possible with a properly stigmated electron probe. Yet, the resolution in the micrograph would be isotropic. However, if a certain overfocus ($C_1 > 0$) is applied, a micrograph is recorded which shows a distinct resolution loss in one particular direction, i.e. in the direction the electron probe is elongated. The resolution loss reflects the impact of the twofold distortion of the electron probe caused by twofold astigmatism. Recording a micrograph in underfocus condition ($C_1 < 0$) would yield a similar micrograph for which, however, the resolution loss is observable in perpendicular direction. This simply reflects the fact that the twofold distortion of the electron probe is opposite for under- and overfocus, respectively.

Let us now assume that we image with this astigmatic STEM probe small spherical particles dispersed on a thin film. In overfocus, the particles would be imaged as ellipses, with their major axes all pointing in one particular direction. In underfocus, the particles would be imaged as ellipses with their major axes perpendicular to the overfocus condition. Comparing these two defocused images with the isotropic micrograph obtained near Gaussian focus, a deconvolution can be applied in order to extract the change in the probe shape by over- and underfocusing the lens. Or, alternatively, the function needs to be founds which blurs the isotropic Gaussian image such that the convolution of the blurring function with the Gaussian image yields the defocused micrograph. The blurring function, obtained from the deconvolution, represents the electron probe for the particular defocus applied. If twofold astigmatism is present, the blurring function is elliptical. The main axis of the ellipse representing the electron probe in overfocus condition is perpendicular to the main axis of the ellipse representing the electron probe in underfocus condition. Hence, this type of deconvolution yields characteristic images of the electron probe for under- and overfocus conditions. Knowing the applied defocus used to record the under- and overfocused images as well as the elongation of the electron probe in under- and overfocus conditions, the twofold astigmatism $A_1$ can be quantified (Uno *et al.*, 2005).

Let us now assume that the microscope is free of twofold astigmatism. Again, recording an image supposedly close to the Gaussian focus yields an isotropic micrograph of the round particles. Since there is no astigmatism, recording an image in over- and underfocus condition yields isotropic micrographs as well. Yet, because of the defocus, the resolution would be lower compared to the micrograph recorded near Gaussian focus. Performing a deconvolution to obtain the probe profiles for under- and overfocus conditions yields two electron probes characteristic for the particular defocus applied. The supposedly Gaussian image represents the reference point of the focus. The over- and underfocused images are recorded in equal focal distances from this reference point. Hence, if the applied under- and overfocus are identical and the supposedly Gaussian image is strictly recorded at the actual Gaussian focus, the probe profiles of the over- and underfocus conditions have equal size. If, however, the supposedly Gaussian image is not recorded at the actual Gaussian focus, the over- and underfocused electron probes have different sizes. Indeed, the difference in probe size of the over- and underfocused electron probe would indicate where to find the real Gaussian focus, and one could determine the actual (finite) focus $C_1$ of the reference point (Uno *et al.*, 2005).

Hence, from comparing a reference image which is recorded close to Gaussian focus with one recorded at overfocus and one recorded at underfocus the actual defocus the reference image was recorded, as well as the twofold astigmatism of the electron probe, can be determined. This method thus allows measuring of $C_1$ and $A_1$ of a STEM (or SEM) microscope (Uno *et al.*, 2005).

Considering that the Zemlin-tableau method is based on measuring the effective defocus $\tilde{C}_1$ and the effective twofold astigmatism $\tilde{A}_1$ under tilted illumination, performing STEM imaging under tilted illumination, the method measuring of the defocus and twofold astigmatism of a STEM microscope can be adapted to measure the variety of geometrical axial aberrations in the same way that this is done in the Zemlin-tableau method for HRTEM. Hence, one needs to record a reference image which is close to Gaussian focus. Then, one records a series of micrographs under well-defined tilted illumination in underfocus condition and a corresponding series of micrographs under tilted illumination in overfocus condition. For each pair of under- and overfocused micrographs recorded under the same tilt, the effective defocus $\tilde{C}_1$ and the effective twofold astigmatism $\tilde{A}_1$ can be determined by comparing them with the reference image recorded under untilted illumination. Similar to the HRTEM method, one ends up with a list of effective defocus values and effective twofold astigmatism values, which, by employing Eqs. (9.4) and (9.5), can be used to determine the geometrical axial aberration coefficients.

The Zemlin-tableau method modified for STEM imaging is used for the aberration diagnosis of the hexapole spherical aberration corrector (Müller *et al.*, 2006)[2]. The crucial point about this method is the deconvolution of the probe intensity profiles from a pair of micrographs — one recorded under defocused, tilted illumination and one recorded in untilted conditions at the reference focus. Yet, compared to HRTEM, two tilt tableaux are necessary — one recorded in underfocus condition and one recorded in overfocus condition.

[2]The method, which is sketched in the text, is patented: J. Zach. *Method for detecting geometrical-optical aberrations.* WO 01.56057, 2001. J. Zach. *Method for detecting geometrical-optical aberrations.* US2003/0001102, 2003.

*Practical aspects of tilt-tableau methods*

The Zemlin-tableau method and the tilt-tableau method for the aberration diagnosis of a TEM and a STEM instrument, respectively, are both based on the assessment of the tilt-induced defocus $\tilde{C}_1$ and the tilt-induced twofold astigmatism $\tilde{A}_1$. Though both methods are robust and can provide aberration coefficients of high precision, the accuracy of the methods depends strongly on their application. The user of the microscope is essentially free in setting the induced beam tilt which is applied to record a tilt tableau. For a given set of parameters (beam tilt, magnification, electron energy), the software measures $\tilde{C}_1$ and $\tilde{A}_1$ for each diffractogram of the tableau. One obtains a list of $\tilde{C}_1$- and $\tilde{A}_1$-values as a function of the (complex) beam tilt $\tau$. The software then determines a set of aberration coefficients on the basis of a fit such that the set of aberration coefficients is in agreement with the tableau characteristics. However, it is in the setup of the tableau acquisition where one has to be a bit careful.

First of all, one has to realize that the impact of lower-order aberrations (like first- and second-order aberrations) is already apparent at small tilt angles, while in general higher-order aberrations (third-order and higher) are only recognizable at increased tilt angles. This is expressed in Eqs. (9.4) and (9.5), where the angular dependence of each aberration on $\tilde{C}_1$ and $\tilde{A}_1$ is shown. Hence, recording a tilt tableau with a small tilt angle (like 5 to 10 mrad), the influence of higher-order aberrations is small. It might happen that their effect is fully overruled by the impact of residual lower-order aberrations. Hence, in order to measure (and optimize) higher-order aberrations, one needs to apply a sufficiently large beam tilt in order to enhance their contribution to $\tilde{C}_1$ and $\tilde{A}_1$. Furthermore, as the impact of residual lower-order aberrations can dominate the effect of higher-order aberrations, it is of crucial importance to minimize lower-order aberrations prior to the precise assessment of higher-order aberration.

For this reason, one should first aim at minimizing lower-order aberrations employing small and medium beam-tilt angles (10–20 mrad). Once the values of the lower-order aberrations reach the achievable measuring precision, which for instance depends on the signal-to-noise ratio of the underlying micrographs, the field of view as well as on the quality of the specimen, the tilt-angle can be increased ($\gtrsim 25$ mrad) such that the effect of higher-order aberrations can be made visible in the tableau characteristics. If possible and needed, action can then be taken to minimize or adjust higher-order aberrations, like, for example, $C_3$, $S_3$ or $A_3$.

As a rule of thumb, we can lay out the following strategy for the fine-tuning of an aberration corrector. Firstly, one needs to minimize the effect of lower-order aberrations such as $A_1$, $B_2$, and $A_2$, employing small and medium beam-tilt angles. In a second step, the tilt angle can be increased such that higher-order aberrations can be measured and adjusted. Repeating these steps, one should iteratively approach the desired state of correction. If the first step is not fulfilled, it is not possible to derive accurate values for the higher-order aberrations.

The important point about a tableau analysis is that one needs to apply tableau settings which actually make it possible to have significant contributions in the tableau of the aberration coefficients one wants to measure. Analyzing a tilt-tableau recorded with a maximum tilt angle of less than 10 mrad in order to assess fifth-order aberrations might provide the desired small values of these coefficients. Yet, the analysis might not be reliable since the impact of the fifth-order aberrations might be negligibly small compared to the impact of the lower-order aberrations on the tableau characteristics.

### 9.1.4 *Ronchigram methods*

Apart from the methods based on the analysis of Zemlin tableaux, for the alignment of STEM imaging systems so-called Ronchigram-based methods are employed. A *Ronchigram* is the shadow image of a specimen, which can be observed in the bright-field diffraction disk of a convergent electron diffraction pattern. The Ronchigram directly reflects the characteristics of the specimen and the characteristics of the electron probe. In fact, it reflects the phase shifts of the illuminating electron beam within the probe forming aperture. The Ronchigram is widely used to manually optimize the electron probe in scanning transmission electron microscopy of conventional (James and Browning, 1999) and aberration-corrected instruments. The phase shifts, which are due to the presence of geometrical axial aberrations, lead to characteristic distortions in the Ronchigram. Indeed, considering the Ronchigram to correspond to a real image of the specimen, the axial geometrical aberrations of the illumination system reveal themselves comparable to the image aberrations due to off-axial aberrations in TEM imaging.

Figure 9.4 shows two Ronchigrams recorded with STEM probes largely underfocused on a silicon crystal in a $\langle 110 \rangle$ zone-axis orientation. Figure 9.4a was recorded on a conventional probe-forming microscope operated at 200 kV with $C_3 \approx 0.6$ mm, and Fig. 9.4b was recorded on an aberration-corrected instrument operated at 300 kV with $C_3 \approx 1\,\mu$m. The impact of the phase shifts caused by the axial geometrical aberrations, effective in the plane of the probe forming aperture, is apparent. While the presence of the spherical aberration leads to severe image distortions of the crystal planes of the silicon specimen in the Ronchigram (Fig. 9.4a), the aberration-corrected microscope enables the formation of a Ronchigram which indeed is very similar to a conventional atomic-resolution TEM micrograph. The absence of substantial phase shifts in the aperture plane of the aberration-corrected microscope reduces the distortions of the specimen information contained in the Ronchigram.

However, what exactly is the difference between the two Ronchigrams shown in Fig. 9.4? Starting in the center of the Ronchigrams, both Ronchigrams reflect the silicon crystal lattice. The crystal lattice seems to be imaged with lower resolution in Fig. 9.4a. Moving in radial direction, the magnification of the Ronchigram in

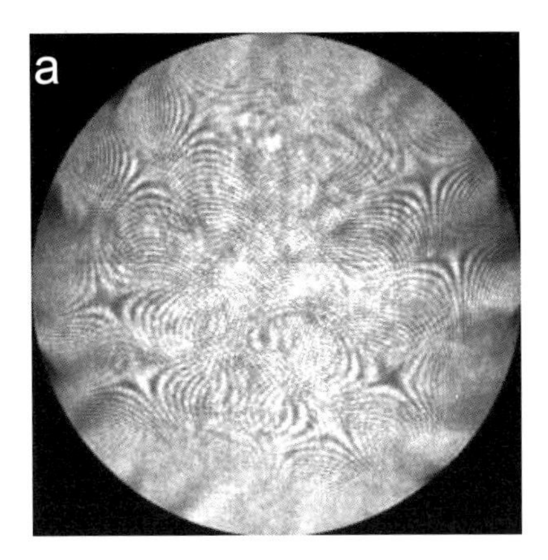

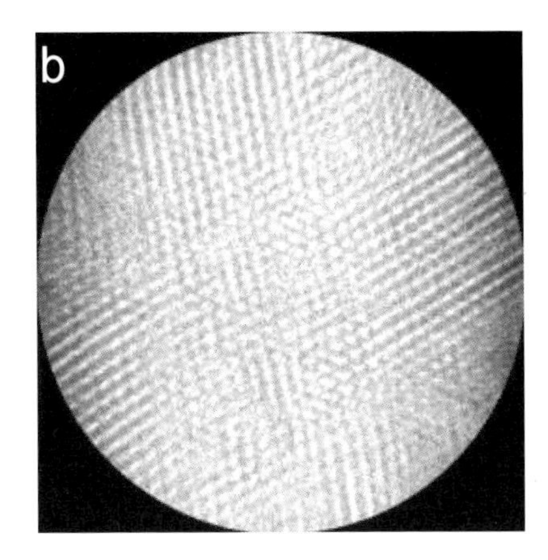


Fig. 9.4 Ronchigrams recorded with an underfocused electron probe on a silicon ⟨110⟩ specimen. (**a**) was recorded on a 200 kV microscope with $C_3$ = 0.6 mm, and (**b**) was recorded on an aberration-corrected STEM instrument operated at 300 kV with $C_3 \approx 0$. The diameter of both diffraction disks is about 50 mrad.

Fig. 9.4a seems to increase, while the magnification of the silicon lattice in Fig. 9.4b seems to be nearly constant across the entire field of view. At a radial distance of about two-thirds of the diffraction disk, the magnification of the silicon lattice in Fig. 9.4b seems to approach a maximum, from where it decays towards the edge of the diffraction disk. The maximum magnification reflects the circle of infinite magnification (see, e.g. James and Browning, 1999), which is typical for a defocused electron probe suffering from spherical aberration.

Hence, without going into the details of the Ronchigram alignment, what can be concluded from comparing the two Ronchigrams shown in Fig. 9.4 is that aberration phase shifts in the aperture plane of the probe-forming instrument lead to distortions in the shadow image observable in the bright-field diffraction disk. These distortions lead to local changes of the magnification. The variation of the local magnification in the shadow image are directly related to the aberration function $\chi(\omega)$ (Dellby *et al.*, 2001). With the complex coordinate in the aperture plane $\omega = \theta_x + \mathrm{i}\theta_y$, the local magnification $\mathbf{M}$ is given by

$$\mathbf{M} = \frac{D}{\lambda}\begin{pmatrix} \dfrac{\partial^2\chi}{\partial\theta_x^2} & \dfrac{\partial^2\chi}{\partial\theta_x\partial\theta_y} \\ \dfrac{\partial^2\chi}{\partial\theta_x\partial\theta_y} & \dfrac{\partial^2\chi}{\partial\theta_y^2} \end{pmatrix}^{-1}, \tag{9.5}$$

where $D$ is the effective camera length employed to record the Ronchigram and $\lambda$ is the wavelength. Equation (9.5) in combination with the aberration function in Eq. (7.14) reveals that by changing the defocus $C_1$, the overall magnification of the Ronchigram changes homogeneously. By defocusing the electron probe on the specimen, the field of view of the shadow image inside the diffraction disk can be adjusted. Other axial aberrations lead to a characteristic change of the local magnification of the shadow image.

Hence, being able to measure the local magnification of the shadow image of the specimen within the Ronchigram makes it possible to deduce the aberration coefficients which define $\chi(\omega)$. However, compared to the Ronchigrams shown in Fig. 9.4, the Ronchigram-based evaluation of the axial aberration coefficients is normally carried out on amorphous specimens which show distinct features, like particles, that can be evaluated in order to measure the local magnification[3]. Indeed, instead of analyzing the local magnification in a single Ronchigram, several Ronchigrams are recorded, each with a well-determined beam shift. By comparing the different Ronchigrams, the shifts of the individual image features can be related to the applied beam shift, from where the local magnification of the Ronchigram can be determined (Dellby *et al.*, 2001).

### 9.1.5 *Other methods*

The methods mentioned above, i.e. the Zemlin-tableau method for HRTEM imaging, the tilt-tableau method for STEM and the Ronchigram-based method, are the most common methods used to measure the geometrical aberration coefficients in transmission electron microscopy. However, in principle, any imaging effect, which in a well definable and distinguishable manner depends on the aberrations, can be employed for an aberration diagnosis. For instance, tilting the illumination in TEM mode also leads to an effective image shift $\tilde{A}_0$, which can be employed to determine the aberration coefficients. In an equivalent way, this method was adopted in the first workable probe aberration corrector on a dedicated scanning transmission electron microscope (Krivanek *et al.*, 1997). Prior to the era of aberration correction, the beam-tilt induced image shift was employed to measure the spherical aberration of transmission electron microscopes (Koster and de Jong, 1991).

A distinctly different method was developed by Ramasse and Bleloch (2005). This method employs the shadow image in the bright field disk of a known crystal in order to measure geometrical axial aberration coefficients of symmetry $N > 0$ of a probe-forming instrument. The method exploits the fact that information contained in a so-called achromatic pattern does not change with defocus. Recording and integrating the Ronchigram over a certain defocus range thus allows for extracting the achromatic pattern as the only feature which is robust against defocus change. The intensity along well defined lines within the achromatic pattern characteristically changes with the geometrical aberrations of the instrument. Measuring the intensity distribution along a few lines allows for quantifying the aberration coefficients with high accuracy (Ramasse and Bleloch, 2005).

[3]The evaluation of the aberrations based on the analysis of the local magnification in Ronchigrams is patented: O.L. Krivanek, N. Dellby, and A.R. Lupini. *Autoadjusting charged-particle probe-forming apparatus.* US6552340, 2003.

The last method mentioned in this context is another Ronchigram-based method, which, as a matter of fact, is employed to align probe-forming instruments (Mitsuishi *et al.*, 2006). In contrast to the Ronchigram method of Ramasse and Bleloch (2005), the method of Mitsuishi *et al.* (2006) enables the measurement of aberrations of symmetry $N = 0$ only, i.e. defocus $C_1$, third-order spherical aberration $C_3$ and fifth-order spherical aberration $C_5$. The method is based on the analysis of the spacing of interference fringes in a Ronchigram recorded on a crystalline specimen. From the spacing of the fringes, information about the local magnification is obtained which can be used to derive the isotropic axial aberrations.

The importance of aberration-diagnosis methods cannot be overestimated. Aberration diagnosis is of crucial importance for practical aberration-corrected electron microscopy. While in an uncorrected electron microscope it is the spherical aberration $C_3$ which is invariant and dominates the information transfer, for spherical aberration-corrected microscopes, in principle, any geometrical aberration can become decisive. If all the controllable aberrations are sufficiently small, what remain are the residual intrinsic aberrations of the instrument. This represents the ideal alignment status of a microscope. However, in practice it is necessary to monitor all possible parasitic and symmetry-allowed aberrations in order to verify that the state of correction fulfills the resolution requirement of the experiment. Though it is of practical importance to know which aberration limits the point resolution, aberration diagnosis software is fundamental for any kind of auto-alignment routines. Moreover, high-performance diagnosis algorithms are essential for precise auto-alignment routines. Such routines enable the daily fine-tuning of correctors, which, as a matter of fact, is fundamental for aberration-'corrected' electron microscopy.

## 9.2 Aberration-Corrected HRTEM

The image formation process in HRTEM imaging was discussed in detail in Chapter 2. This was done for the case of conventional transmission electron microscopes which are not equipped with a spherical aberration corrector. With some exceptions, which are mostly related to the fact that setting $C_3 = 0$ can lead to meaningless expressions, most of the concepts described in Chapter 2 can be applied to aberration-corrected phase contrast imaging as well. The crucial point is that the aberration function given in Eq. (2.18) needs to be replaced by the generalized axial aberration function given in Eq. (7.14).

The figure of merit of phase contrast transmission electron microscopy is the phase contrast transfer function. For the case of conventional electron microscopes, its coherent part is determined by the wavelength of the electrons and, most importantly, by the aberration function given by defocus $C_1$ and third-order spherical aberration $C_3$. The first zero-crossing of this transfer function at Scherzer focus provides a measure for the point resolution of the microscope. The point resolution of a conventional microscope is determined by the wavelength and the constant of spherical aberration $C_3$. The incoherent contributions to the phase contrast transfer function, namely the envelope functions due to partial temporal and partial spatial coherence, determine the information limit. While the envelope function due to partial temporal coherence depends on the defocus spread, which is proportional to the constant of chromatic aberration, the envelope function due to partial spatial coherence depends on a derivative of the aberration function $\chi$. Since both the coherent part of the phase contrast transfer function and the envelope function due to partial spatial coherence depend on the aberration function, it can be expected that these are the quantities which strongly depend on whether or not an instrument is equipped with a spherical aberration corrector.

### 9.2.1 *The coherent phase contrast transfer function*

The coherent part of the phase contrast transfer function is given by

$$t_c(\omega) = \Im\left[\exp\left\{-\frac{2\pi i}{\lambda}\chi(\omega)\right\}\right] = \sin\left\{-\frac{2\pi}{\lambda}\chi(\omega)\right\}. \tag{9.6}$$

This expression is identical to Eq. (2.19), with the small exception that instead of using the isotropic aberration function given in Eq. (2.18), the general expression for the axial aberration function in Eq. (7.14) needs to be employed, i.e. for the case that the third-order spherical aberration is small, additional axial aberrations need to be considered in the information transfer process.

However, though the similarity between the corrected and the uncorrected setup is obvious, a fundamental question needs to be resolved: what does it actually imply to do phase contrast imaging with an aberration-corrected microscope? Ignoring the residual intrinsic aberrations of the instrument, which, under the assumption that incoherent effects are sufficiently small, would be limiting the point resolution, aberration correction simply means that the aberration function is vanishingly small, i.e. $\chi \to 0$. Regardless of the characteristics of incoherent effects, with $\chi = 0$, the coherent phase contrast transfer function given in Eq. (9.6) becomes zero. Unless special phase plates are employed in the path of the beam (Majorovits *et al.*, 2007; Gamm *et al.*, 2008; Danev *et al.*, 2009), phase contrast imaging is not feasible in an aberration-*free* microscope. The only contrast mechanism that enables the formation of an image with $\chi = 0$ is amplitude contrast. However, particularly true for thin specimens, it is the phase of the electron wave at the exit plane of the specimen which carries essential information about the specimen (see, e.g. Hawkes and Kasper, 1994). The phase contrast mechanism is important for atomic-resolution TEM imaging. Therefore, in order to translate the phase information of the exit-plane wave into an image, one has to 'design' a phase contrast transfer function in an aberration-corrected microscope. This can be done by a suitable adjustment of some of the geometrical aberrations. Hence, $\chi$ becomes finite again. The goal, however, is to modulate $\chi$ in such a way that the corresponding phase contrast transfer function provides an image contrast which directly reflects the main characteristics of the specimen with maximum resolution. Of course, one could in principle use any of the axial aberrations in order to induce phase contrast. It is, however, of obvious advantage to warrant an isotropic information transfer. Therefore, the aberrations which should be employed to design the phase contrast transfer function are the isotropic axial aberrations: defocus $C_1$ and any of the $n$th-order spherical aberrations $C_n$ $(n = 3, 5, 7, ...)$.

Of course, $\chi = 0$ implies that the defocus $C_1$ is annulled. Hence, by simply applying a certain defocus, $\chi$ becomes finite. This *re*activates the phase contrast mechanism, even if the spherical aberration $C_3 = 0$ (Chen *et al.*, 2004). Adjusting the defocus in order to induce phase contrast is certainly a viable method. Yet, we have to consider that the actual goal of designing the transfer function is to form a phase contrast transfer function which allows for transferring the widest possible passband warranting a simple transfer for a large spatial frequency range. With an aberration-corrected microscope, $C_3$ can be annulled. However, the third-order spherical aberration $C_3$ can also be set to finite values, positive and negative. Hence, apart from defocus $C_1$, the third-order spherical aberration $C_3$ represents a second parameter which can be used to suitably model the phase contrast transfer function. The adjustment of $C_3$ is particularly useful for the case that the fifth-order spherical aberration $C_5$ cannot be corrected. As we saw in the previous chapter, $C_5$ is a potential residual intrinsic aberration of a third-order aberration-corrected instrument. Normally, the residual intrinsic fifth-order spherical aberration of a third-order aberration-corrected instrument is of the order of a few millimeters. It can only be adjusted (or even annulled) at the expense of increasing the off-axial coma. This reduces the number of equally well resolved image points[4]. Therefore, the main idea of designing an optimized phase contrast transfer function on a third-order spherical aberration-corrected microscope is to suitably adjust defocus $C_1$ and the third-order spherical aberration $C_3$ such that, in combination with the intrinsic fifth-order spherical aberration $C_5$, a broad passband can be realized which enables a simple visualization of the object over a large spatial frequency range. The fact that it is advantageous for phase contrast imaging to set the third-order spherical aberration $C_3$ to a finite value in a third-order spherical aberration-corrected

[4]An aberration corrector is tuned for a certain nominal excitation of the objective lens. The location of the coma-free point of the objective lens is determined by the excitation of the lens. Changing the excitation of the lens to focus the specimen thus also implies that the coma-free point of the lens is moved. As a consequence of the focus adjustment, the coma-free point no longer matches with the coma-free point of the corrector. This means that the instrument is no longer optimized in respect of coma. Furthermore, since by defocusing the coma-free point of the objective lens in respect to the coma-free point of the corrector, the fifth-order spherical aberration can be adjusted, a significant deviation from the objective lens' nominal focus leads to a change of $C_5$. Hence, when working with an aberration corrector it is important to work close to the objective lens' nominal excitation for which the coupling optics to the corrector is optimized. Focusing of the specimen should primarily be performed by changing the height of the specimen. Only minor focus adjustments (clearly less than 1 $\mu$m) should be done by changing the excitation of the lens. This applies to aberration-corrected TEM and STEM.

microscope has already been noticed by Scherzer (1949). Similar to the Scherzer focus in Chapter 2, there are conditions that allow for optimizing the phase contrast transfer function of an instrument whose $C_3$ is adjustable and whose $C_5$ is fixed. An optimized phase contrast transfer function is obtained if defocus and third-order spherical aberration are set to (Scherzer, 1970)

$$C_{1\,\mathrm{Scherzer,\,corr}} = a_1 \sqrt[3]{\lambda^2 C_5}$$
$$C_{3\,\mathrm{Scherzer,\,corr}} = -a_2 \sqrt[3]{\lambda C_5^2}\,. \tag{9.7}$$

According to Scherzer (1970), the numerical parameters $a_1$ and $a_2$ are $a_1 \approx 2$ and $a_2 \approx 3.2$. Similar values were deduced by Chang *et al.* (2006), who found $a_1 = 1.56$ and $a_2 = 2.88$. Though the numerical difference between the two sets of $(a_1, a_2)$ parameters is quite small, its impact on the coherent phase contrast transfer function is clearly noticeable (see Fig. 9.5). Scherzer (1970) also deduced a relation for the geometrical resolution limit associated with a phase contrast transfer function optimized according to Eq. (9.7)

$$\rho_{\mathrm{r}\,C_5} = \frac{4}{7}\sqrt[6]{C_5 \lambda^5}. \tag{9.8}$$

The optimization of the phase contrast transfer function according to the conditions given in Eq. (9.7) is carried out in one dimension along a line profile of the two-dimensional phase contrast transfer function. Lentzen (2008) employed an optimization which takes into account the two-dimensional information transfer in the microscope. He derives the following conditions for an optimized two-dimensional transfer of a microscope limited by the fifth-order spherical aberration $C_5$

$$C_{1\,\mathrm{Lentzen}} = \frac{2}{\lambda q_{\mathrm{max}}^2} + \frac{2}{15} C_5 \lambda^4 q_{\mathrm{max}}^4$$
$$C_{3\,\mathrm{Lentzen}} = -\frac{10}{3\lambda^3 q_{\mathrm{max}}^4} - \frac{8}{9} C_5 \lambda^2 q_{\mathrm{max}}^2. \tag{9.9}$$

In the above relations, $q_{\mathrm{max}}$ is the maximum spatial frequency that contributes to the image formation. This value is determined either by the information limit of the microscope or by the objective aperture.

The calculated coherent phase contrast transfer functions $t_\mathrm{c}$ in Fig. 9.5 reveal that the optimization criteria lead to similar results. Though the difference between the optimization according to Chang *et al.* (2006) shows a distinctly different transfer at scattering angles $\theta > 30$ mrad, the general characteristics of the optimization according to Scherzer (1970) and the optimization according to Lentzen (2008) are nearly equivalent. While the optimized transfer function according to the relations in Eq. (9.7) has higher values near the first zero crossing, optimization according to Eq. (9.9) shows a more even transfer throughout the main passband.

There is a distinct difference between an optimized phase contrast transfer function of a conventional, i.e. uncorrected, electron microscope and the corresponding transfer function of a microscope which is essentially limited by fifth-order spherical

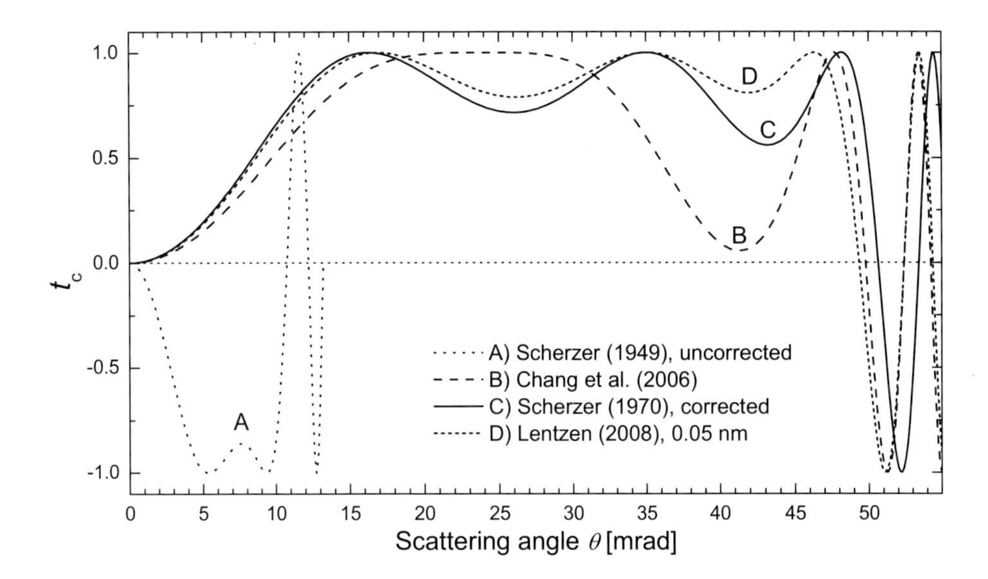


Fig. 9.5 Coherent phase contrast transfer functions $t_c$. The curves are calculated for 200 kV according to the following parameters: (A) $C_1 = -58$ nm, $C_3 = 1$ mm, $C_5 = 0$ mm, i.e. for Scherzer focus of an uncorrected microscope according to Scherzer (1949); curves (B) through (D) are representative for a $C_3$-correctable microscope which is limited by $C_5$; (B) $C_1 = +4.9$ nm, $C_3 = -11.4$ $\mu$m, $C_5 = 5$ mm, according to Chang *et al.* (2006); (C) $C_1 = +6.3$ nm, $C_3 = -12.7$ $\mu$m, $C_5 = 5$ mm, according to Scherzer (1970); (D) $C_1 = +6.0$ nm, $C_3 = -12.2$ $\mu$m, $C_5 = 5$ mm, according to Lentzen (2008) for $q_{\max} = 20$ nm$^{-1}$, which corresponds to the point resolution of a $C_5$-limited transmission electron microscope with $C_5 = 5$ mm. The curves are plotted as functions of the scattering angle $\theta$, which relates to the real space distance $d$ by $d \approx \lambda/\theta$ with the wavelength $\lambda = 2.5$ pm for 200 keV electrons. For clarity, curve (A) is only plotted up to about 13 mrad.

aberration $C_5$ (see curve A vs. curves B, C and D in Fig. 9.5). While optimizing the phase contrast transfer function on a conventional electron microscope according to Scherzer (1949) leads to positive phase contrast, i.e. dark atomic columns on a bright background, the equivalent optimization according to Eqs. (9.7) gives rise to negative phase contrast and thus to bright atoms on a dark background (see Fig. 9.6) (Scherzer, 1970; Chang *et al.*, 2006; Lentzen, 2008). Using a negative third-order spherical aberration $C_3$ in combination with a fixed positive fifth-order spherical aberration $C_5$ in order to form a broad passband in the phase contrast transfer function is known as *negative* spherical aberration imaging (NCSI) (Urban *et al.*, 2009). This imaging mode has made it possible to directly image light atoms such as individual oxygen atoms or individual carbon atoms (see, e.g. Jia *et al.*, 2003, 2008; Girit *et al.*, 2009).

Lentzen (2008) also derived expressions for a microscope whose fifth-order spherical aberration $C_5$ is adjustable. By making use of the combination aberration

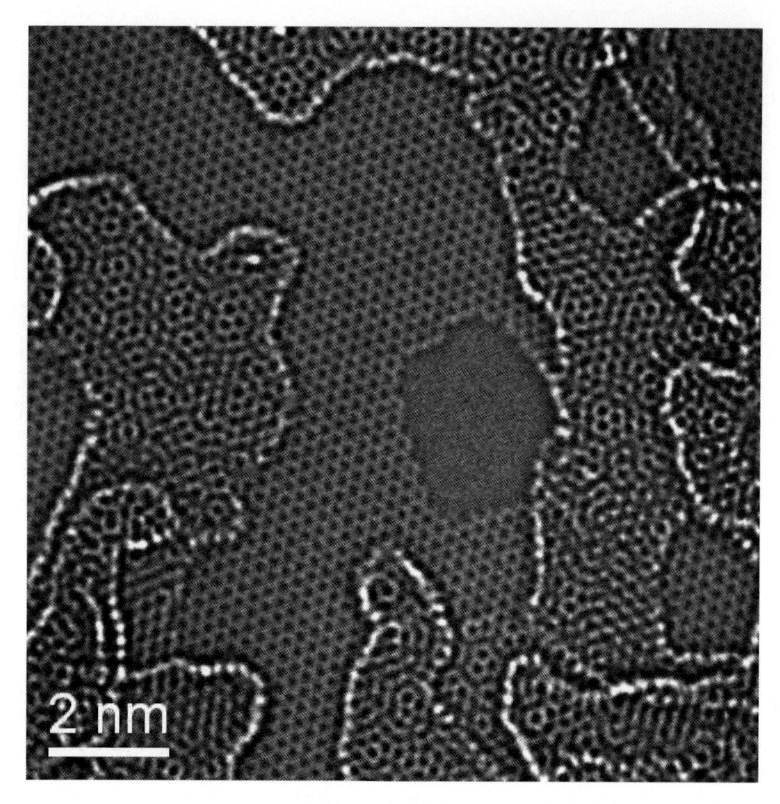


Fig. 9.6 HRTEM micrograph of graphene: a single layer of carbon atoms arranged in a honeycomb lattice. The micrograph was recorded at 80 kV on a microscope whose $C_5$ is 5.2 mm. With an information limit of 0.08 nm ($q_{\text{max}} = 12.5$ nm$^{-1}$), $C_1$ and $C_3$ were optimized according to Eqs. (9.9) yielding $C_1 = +8$ nm and $C_3 = -14$ $\mu$m. The carbon atoms are imaged with negative phase contrast, i.e. they appear as bright spots on a dark background. Besides the clean honeycomb structure, a hole in the graphene and amorphous surface layers are also observable (Girit *et al.*, 2009).

between the (negative) $C_3$ of the corrector and the (positive) $C_3$ of the objective lens, $C_5$ is in principle adjustable on a third-order aberration-corrected microscope as well. Yet, the adjustment of the fifth-order spherical aberration $C_5$ increases the off-axial coma and thus significantly reduces the field of view. This was explained in the previous chapter. Nonetheless, provided a $C_5$-correctable HRTEM instrument is available, the conditions for the adjustment of defocus, third- and fifth-order spherical aberration are (Lentzen, 2008)

$$C_{1\,\text{Lentzen}} = \frac{15}{4\lambda q_{\text{max}}^2}, \quad C_{3\,\text{Lentzen}} = -\frac{15}{\lambda^3 q_{\text{max}}^4} \quad \text{and} \quad C_{5\,\text{Lentzen}} = \frac{105}{8\lambda^5 q_{\text{max}}^6}. \tag{9.10}$$

In the above relations, it is assumed that the seventh-order spherical aberration

$C_7$ is negligibly small[5]. For a 200 kV electron microscope with an information limit of 0.05 nm ($= 1/q_{\max}$), one obtains for $C_1 = 3.75$ nm, $C_3 = -6$ $\mu$m and for $C_5 = 2.1$ mm. Figure 9.7 plots the coherent phase contrast transfer function of a 200 kV electron microscope optimized according to the relations in Eq. (9.10), and the transfer function of a microscope with an intrinsic fifth-order spherical aberration $C_5$ of 5 mm optimized according to Eq. (9.9). Both transfer functions were optimized for a theoretical information limit corresponding to $q_{\max} = 20$ nm$^{-1}$. The curves in Fig. 9.7 reveal that the adjustability of $C_5$ improves the achievable phase contrast function. The transfer function optimized according to Eq. (9.10) shows fewer modulations in the main passband and, furthermore, the actual decay of the passband becomes significant only at the point of the information limit (50 mrad). In contrast, the phase contrast transfer function optimized for a fixed $C_5$ of 5 mm shows a first zero-crossing already at $q_{\max}$, which corresponds to 50 mrad.

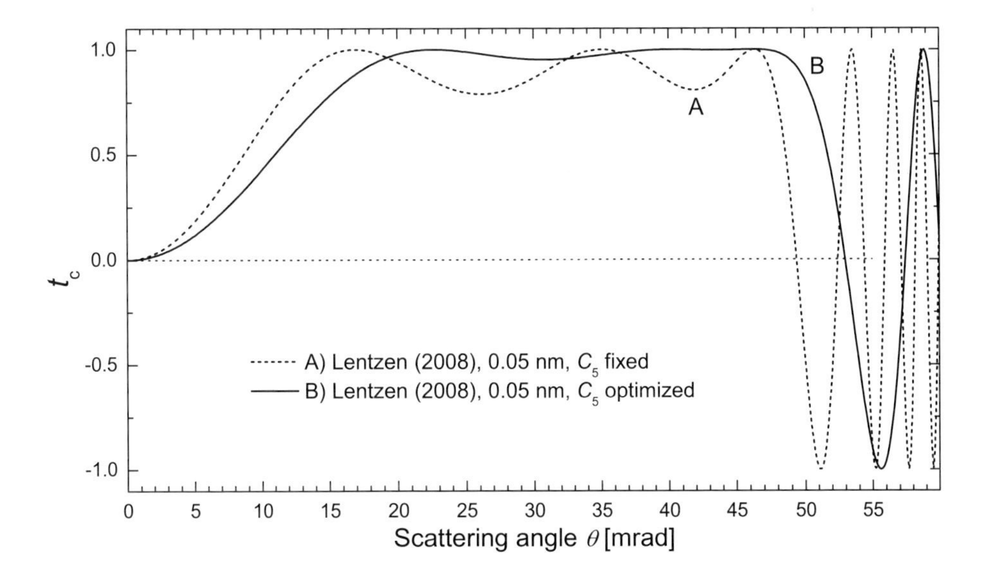


Fig. 9.7 Coherent phase contrast transfer functions $t_c$. The curves are calculated for 200 kV according to the following parameters: (A) $C_1 = +6.0$ nm, $C_3 = -12.2$ $\mu$m, $C_5 = 5$ mm, according to Lentzen (2008) for $q_{\max} = 20$ nm$^{-1}$, which corresponds to the point resolution of a $C_5$-limited transmission electron microscope (identical to curve D in Fig. 9.5), and (B) $C_1 = +3.8$ nm, $C_3 = -6.0$ $\mu$m, $C_5 = 2.1$ mm, according to Eqs. (9.10) (Lentzen, 2008) for $q_{\max} = 20$ nm$^{-1}$ and variable $C_5$. The curves are plotted as functions of the scattering angle $\theta$, which relates to the real space distance $d$ by $d \approx \lambda/\theta$ with the wavelength $\lambda = 2.5$ pm for 200 keV electrons.

[5] Lentzen (2008) also gives expressions for the case $C_7$ needs to be taken into account. This yields

$$C_1 = \frac{15}{4\lambda q_{\max}^2} - \frac{5C_7\lambda^6 q_{\max}^6}{112},\ C_3 = -\frac{15}{\lambda^3 q_{\max}^4} + \frac{15C_7\lambda^4 q_{\max}^4}{28},\ C_5 = \frac{105}{8\lambda^5 q_{\max}^6} - \frac{45C_7\lambda^2 q_{\max}^2}{32}.$$

The impact of the fixed $C_5$ of a third-order spherical aberration-corrected TEM imaging instrument, which in some way reflects the optimization of the size of the field of view, does not have a deleterious effect on the achievable phase contrast transfer function. However, the flexibility of adjusting the fifth-order spherical aberration opens the way to obtain equally well positive phase contrast. This inversion can be realized simply by reversing all the signs in the relations in Eq. (9.10). This reverses the sign of all three aberration coefficients. Hence, instead of having the atoms appearing bright on a dark background (negative phase contrast), the adjustable $C_5$ enables equally well positive phase contrast, such that the atoms appear as dark spots on a bright background.

In conclusion, one can say that the limitation imposed by the finite and fixed fifth-order spherical aberration on third-order aberration corrected instruments is not critical. Employing phase contrast implies having finite values of the isotropic axial aberrations. Though essentially limited to negative phase contrast under optimized imaging conditions, defocus $C_1$ and third-order spherical aberration $C_3$ can be used to engineer a suitable phase contrast transfer function which enables a broad passband and actually makes use of the finite positive value of $C_5$.

### 9.2.2 *Incoherent contributions*

The coherent phase contrast transfer function $t_c$ provides a means to determine the geometrical resolution limit of a microscope, i.e. the resolution limit under ideal imaging conditions. *Ideal* is meant in the sense that incoherent effects do not reduce the information transfer. In *real* electron microscopes, incoherent effects such as mechanical and electrical instabilities, partial spatial and partial temporal coherence, as well as the point spread function of the recording device, are often the factors which impose the limit on the achievable resolution (De Jong and Van Dyck, 1993; Van Dyck *et al.*, 2003). This is particularly true for electron microscopes whose coherent transfer function reflects the benefit of aberration-corrected optics, independent of whether this is TEM or STEM.

#### 9.2.2.1 *Enhanced instrument stability*

The factors discussed in a qualitative way in Chapter 2 for the case of conventional electron microscopes are also valid for aberration-corrected microscopes. As a matter of fact, the stability requirements and the possibility that incoherent optical effects become resolution-limiting strongly increase with the target resolution of the microscope. It is for this reason that special care has to be taken in the design of the microscope room (Muller *et al.*, 2006) and also in the design of high-performance low-noise electronics in ultra-high stable microscope platforms (von Harrach, 1995; O'Keefe *et al.*, 2001b; Haider *et al.*, 2008b). Indeed, the importance of the instrument stability has already been pointed out, in the early years of electron microscopy (Hillier, 1946). Before electron microscopes were actually limited by the optics, i.e. by the third-order spherical aberration, instruments needed to be built whose stability would allow to touch the optical limit.

It is not the goal of the present text to discuss in more detail the stability issues of high-end electron microscopes. However, it is a simple fact that in order to improve the resolution of a microscope, it is not only the optical performance that needs to be enhanced, but it is of equal importance (and complexity) to enhance the overall stability of the instrument. Considering that an aberration-corrected electron microscope contains many more optical elements, it is of paramount importance to implement all these optical elements in such a way that degradation of the overall stability is avoided. In contrast, the stability even needs to be enhanced in order to make use of the improved optical performance. Therefore, it is certainly an advantage, if not to say a requirement, to make available dedicated aberration-corrected microscope platforms which warrant the necessary stability and thus simplify the implementation of aberration correctors. Apart from the actual stability of the instrument, another important factor that can impact the stability of a microscope is the presence of the operator of the microscope. In order to avoid disturbances, it is beneficial to be able to control the microscope remotely, from a separate room or from a workstation which is detached from the microscope.

The optical performance dictates the stability: the stability of an electron microscope needs to be good enough to be limited by the optics. If the stability of the microscope imposes a limit on the observable resolution, the optical components or the optical setup is of too high quality. There is no reason for implementing an aberration corrector in a microscope which does not previously show an information limit which is beyond the point resolution of the (conventional) microscope. If the resolution and the information limit of a microscope is limited by stability issues, the first step towards improving the resolution is not the implementation of an aberration corrector but tackling the instabilities. Furthermore, a spherical aberration corrector does not improve the information limit; in contrast, a spherical aberration corrector adds chromatic aberration to the system and as such tends to decrease the temporal information limit.

#### 9.2.2.2 *Overcoming the limitation imposed by partial coherence*

Provided that all mechanical stability issues are resolved, the two optical effects that can alter the information transfer in an electron microscope are the damping envelope functions due to partial temporal $E_{\mathrm{t}}$ and partial spatial coherence $E_{\mathrm{s}}$. Although the focus spread caused by the chromatic aberration depends on the stability of the high tension as well as on the stability of the current of the objective lens (see, e.g. Eq. (2.23)), we shall call the effects due to partial coherence optical effects. The general expressions of these two damping envelope functions are given in Eqs. (2.24) and (2.26) as

$$E_{\mathrm{t}}(\omega) = \exp\left\{-\frac{1}{2}\frac{\pi^2}{\lambda^2}\Delta C_1^2\,(\omega\overline{\omega})^2\right\},$$

and with $\omega = \theta_x + \mathrm{i}\theta_y$,

$$E_{\mathrm{s}}(\omega) = \exp\left\{-\frac{\pi^2\theta_{\mathrm{s}}^2}{\lambda^2}\left[\nabla\chi(\omega)\right]^2\right\} = \exp\left\{-\frac{\pi^2\theta_{\mathrm{s}}^2}{\lambda^2}\left|\frac{\partial\chi}{\partial\theta_x} + \mathrm{i}\frac{\partial\chi}{\partial\theta_y}\right|^2\right\}.$$

These two equations reveal that it is primarily the envelope function due to partial spatial coherence which is influenced by aberration correction. In order to properly derive the envelope function $E_{\mathrm{s}}$ for the case of an aberration-corrected microscope, one would have to employ the expression for $\chi$ given in Eq. (7.14). In any case, aberration correction leads to a more flat aberration function — this is the main purpose of an aberration corrector. Hence, the gradient of the aberration function in the expression for $E_{\mathrm{s}}$ becomes smaller and, as a direct consequence, the deleterious effect due to partial spatial coherence is reduced. As a rule of thumb, one can state that the envelope function due to partial spatial coherence is not usually critical in aberration-corrected TEM imaging. Of course, this statement presumes that all geometrical axial aberrations are sufficiently small. A crucial point about the envelope function due to partial spatial coherence is that in the case of aberration correction, the beam divergence angle $\theta_{\mathrm{s}}$ is no longer as decisive as it is for conventional TEM imaging (see Eq. (2.34)). Indeed, as long as the aberration function is sufficiently flat, the impact of a slightly convergent illumination does not substantially affect the information transfer of the instrument.

Under normal TEM imaging conditions, it is the envelope function due to partial temporal coherence $E_{\mathrm{t}}$ which limits the information transfer in aberration-corrected microscopes. There are essentially two strategies to reduce the negative impact of the partial temporal coherence on the imaging process: one can either employ a microscope of low, or even corrected, chromatic aberration, or one can try to reduce the focal spread caused by the chromatic aberration by improving the high tension stability and the current stability of the objective lens, and by reducing the energy spread of the electron beam. Indeed, on a thermally assisted field-emission electron microscope with $\Delta E$ of about 1 eV, the dominant term in Eq. (2.23) is the energy spread of the electron beam. Assuming that the energy spread is independent of the applied high tension, a crucial point about the impact of the energy spread of the beam is that its negative effect increases with decreasing high tension of the microscope.

Employing an electron monochromator in the gun area of the microscope makes it possible to reduce the energy spread of the beam by roughly one order of magnitude, from about 1 eV to less than 100 meV (Tiemeijer, 1999b; Benner *et al.*, 2004). Neglecting high tension instabilities as well as current fluctuations, the focus spread $\Delta C_1$ is approximately given by (see Eq. (2.23))

$$\Delta C_1 \approx C_{\mathrm{C}}\frac{\Delta E_{\mathrm{rms}}}{E_0}, \tag{9.11}$$

and the information limit due to partial temporal coherence is (see Eq. (2.33))

$$\rho_t = \left(\frac{\pi \Delta C_1 \lambda}{2}\right)^{\frac{1}{2}}$$

Recalling that $\Delta E_{\rm rms}$ is related to the full width at half maximum of the energy spread $\Delta E$ by $\Delta E = 2.355 \Delta E_{\rm rms}$, we obtain for a 200 kV microscope ($\lambda = 2.5$ pm) with an energy spread of $\Delta E = 1$ eV and a constant of chromatic aberration[6] $C_{\rm C} = 2$ mm a focus spread of $\Delta C_1 = 4.2$ nm, which enables an information limit of $\rho_t = 0.13$ nm. Decreasing the energy spread to $\Delta E = 0.1$ eV with the aid of an electron monochromator, the focus spread decreases by the same factor and the information limit becomes $\rho_t = 41$ pm.

From these values we can conclude that the temporal information limit $\rho_t$ of a 200 kV microscope with $\Delta E = 1$ eV allows for an information transfer which is about 0.1 nm smaller than the point resolution of a conventional instrument with $C_3 \approx 1$ mm (see Fig. 9.5). Hence, if such a microscope is equipped with a spherical aberration corrector, the point resolution can be improved from about 0.23 nm to 0.13 nm. Provided that $C_1$ and $C_3$ are optimized according to Eq. (9.9) for a $C_5$ of 5 mm, the geometrical resolution limit of this aberration-corrected 200 kV microscope is about 50 pm (see Fig. 9.5). This is clearly smaller than the limitation imposed by the partial temporal coherence. Adding a monochromator to this microscope, the temporal information limit can be improved to about 41 pm, which allows for transferring information just a bit beyond the geometrical resolution limit of 51 pm. Hence, provided the microscope's overall stability enables a transfer to about 50 pm, the monochromator makes it possible to transfer information down to the geometrical resolution limit of this instrument. The combination of an electron monochromator and a spherical aberration corrector provides one way of achieving sub-Ångström resolution at 200 kV.

Freitag *et al.* (2004, 2005) confirmed that an improvement of the information limit on a spherical aberration-corrected microscope is feasible by employing an electron monochromator, and that an information transfer of about 50 pm becomes feasible at 300 kV (Tiemeijer *et al.*, 2008). Similarly, the micrograph shown in Fig. 9.6 was recorded by employing an electron monochromator in combination with a spherical aberration corrector. While the expected information limit at 80 kV is around 0.16 nm, the application of an electron monochromator makes it possible to shift the information limit beyond 0.1 nm. This resolution improvement is needed to resolve the interatomic distance of 0.14 nm between two carbon atoms in a graphene lattice. However, instead of making use of an electron monochromator, it is also possible to employ a cold field-emission electron source of $\Delta E \approx 0.3$ eV or — under certain circumstances — to reduce the extractor voltage of a thermally assisted field-emission electron source in order to reduce the energy spread of the beam to about $\Delta E \approx 0.2$ eV. This enables a resolution which is comparable to what is achievable with an electron monochromator at 80 kV (Meyer *et al.*, 2009).

[6]As mentioned in Chapter 8, a spherical aberration corrector increases the chromatic aberration of the system. A value of about 2 mm is typical for aberration-corrected microscopes.

As mentioned above, there is an alternative approach to reducing the impact of the partial temporal coherence on the imaging process in TEM. Instead of decreasing the energy spread of the beam, one could think about an aberration corrector which, apart from the geometrical aberrations, also corrects for the main chromatic aberration. Although an optical unit which corrects the chromatic aberration is very demanding in respect of alignment and (electrical) stability (Haider *et al.*, 2008b), this idea was realized roughly ten years after the first working spherical aberration corrector was brought to application in a transmission electron microscope. First results reveal that the combined correction of spherical and chromatic aberration is feasible and that it is beneficial for various applications (Kabius and Rose, 2008; Kabius *et al.*, 2009). As already mentioned in Chapter 8, the chromatic and spherical aberration corrector is based upon electric-magnetic quadrupole fields. The design of this corrector is described in detail by Rose (2009a). Considering only elastic electron scattering within the specimen, a monochromatic beam with $\Delta E \to 0$ combined with a spherical aberration corrector is, in principle, equivalent to a chromatic and spherical aberration-corrected microscope without monochromator. This equivalence is not fully realized in practice where a monochromator provides a finite energy resolution of $\Delta E \approx 100$ meV. Furthermore, a corrector for $C_C$ makes the imaging process insensitive to current fluctuations of the objective lens and high tension instabilities (see Eq. (2.23)).

The fundamental advantage of a chromatic aberration corrector is that electrons of varying energy have the same imaging property. This makes a chromatic aberration corrector a very powerful tool for energy-filtered imaging where, particularly for elemental mapping, electrons are collected to form images that span over large energy windows (about 5 to 50 eV). On a microscope of finite $C_C$, the width of the energy window leads to a focus blur given by Eq. (9.11) and thus to a loss of resolution (Krivanek *et al.*, 1995). On the other hand, a corrector for the chromatic aberration warrants that electrons of different energies are focused onto the same plane. Hence, a spherical and chromatic aberration-corrected microscope can, in principle, provide elemental maps at atomic resolution. However, one has to be careful in the interpretation of such images, since combined inelastic and elastic scattering can lead to elastic image contributions in energy-filtered images which are supposedly formed only by incoherently, i.e. inelastically scattered electrons (see, e.g. Krivanek *et al.*, 1992). Hence, apart from advanced HRTEM imaging, a spherical and chromatic aberration-corrected microscope opens the way for many new applications, including the expansion of high-resolution imaging at reduced microscope high tensions, like 60 or 50 kV and even lower (Rose, 1971b; Kabius and Rose, 2008).

### 9.2.3 *Summary*

Provided no physical phase plate is employed, phase contrast imaging is not feasible in an aberration-free microscope. In order to do phase contrast TEM imaging, one needs to adjust the isotropic geometrical aberrations in such a way that the coherent phase contrast transfer function shows a broad passband which enables a direct image interpretation, in principle, up to the geometrical resolution limit. Besides potential instabilities, the limited coherence of the imaging process also needs to be considered. From the two damping envelope functions due to partial coherence, it is primarily the envelope function due to partial temporal coherence which can limit the information transfer. There are two strategies to overcome the limitation imposed by partial temporal coherence: reducing the energy spread of the beam (for instance by making use of an electron monochromator), or by employing a chromatic aberration corrector.

Table 9.1 summarizes the optimum settings for phase contrast TEM imaging according to Eqs. (2.21), (9.9) and (9.10).

Table 9.1 Optimum settings for phase-contrast TEM imaging.

| Microscope | Optimum $C_1$ | Optimum $C_3$ | Optimum $C_5$ |
|---|---|---|---|
| $C_3$-limited* | $-\sqrt{\frac{4}{3}\lambda C_3}$ | fixed, positive | n/a |
| $C_5$-limited** | $\frac{2}{\lambda q_{max}^2}+\frac{2}{15}C_5\lambda^4 q_{max}^4$ | $-\frac{10}{3\lambda^3 q_{max}^4}-\frac{8}{9}C_5\lambda^2 q_{max}^2$ | fixed, positive |
| $C_5$ adjustable*** | $\pm\frac{15}{4\lambda q_{max}^2}$ | $\mp\frac{15}{\lambda^3 q_{max}^4}$ | $\pm\frac{105}{8\lambda^5 q_{max}^6}$ |

(*) see Eq. (2.21) from Scherzer (1949); (**) Eq. (9.9) from Lentzen (2008); (***) Eq. (9.10) from Lentzen (2008).

## 9.3 Aberration-Corrected STEM

In contrast to TEM imaging, where the impact of geometrical aberrations can be separated to a large extent from the incoherent effects, like the ones caused by partial temporal and partial spatial coherence, in STEM imaging this is, in general, not possible. The figure of merit in STEM imaging is the size and the characteristics of the focused electron beam, i.e. the STEM probe. The individual effects that define the STEM probe are entangled. We can distinguish between four main effects.

### 9.3.1 *Illumination aperture*

The first one is the size of the aperture. As shown in Chapter 3, the presence of a beam-defining aperture leads to a characteristic Airy pattern which corresponds to the ideal STEM probe. If the size of the STEM probe is limited by the aperture, the probe and the STEM resolution are said to be diffraction-limited. An Airy function does not consist simply of a peak — there are side lobes present which contribute to the shape of the STEM probe (see Fig. 3.2). The size of the aperture and the resulting Airy function provide the skeleton upon which the other three effects are superimposed.

The central peak of the Airy function decreases with increasing size of the aperture, i.e. with increasing illumination semi-angle. Hence, in order to form a small electron probe, it is advantageous to work with the largest feasible illumination semi-angle. This is essentially described by the diffraction limit expressed in Eq. (3.1). However, for the case of conventional probe-forming instruments we saw that there is an optimum illumination semi-angle which considers the influence of the spherical aberration and the way its impact can be balanced by a suitable adjustment of the defocus. This is similar in aberration-corrected probe-forming instruments. Instead of the third-order spherical aberration, other residual intrinsic aberrations, including the chromatic aberration, define the optimum illumination semi-angle. Hence, in principle one aims to work with the largest possible illumination semi-angle $\alpha$, yet one needs to consider the influence of residual geometrical aberrations as well as of the chromatic aberration on the characteristics of the STEM probe. This will be discussed in the following subsections.

First, however, we draw our attention to another effect. While the illumination semi-angle is crucial for the probe's lateral extension, the size of the aperture also influences the characteristics of the electron probe in a different way. Imagine the specimen is illuminated with a parallel beam of finite lateral extension. Hence, there is no focused electron probe on the specimen. From a different point of view, one can say that a parallel beam can be considered as a (broad) electron probe which has an infinite extension along the axis of the microscope[7]. The *focal depth* or the *depth of field* is in principle infinite. However, with increasing illumination angle of the electron beam, the depth of field $\Delta_{C_1}$ decreases. For a convergent electron beam, it is the probe-forming aperture which defines the illumination semi-angle $\alpha$. The dependence of the depth of field $\Delta_{C_1}$ on the illumination semi-angle $\alpha$ can be expressed as (Born and Wolf, 2001)[8]

$$\Delta_{C_1} \approx \frac{\lambda}{\alpha^2}. \tag{9.12}$$

Since the size of a diffraction-limited electron probe decreases with increasing illumination angle, the depth of field can also be expressed in terms of the STEM resolution $\rho_r$, which is directly related to the size of the probe. Rose (1975) derives the following approximate formula as a measure for a quantity which is called the

[7]Indeed, parallel electron probes are used for coherent diffraction of nanoparticles (see e.g. Zuo *et al.*, 2003; Huang *et al.*, 2008).

[8]The depth of field can be considered to correspond to the total focal tolerance of a light optical system (see Born and Wolf, 2001, 491).

*depth of contrast* $\Delta$

$$\Delta \approx \frac{9}{2}\frac{\rho_{\mathrm{r}}^{2}}{\lambda}. \tag{9.13}$$

The depth of contrast $\Delta$ can be explained as follows. Imagine an atom in vacuum located in the specimen plane of a microscope. The electron probe is now focused onto the plane of this atom and a STEM image is recorded. Indeed, what we would like to record is a whole focal series of STEM images showing this atom at different defoci. Because of the defocused electron probe at the plane of the atom, the image of the atom becomes wider with increasing defocus $|C_1|$ and its contrast decreases. Comparing all the images of the focal series, at a certain defocus $C_{1\,50\%}$, the contrast is dropped to half the value it has in the image recorded with the properly focused electron probe. The defocus $C_{1\,50\%}$ where the contrast dropped to 50% corresponds to half the depth of contrast, i.e. $\Delta/2$ (Rose, 1975).

Both the depth of field $\Delta_{C_1}$ and the depth of contrast $\Delta$ provide a measure for the vertical extension of the electron probe. How are these quantities related to each other? Let us assume the electron probe is diffraction-limited. The electron probe is thus an Airy pattern and the corresponding STEM resolution $\rho_{\mathrm{r}}$ shall be of the order of the full width at half maximum of the central peak of the Airy pattern, which is given by

$$\delta_{\mathrm{FWHM}} \approx 0.515\frac{\lambda}{\alpha}. \tag{9.14}$$

Under the approximation $\rho_{\mathrm{r}} \approx \delta_{\mathrm{FWHM}}$, we can substitute $\alpha$ from Eq. (9.14) in Eq. (9.12) and obtain the following (modified) expression for the depth of field

$$\Delta_{C_1}(\delta_{\mathrm{FWHM}}) \approx 3.8\frac{\delta_{\mathrm{FWHM}}^{2}}{\lambda}. \tag{9.15}$$

This expression is comparable to the approximate expression for the depth of contrast in Eq. (9.13). Though the numerical pre-factor is different, the above modified expression for the depth of field shows that the relations given in Eqs. (9.12) and Eq. (9.13) have the same dependency: with increasing $\alpha$ and with decreasing $\lambda$ the vertical extension of the electron probe decreases.

### 9.3.2 *Geometrical aberrations*

The second effect affecting the electron probe is determined by the (residual) geometrical aberrations. These aberrations lead to a variation of the electron wave's phase within the aperture opening. The modulation of the phase within the illumination aperture translated to the specimen plane leads to deviations of the shape of the electron probe from a simple Airy pattern; geometrical aberrations deform the Airy pattern. The shape of the electron probe reflects the symmetry of the dominant geometrical aberration. Figure 9.8 shows a series of calculated electron probes which illustrate the effects of axial coma $B_2$, threefold astigmatism

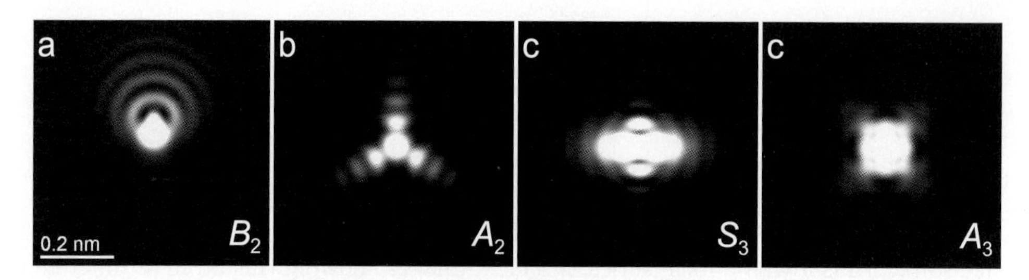


Fig. 9.8 Calculated electron probe intensity distributions for 200 keV electrons. The electron probe in (**a**) was calculated for $B_2 = 100$ nm, (**b**) for $A_2 = 100$ nm, (**c**) for $S_3 = 1$ $\mu$m and (**d**) for $A_3 = 2$ $\mu$m. All probes show in addition to the indicated aberration a defocus $C_1$ of -3 nm. The imaginary part of the aberrations are set to zero. The illumination semi-angle is 30 mrad.

$A_2$, third-order star aberration $S_3$ and fourfold astigmatism $A_3$. The probe parameters ($\alpha = 30$ mrad and $E_0 = 200$ keV) are randomly chosen, however the residual aberration coefficients for $B_2$, $A_2$, $S_3$ and $A_3$ are typical for a slightly misaligned aberration corrector. While Fig. 9.8 illustrates the effect of individual aberrations, in general, the actual shape of the electron probe is determined by the sum of all the residual geometrical aberrations.

The aperture and the geometrical aberrations are the coherent probe contributions. Employing the complex aperture coordinate $\omega$, the coherent electron wave associated with the electron probe can be expressed in analogy to Eq. (3.16) by

$$\psi_0(\mathrm{w_o}) = \int_{-\infty}^{\infty} \frac{\exp\left\{-\dfrac{2\pi \mathrm{i}}{\lambda}\chi(\omega)\right\}}{1+\exp\left\{\dfrac{\omega\overline{\omega}-\alpha^2}{\delta_\mathrm{a}^2}\right\}} \exp\left\{-\frac{2\pi \mathrm{i}}{\lambda}\Re\left[\mathrm{w_o}\overline{\omega}\right]\right\} d\omega. \tag{9.16}$$

The (complex) position of the electron probe in the specimen plane is $\mathrm{w_o}$. Similar to Eq. (3.16), the Fermi function is used to describe the aperture. However, in contrast to Eq. (3.16) one has to employ the generalized aberration function $\chi$ given in Eq. (7.14), considering all relevant and measured aberration coefficients. This is particularly true for the case of an aberration-corrected microscope, where in general there is no distinct dominant aberration limiting the size and shape of the electron probe. The intensity of the (coherent) electron probe, as for instance depicted in Fig. 9.8, is then given by Eq. (3.17).

In contrast to phase contrast TEM, the optimum electron probe is obtained by having an aberration function $\chi = 0$. This is the primary goal of the optical setup. Adjustments for the isotropic aberrations are not necessary in the way this is done in order to induce phase contrast in TEM imaging. However, since a third-order aberration-corrected microscope can be limited by fifth-order aberrations, it can be of help to adjust the third-order spherical aberration $C_3$ such that it balances the effect of the fifth-order spherical aberration $C_5$, provided that $C_5$ cannot be annulled (see, e.g. Dellby *et al.*, 2001). Using symmetry-equivalent lower-order aberrations, similar adjustments are feasible for other higher-order aberrations which are not correctable. For instance, threefold astigmatism $A_2$ can be set to a value which allows for compensating the impact of fourth-order three-lobe aberration $D_4$.

On a conventional scanning transmission electron microscope, the defocus $C_1$ needs to be adjusted such that it balances the effect of the third-order spherical aberration $C_3$ within a limited aperture opening. Because of this restriction, the illumination semi-angle of conventional probe-forming instruments is typically of the order of 10 mrad. On an aberration-corrected instrument we are no longer limited by the spherical aberration $C_3$. Indeed, we can correct for the geometrical aberrations such that the aberration function becomes a flat function within a significantly expanded angular area. Let us assume we are limited by a fifth-order aberration, like $A_5$ or $C_5$, whose uncorrectable value shall be 2 mm. Let us further assume that a phase shift $\gamma = 2\pi\chi/\lambda$ of $\pi/4$ is tolerable within the radius of the illumination aperture. Hence, according to Table 7.2, the maximum tolerable illumination semi-angle is given by

$$\alpha_{C_5} = \sqrt[6]{\frac{3}{4}\frac{\lambda}{C_5}}, \tag{9.17}$$

which yields for a 200 kV microscope $\alpha = 31$ mrad. Hence, instead of working with a 200 kV electron probe employing an illumination semi-angle of about 10 mrad whose diffraction limit $\delta_{\mathrm{D}}$ is 0.15 nm, one can, in principle, triple the illumination semi-angle and obtain a diffraction limit of 49 pm. This estimation of the resolution is based on the diffraction limit given in Eq. (3.1) and on the derivation of a maximum illumination semi-angle, which, for a given aberration coefficient, does not tolerate a phase shift exceeding $\pi/4$. With the limited illumination semi-angle of Eq. (9.17), the following expression for the size of a $C_5$-limited electron probe can be obtained:

$$\delta_{C_5} = a_1\sqrt[6]{\lambda^5 C_5}. \tag{9.18}$$

Here, the numerical parameter $a_1$ is 0.64 ($\approx 0.61(4/3)^{1/6}$). However, similar to the incoherent Scherzer conditions, the defocus $C_1$ can be adjusted to counterbalance the effect of $C_5$. Optimizing the defocus $C_1$ according to the fixed value of $C_5$ yields

$$C_1 = -1.3\frac{\lambda}{\alpha_{C_5}^2}. \tag{9.19}$$

Furthermore, the optimized illumination semi-angle is then given by

$$\boxed{\alpha_{C_5} = \sqrt[6]{a_2\frac{\lambda}{C_5}},} \tag{9.20}$$

with $a_2 = 3.9$. Employing these settings, Crewe and Salzman (1982) show that the constant $a_1$ in Eq. (9.18) can be reduced to 0.48. Optimizing defocus $C_1$ *and* third-order spherical aberration $C_3$ according to

$$C_1 = -1.56\sqrt[3]{\lambda^2 C_5} \tag{9.21}$$

$$C_3 = -2.88\sqrt[3]{\lambda C_5^2},$$

Intaraprasonk *et al.* (2008) show that for a $\pi/4$ limit one can further reduce the constant $a_1$ in Eq. (9.18) to 0.4 (Krivanek *et al.*, 2003)[9]. The optimized illumination semi-angle is then given by setting in Eq. (9.20) $a_2 = 12$, yielding for the parameters mentioned above $\alpha_{C_5} = 50$ mrad.

Hence, one can conclude that a microscope which is limited by a fifth-order aberration like $C_5$ enables a probe size and an achievable STEM resolution as given in Eq. (9.18), with $a_1 \approx 0.4$.

For an electron microscope whose illumination optics corrects fifth-order aberration as well, seventh-order geometrical aberrations become, in principle, limiting. For this case, the geometrical probe size $\delta_{C_7}$ can be estimated by

$$\delta_{C_7} \approx 0.36\sqrt[8]{\lambda^7 C_7}, \tag{9.22}$$

with $C_7$ the spherical aberration of seventh order (Krivanek *et al.*, 2003; Intaraprasonk *et al.*, 2008). However, at this point it is crucial to mention that apart from geometrical effects, i.e. aperture size and geometrical aberrations, incoherent aberrations need to be considered in the evaluation of the performance of aberration-corrected probe-forming instruments. Hence, while incoherent effects can largely be neglected in conventional probe-forming systems, they become important once electron probes can be formed which are no longer limited by the third-order spherical aberration. The limitation imposed by incoherent aberrations, particularly by the chromatic aberration, are discussed in the following subsection.

In any case, an important point about this geometrical analysis of electron probes is that probe-forming optics, which corrects the spherical aberration $C_3$, enables the formation of an electron probe whose diffraction limit is significantly improved compared to conventional microscopes. On a third-order aberration-corrected microscope, one can roughly gain a factor three in resolution provided that only geometrical effects contribute to the size of the electron probe.

Aside from the resolution improvement, the increased illumination angle leads to another very beneficial effect. Let us assume that a conventional and an aberration-corrected microscope are operated with the same type of electron source and that the source demagnification is also equal in both systems. On a conventional microscope one would employ an optimized illumination semi-angle of 10 mrad, while on the corrected one $\alpha$ is set to 30 mrad. As the intensity of the electron wave can be considered to be constant across the aperture plane, by choosing a larger aperture the intensity of the electron probe increases. Indeed, since the probe current scales with the area of the aperture opening, an increase of $\alpha$ by a factor of 3 enhances the probe current by a factor of 9.

A large beam current in a small electron probe is firstly advantageous to increase the signal-to-noise ratio in STEM micrographs. However, a large probe current is also of crucial importance for analytical techniques such as electron energy-loss

[9]Similar values were found by Rose (1974) for phase-contrast STEM. A $C_3$-limited instrument would yield a resolution limit of $0.36\sqrt[4]{\lambda^3 C_3}$ and for a $C_5$-limited microscope the resolution limit for phase-contrast STEM would be $0.31\sqrt[6]{\lambda^5 C_5}$.

spectroscopy or energy-dispersive X-ray analysis. The large probe current enables shorter acquisition times for spectra, which are then less prone to specimen drift. Alternatively, one obtains with an equal acquisition time a spectrum which has a higher signal-to-noise ratio. Hence, the increased probe current of aberration-corrected probe-forming instruments enhances the sensitivity and the spatial resolution of analytical techniques.

### 9.3.3 *Partial coherence*

#### 9.3.3.1 *Effective source size*

Besides the coherent contribution determined by the illumination aperture and the geometrical aberrations, there are also incoherent effects which need to be considered in the evaluation of the electron probe. Provided that instabilities are sufficiently small, the two incoherent optical contributions are the effects due to partial temporal and partial spatial coherence. As already outlined in Chapter 3, the effect of partial spatial coherence is described by an effective source distribution function which essentially leads to a incoherent broadening of the electron probe. Convolution of the intensity of the (coherent) probe wave field with the effective source distribution yields the incoherently broadened electron probe (Dwyer *et al.*, 2008). Of course, as long as the effective source distribution function is clearly narrower than the coherent probe intensity distribution, the impact of partial spatial coherence can be considered to be marginal. The effect of the effective source distribution is similar to a damping envelope function in the phase contrast transfer function of HRTEM imaging. The broadening of the electron probe causes object information of high spatial frequencies to be imaged with less (or no) contrast.

Though the effect of the effective source size is conceptually identical to the case of conventional probe-forming instruments, there is a quantitative difference. As illustrated in the previous section, the geometrical resolution limit of an aberration-corrected probe-forming instrument can be more than a factor three better than what is achievable on a conventional microscope. While for a (conventional) STEM probe of 0.2 nm diameter an effective source size of the order of 0.1 nm is not critical, it certainly becomes the limiting factor if the geometrical resolution limit is smaller than 0.1 nm. This simply implies that in order to be able to benefit from the improved geometrical resolution limit one needs to have a smaller effective source. This can be achieved by increasing the demagnification of the electron source. Increasing the demagnification implies that the probe current drops. Hence, instead of having a probe current which is almost an order of magnitude greater than the probe current of a conventional microscope, this factor is normally between 1 and 5, provided that on both systems the currents of the smallest resolution-carrying electron probes are compared.

Hence, due to the enhanced geometrical resolution limit in aberration-corrected microscopes, the requirement on the effective source size increases. This makes it necessary to have electron sources of highest brightness available. While a high-brightness electron source makes it generally possible to reduce the demagnification in order to achieve a certain effective source size, an electron source of high brightness also warrants a certain probe current for a small effective source size.

#### 9.3.3.2 *Chromatic aberration*

The effect of partial temporal coherence due to the finite energy spread and the chromatic aberration is more intricate. As illustrated in Fig. 3.5, the finite energy spread of the electron beam in combination with the chromatic aberration of the probe-forming lens leads to a blurring of the electron probe. The blurring caused by the chromatic aberration increases the lateral extension of the electron probe, and it increases the extension of the electron probe in direction parallel to the optical axis. The crucial factor in describing the effect of the chromatic aberration on the electron probe is the *energy length* $\ell_C$, which is the product of the coefficient of the chromatic aberration $C_C$ and the energy spread $\Delta E$, i.e. $\ell_C = C_C\,\Delta E$. The usage of the energy length essentially expresses the importance of having a narrow energy spread in combination with a small chromatic aberration.

However, as illustrated in Fig. 3.9, the lateral blurring of the electron probe due to partial temporal coherence does not predominantly affect the central maximum of the probe's intensity profile. The lateral blurring due to the chromatic aberration essentially affects the side lobes of the probe: they gain intensity. Indeed, the entire tail of the probe becomes more intense. As a consequence, it is not primarily the STEM resolution which is affected by the chromatic aberration but the achievable image contrast. Fertig and Rose (1979) show that it is possible to fulfill the Rayleigh criterion for a certain resolution with an electron probe whose side lobes would imply a vanishingly small image contrast. Resolution without contrast — this is the ultimate effect of the chromatic aberration. Hence, in order to keep a certain level of image contrast, one needs to find a setting for the electron probe which provides optimized spatial resolution and adequate image contrast. A compromise is often necessary.

The side lobes of the electron probe caused by the chromatic aberration strongly increase with increasing illumination angle. Provided that the energy spread and the coefficient of chromatic aberration are fixed, one way to reduce the impact of the chromatic aberration is to limit the illumination semi-angle $\alpha$. Krivanek *et al.* (2008a) show that an electron probe optimized in regard to the chromatic aberration has a limited illumination semi-angle $\alpha$, given by

$$\boxed{\alpha_{\mathrm{chrom}} = 1.2\sqrt{\lambda\frac{E_0}{\ell_C}}\,.} \tag{9.23}$$

For a $E_0$=200 keV electron microscope with $\Delta E = 1$ eV and a constant of chromatic aberration $C_C$ of 2 mm, i.e. with an energy length $\ell_C = 2$ mm eV, the optimal illumination semi-angle $\alpha_{\mathrm{chrom}}$ is 19 mrad. Provided this microscope is aberration corrected with its geometrical resolution limited by $C_5 = 2$ mm, the restriction on the illumination semi-angle imposed by the chromatic aberration is significantly stronger than the $\pi/4$–limitation due to $C_5$ (see Eq. (9.20)). While $\alpha_{C_5} = 50$ mrad, $\alpha_{\rm chrom} = 19$ mrad. The limited illumination semi-angle $\alpha_{\rm chrom}$ restricts the achievable diffraction-limited resolution to (see Eq. (3.1))

$$\boxed{\delta_{\rm chrom} = 0.51\sqrt{\lambda\frac{\ell_{\rm C}}{E_0}}}\,, \tag{9.24}$$

which for the values above yields $\delta_{\rm chrom}$ =81 pm. Hence, for the instrument discussed, the limitation of the illumination angle due to the chromatic aberration imposes a limit on the achievable resolution which is roughly a factor two larger than the geometrical resolution limit.

Intaraprasonk *et al.* (2008) derive for the maximum illumination semi-angle $\alpha_{\rm chrom}$ an expression which is similar to Eq. (9.23):

$$\alpha_{\rm chrom} = a_3\sqrt{\lambda\frac{E_0}{\ell_{\rm C}}}\,. \tag{9.25}$$

The numerical parameter $a_3$ depends on the intensity of the side lobes caused by the chromatic aberration. For $a_3 = 1.4$, the side lobes contain 50% of the total intensity of the probe, and for $a_3 = 1.2$, it is 25%. This explains the numerical factor in Eq. (9.23). From the value of $a_3$ and its impact on the intensity of the side lobes, it becomes clear that a slight increase of the illumination semi-angle leads to a drastic increase of the intensity of the side lobes. From this we can conclude that in order to optimize the contrast in STEM imaging, it is of crucial importance to evaluate the effect of the chromatic aberration. Yet, because residual geometrical aberrations also contribute to the side lobes of the electron probe (see Fig. 9.8), it is in general not possible to treat these effects independently from each other. Still, the above analysis of partial temporal coherence shows that the finite chromatic aberration and the finite energy spread of the electron source can have a significant impact on an aberration-corrected electron probe. This is distinctly different from the case of a conventional probe-forming microscope, where the illumination semi-angle $\alpha$ has to be set to a value where the chromatic aberration is in principle irrelevant for typical operation voltages of scanning transmission electron microscopes (100–300 kV).

The lateral blurring of the electron probe due to the chromatic aberration is usually analyzed by integrating the electron probe over a certain defocus range. This defocus range reflects the magnitudes of the energy spread and the chromatic-aberration coefficient. Hence, the primary effect of the chromatic aberration is a blurring of the electron probe along the optical axis of the microscope. Instead of working with an electron probe at a well defined focus, which indeed would only be possible if $\varDelta_{C_1} \rightarrow 0$, the blurring due to the chromatic aberration leads to an addition elongation of the electron probe along the optical axis. This elongation is insignificant for electron probes formed on conventional electron microscopes with

$\alpha \approx 10$ mrad, but it becomes a significant contribution to the vertical extension of the electron probe in aberration-corrected probe-forming instruments which are operated with larger illumination semi-angles. Hence, the crucial point is that while a large illumination semi-angle reduces the depth of field of an electron probe (see Eq. (9.12)), the chromatic aberration blurs the electron probe along the optical axis. These two effects thus work in opposite directions.

### 9.3.4 *Considerations about small electron probes*

In a conventional probe-forming instrument, the elongation of the electron probe along the optical axis is of the order of the thickness of a high-resolution STEM specimen. The limited focal depth of such an electron probe is not of importance — it is the lateral extension of the electron probe which is solely decisive for the imaging process.

On an aberration-corrected probe-forming instrument, the illumination semi-angle can be increased in order to reduce the impact of the diffraction limit on the lateral resolution. However, on opening the illumination angle, the depth of field of the electron probe decreases. Typically, the depth of field of an aberration-corrected electron probe is smaller than the thickness of a high-resolution specimen, even if the focus blur due to the chromatic aberration is taken into account. For this case, the electron probe, which rasters the specimen, needs to be regarded as a three-dimensional object which *locally* gathers structural information within the finite volume of the specimen. The information depends on the lateral position of the electron probe as well as on the vertical position within the specimen the probe is focused onto. The spatial extension of such an electron probe is determined by the parameters mentioned above, i.e. the illumination semi-angle $\alpha$, geometrical aberrations expressed by $\chi$, the size of the effective source and the chromatic aberration $C_C$. An electron probe does not have sharp borders. Figuratively, an electron probe is a local accumulation of electron intensity. Figure 9.9 illustrates the three-dimensional structure of a typical electron probe of an aberration-corrected microscope operated at 300 kV with an energy length $\ell_C$ of 1.7 mm eV, employing an illumination semi-angle $\alpha$ of 29 mrad.

The specimen position, which experiences high electron intensity, dominates the scattered intensity and thus also the structural information contained in a STEM micrograph. Of course, this is in general more complicated, particularly for the case in which channelling of the electron probe along the atomic columns takes place. Still, if the extension of the electron probe is smaller than the thickness of the specimen, the structural information contained in a STEM micrograph can depend on the height of the specimen onto which the electron beam is focused.

In principle, the electron probe can be focused onto the top or bottom surface of the specimen or onto any other plane inside the specimen. By recording focal series of STEM micrographs, three-dimensional information of the specimen can be

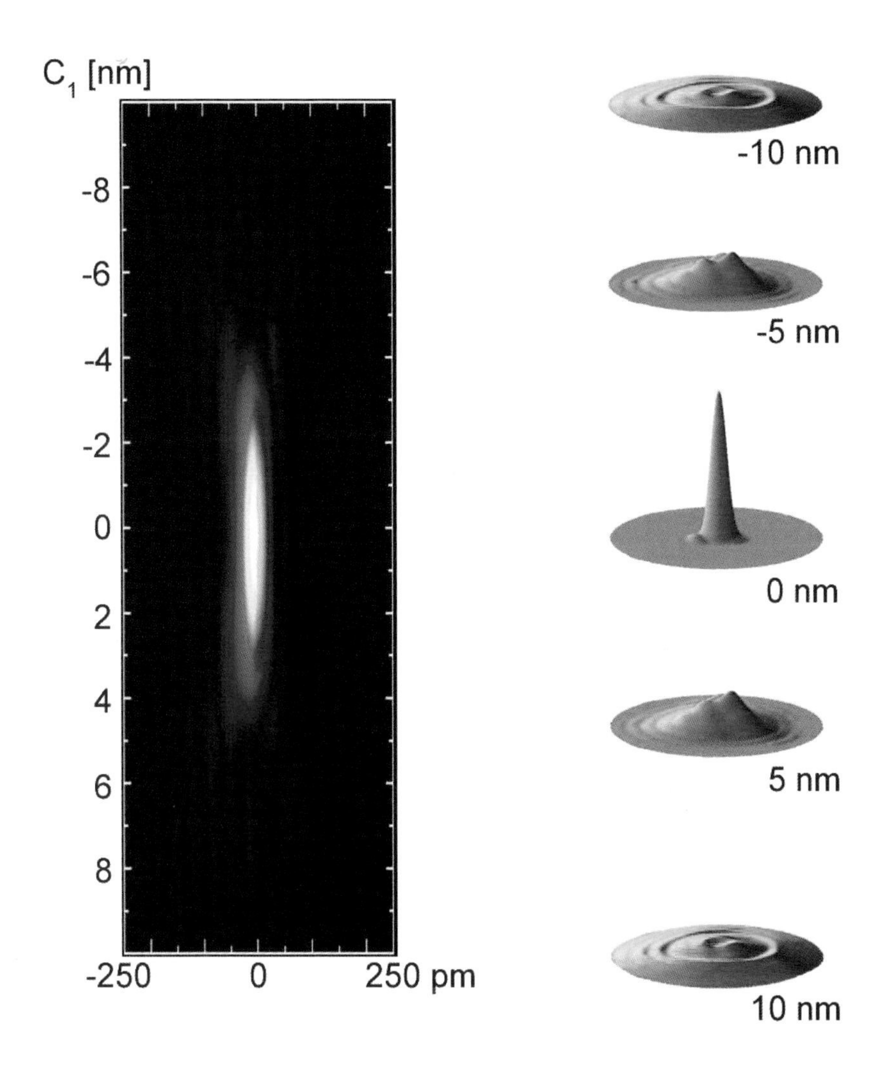


Fig. 9.9 Three-dimensional structure of a 300 keV electron probe formed with an illumination semi-angle of 29 mrad. The effective (Gaussian) source size is 25 pm and the residual geometrical aberrations considered in the probe calculation are $A_2 = 24$ nm, $B_2 = 8$ nm, $C_3 = -149$ nm, $A_3 = 97$ nm, $S_3 = 90$ nm, $A_4 = 10.2\ \mu$m, $D_4 = 7.2\ \mu$m, $B_4 = 5.5\ \mu$ m, $C_5 = 509\ \mu$m, $A_5 = 221\ \mu$m, $S_5 = 7\ \mu$m and $R_5 = 24\ \mu$m. The energy length $\ell_C$ is 1.7 mm eV.

gathered. This type of depth sectioning by STEM imaging was proposed by Rose (1975). However, aberration-corrected probe-forming instruments providing small electron probes of small depth of field were needed to carry out suitable experiments. Indeed, recent experiments have shown that similar to tilt electron tomography (see, e.g. Li *et al.*, 2008), it is feasible to access three-dimensional specimen information on the atomic scale by depth sectioning or confocal STEM imaging (van Benthem *et al.*, 2005, 2006). The interpretation of such sets of three-dimensional data is particularly complex because of dynamic scattering of the electron probe within a crystalline specimen and the resulting channelling effect which arises if a crystalline specimen is aligned along a specific zone axis (Cosgriff *et al.*, 2008; D'Alfonso *et al.*, 2008). In general, there is no linear contrast transfer that relates the specimen information to the image. However, assuming purely incoherent image formation, i.e. where the image can be considered as a convolution of the electron probe with the self-luminous specimen response function, the calculated electron probe shown in Fig. 9.9 would yield a depth of field of about 5 nm, which would provide a depth resolution of about twice this value (van Benthem *et al.*, 2006; Intaraprasonk *et al.*, 2008). The focal range, which contains most of the electron probe's intensity, is of the order of the estimated depth of field. Hence, when scanning the probe across a specimen, only a specimen layer of about 5 nm would contribute significantly to the image information. Though this is clearly larger than what is necessary for atomic-resolution imaging, by collecting a focal series of STEM images, the limited depth resolution enables the assessment of three-dimensional information of the specimen which benefits from the high lateral resolution.

Undoubtedly, being able to derive three-dimensional information of a specimen instead of seeing only projected, two-dimensional information is advantageous. However, what does the limited depth of field imply for STEM imaging, which does not intend to assess three-dimensional information but which aims solely at observing a projection of a crystal in one particular orientation with highest lateral resolution? The electron probe illustrated in Fig. 9.9 has a lateral extension of about 50 pm at the position where $C_1 \approx 0$. Hence, in principle the electron probe's lateral extension enables a lateral resolution of about 50 pm. Yet, one needs to consider that the lateral extension of the electron probe increases significantly with increasing $|C_1|$. Hence, it is only within the limited depth of field of about 5 nm where the lateral extension of the electron probe remains comparable to the value at $C_1 \approx 0$. If such an electron probe is focused in the middle of a 10 nm thick crystal, the areas above and below the depth of field of the probe would experience a much wider electron probe. Hence, instead of actually contributing to the high-resolution information, the focus-dependent blurring of the electron probe would add specimen information of reduced resolution to the final STEM micrograph (see Fig. 9.9). These 'low'-resolution contributions do not dominate the STEM micrograph but they reduce the achievable image contrast.

Given these considerations, we can conclude that the gain in lateral resolution which is feasible by increasing the illumination semi-angle on an aberration-corrected microscope comes at the expense of a reduced depth of field. The reduced depth of field opens the way for depth section and confocal STEM imaging. These techniques make it possible to access three-dimensional specimen information. For a single micrograph, however, it can happen that the increased illumination angle and the reduced depth of field impair the achievable image contrast. This effect becomes important if the thickness of the specimen clearly exceeds the depth of field of the electron probe. The reduced depth of field that comes with the highest achievable lateral resolution does not limit the application of (aberration-corrected) STEM imaging to very thin specimens. In contrast, at the expense of a slight reduction in lateral resolution, it is equally well possible to form an electron probe of a somewhat reduced illumination angle. By increasing the depth of field, this strategy can boost the contrast in single acquisition STEM micrographs which are intended to provide projected structural information of a certain specimen in zone-axis orientation and do not require the ultimate STEM resolution of the instrument (see, e.g. Fig. 9.10).

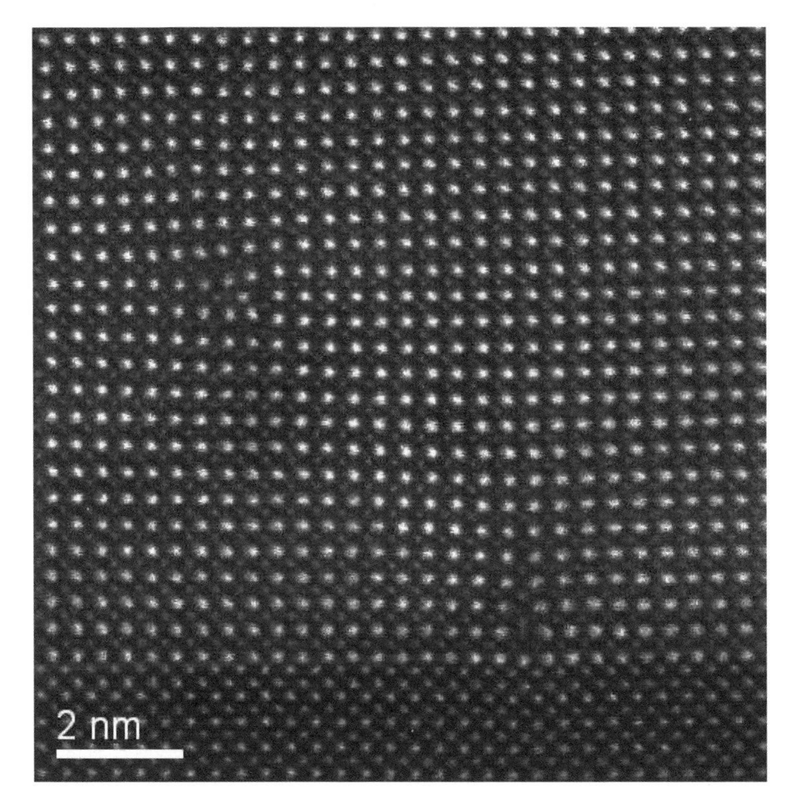


Fig. 9.10 HAADF-STEM micrograph of two edge dislocations in a $BiFeO_3$ crystal recorded in a pseudo [001] zone-axis orientation. While one dislocation is in the bulk of $BiFeO_3$, the other is related to the interface to the adjacent $SrRuO_3$ phase. In order to increase the contrast as well as the bulk information contained in the micrograph, the depth of field of the aberration-corrected electron probe was increased by employing a reduced illumination semi-angle of 18 mrad (at 300 keV), which is sub-optimal in terms of the maximal resolution. With an estimated depth of field of about 8 to 10 nm (considering the chromatic defocus blur as well), the lateral diffraction-limited resolution of about 70 pm is still sufficient to resolve the atomic structure of the dislocation cores. The upper part of the micrograph shows $BiFeO_3$ with the Bi-atomic columns appearing brightest and the FeO columns with lower intensity. Oxygen columns are not visible. The lower part of the micrograph shows $SrRuO_3$, where Sr and RuO-columns have about the same image intensity. (Image courtesy of Dr. M.D. Rossell.)

Furthermore, as the depth of field $\Delta_{C_1}$ is proportional to $\alpha^{-2}$ (Eq. (9.12)) while the size of a diffraction-limited electron probe $\delta_{\mathrm{D}}$ is proportional to $\alpha^{-1}$, a small change of the illumination semi-angle $\alpha$ has a larger effect on the depth of field $\Delta_{C_1}$ than it has on the lateral resolution. However, in the presence of the chromatic focus spread, the relative change of lateral resolution and depth of field do not reveal this exact dependency. Indeed, for small illumination angles, the vertical extension of the electron probe is dominated by the depth of field $\Delta_{C_1}$, while with increasing illumination angle the depth of field starts to be limited by the chromatic focus blur. For this reason it is useful to introduce an empirical parameter which provides a measure for the vertical extension of the electron probe. The *effective* depth of field considers both the depth of field given in Eq. (9.12) and the blur of focus due to the chromatic aberration. In principle, this parameter would need to be evaluated by three-dimensional electron-probe calculations, as depicted in Fig. 9.9. However, we can derive an approximate expression for the effective depth of field $\Delta_{\mathrm{eff}}$ by

$$\Delta_{\mathrm{eff}} \approx \sqrt{\Delta_{C_1}^2 + \Delta C_1^2} \approx \sqrt{\frac{\lambda}{\alpha^2} + \left(C_{\mathrm{C}}\frac{\Delta E}{E_0}\right)^2}, \tag{9.26}$$

where the first term of the geometrical mean reflects the geometrical depth of field from Eq. (9.12) and the second term is the spread of focus due to the chromatic aberration. For a 300 keV electron probe with $\alpha = 29$ mrad and an energy length $\ell_{\mathrm{C}} = C_{\mathrm{C}}\Delta E$ of 1.7 mm eV one obtains for the effective depth of field $\Delta_{\mathrm{eff}}$ about 6 nm. This is in agreement with Fig. 9.9, which illustrates the three-dimensional extension of an electron probe corresponding to this set of parameters. The diffraction limit $\delta_{\mathrm{D}}$ of this electron probe is 42 pm. Reducing the illumination semi-angle to 18 mrad increases the effective depth of field to 8.5 nm, while the diffraction limit of the electron probe becomes 68 pm. Employing the electron probe with $\alpha = 18$ mrad, one would thus expect an increased contrast which is due to the increased focal depth. However, another aspect is the angle $\alpha_{\mathrm{chrom}}$, which, for a 300 keV electron probe with an energy length of $\ell_{\mathrm{C}}$, is 22 mrad (see Eq. (9.23)). Hence, by reducing $\alpha$ from 30 to 18 mrad one primarily gains contrast due to the reduction of the probe tails which are due to the chromatic aberration. Because of the increased depth of field and because of the increased contrast due to the limited illumination semi-angle, the 18 mrad-electron probe enables direct atomic-resolution imaging with optimized contrast. The corresponding micrographs provide direct structural data of the specimen with a slight loss of resolution that comes with the reduced illumination semi-angle (see Fig. 9.10). This type of compromise between resolution and contrast is unavoidable when working with electron probes of limited depth of field. However, it is only the application of an aberration corrector which enables the flexibility of balancing and optimizing STEM probes according to the requirements of the specimen. It all comes back to the fact that one needs to know the STEM probe.

Table 9.2 Optimum settings for electron probes.

| Microscope | Optimum $C_1$ | Optimum $C_3$ | Optimum $C_5$ | Optimum $\alpha$ | Resolution |
|---|---|---|---|---|---|
| $C_3$-limited* | $C_1 = -\sqrt{\lambda C_3}$ | fixed, positive | n/a | $\sqrt[4]{4\frac{\lambda}{C_3}}$ | $0.43\sqrt[4]{\lambda^3 C_3}$ |
| $C_5$-limited** | $-1.56\sqrt[3]{\lambda^2 C_5}$ | $-2.88\sqrt[3]{\lambda C_5^2}$ | fixed, positive | $\sqrt[6]{12\frac{\lambda}{C_5}}$ | $0.40\sqrt[6]{\lambda^5 C_5}$ |
| $C_7$-limited** | $2.38\sqrt[4]{\lambda^3 C_7}$ | $7.07\sqrt[4]{\lambda^2 C_7^2}$ | $-5.05\sqrt[4]{\lambda C_7^3}$ | $\sqrt[8]{64\frac{\lambda}{C_7}}$ | $0.36\sqrt[8]{\lambda^7 C_7}$ |
| $C_C$-limited** | n/a | n/a | n/a | $1.2\sqrt{\lambda\frac{E_0}{\ell_C}}$ | $0.51\sqrt{\lambda\frac{\ell_C}{E_0}}$ |

(*) See Eqs. (3.9) and (3.10) from Scherzer (1949); Crewe and Salzman (1982).
(**) Intaraprasonk *et al.* (2008). For the $C_7$-limited microscope, $C_7 > 0$.

In order to fully derive the three-dimensional shape of the electron probe, calculations based on the strategies and formulas given in Chapter 3 are unavoidable (see, e.g. Lupini *et al.*, 2009). Nonetheless, estimations about the structure of the electron probe are feasible by considering the incoherent effects and the geometrical effects separately. The basis of such probe estimations are the formulas dispersed in the above text. They are summarized in Table 9.2.

The values of the resolution given in the last column of Table 9.2 reflect the diffraction limit for the corresponding optimum illumination semi-angle $\alpha$. It has to be noted that the limitation due to the chromatic aberration as expressed in Table 9.2 is independent from the geometrical aberrations. Similarly, the limitation due to the residual geometrical aberrations do not take into account contributions which are caused by the chromatic effect. The transition between a $C_5$-limited probe-forming instrument and an instrument that is limited by the chromatic aberration occurs when the energy length $\ell_C$ reaches (Intaraprasonk *et al.*, 2008)

$$\ell_C \geq 0.61 E_0 \sqrt[3]{\lambda^2 C_5}. \tag{9.27}$$

For a relatively large energy length $\ell_C$ ($> 2$ mm eV), the limitation on the illumination semi-angle and thus on the achievable resolution can be rather restrictive. Therefore, one aims at finding a compromise between the gain in resolution which one can expect from an illumination semi-angle $\alpha > \alpha_{\text{chrom}}$, and the associated loss of image contrast which is due to the chromatic probe tails.

The depth resolution achievable in depth sectioning or confocal STEM is usually estimated to be of the order of (van Benthem *et al.*, 2006; Lupini *et al.*, 2009)

$$\Delta_R \approx 2\frac{\lambda}{\alpha^2}, \tag{9.28}$$

i.e. about twice the depth of field as given in Eq. (9.12). The actual depth resolution and depth sensitivity in crystalline specimens depends on dynamic scattering effects such as the channelling of the electron probe along the atomic columns (Intarapra-sonk *et al.*, 2008).

## 9.4 New Possibilities and New Limits

A third-order spherical aberration-corrected microscope is geometrically limited by fifth-order aberrations. The limiting fifth-order aberrations are of the order of a few millimeters (Uhlemann and Haider, 1998; Dellby *et al.*, 2001). For a 300 kV electron microscope one would thus expect that in phase contrast TEM imaging, a geometrical resolution of 30 to 40 pm is feasible (Eq. (9.18)), while in STEM mode, the resolution should be around 20 to 30 pm (Table 9.2). The resolution measurements reported of (advanced) third-order aberration-corrected instruments operational at 200 or 300 kV are about a factor two larger than what one would expect if the resolution was limited solely by geometrical aberrations. Hence, the resolution in third-order spherical aberration-corrected microscopes seems not to be limited by the residual intrinsic aberrations but by incoherent effects. As explained above, it is often the partial temporal coherence which defines the information limit. In principle, this barrier can be overcome by employing an electron monochromator (Kisielowski *et al.*, 2008), a chromatic aberration corrector (Kabius *et al.*, 2009), or by designing experiments which can enhance the information transfer due to a special geometrical setup (Haigh *et al.*, 2009a). Hence, after the successful implementation of aberration correctors, which enable the correction of the resolution limiting geometrical aberrations of conventional electron microscopes, the next generation of aberration correctors aims at reducing the chromatic aberration (Haider *et al.*, 2008b; Krivanek *et al.*, 2009a; Zach, 2009; Rose, 2010) and at increasing the field of view by tackling the anisotropic off-axial coma.

Increased resolution — this is the main benefit of spherical-aberration corrected imaging. This characteristic translates into various fields of applications and enables novel imaging and analysis techniques. One particularly important side effect of aberration-corrected imaging is the enhanced sensitivity that comes with the increased resolution. Yet, higher sensitivity, higher signal-to-noise-ratios and enhanced resolution, all these aspects characteristic for an aberration-corrected microscope require a higher electron dose. In order to achieve the same signal-to-noise ratio in a micrograph, which is recorded at twice the magnification of a reference image, an electron dose is required which is four times larger. In order to image individual carbon atoms at 80 kV, the micrograph shown in Fig. 9.6 was recorded with an electron dose of about $28 \cdot 10^5\,\mathrm{e}^-/\text{Å}^2\mathrm{s}$. Each carbon atom thus interacts with thousands of electrons within a fraction of a second. One has to be aware of the fact that electrons can influence the specimen in an electron microscope. However, to quote Otto Scherzer (1970), 'we strongly hope that the irradiated specimen resembles the pristine specimen the same way a grilled chicken is supposed to resemble rather a healthy chicken than a carbonized one'. Hence, with the increased electron dose which is often necessary in order to fully exploit the enhanced optical capabilities and, in particular, the improved sensitivity of aberration-corrected microscopes, the question of radiation damage becomes crucial.

Electrons interact elastically and inelastically with the atoms in the specimen. By inelastic electron–electron interactions, electrons can cause ionization of atoms. Ionization damage can lead to the breakage of atomic bonds. Furthermore, by inelastic electron–electron interaction and by elastic electron–nucleus interactions, electrons can deposit energy to the specimen (Jouffrey and Karlik, 1992). If the deposited energy cannot be dissipated by thermal conductivity, it simply causes the specimen to heat locally. This can result in its thermal destruction. These two types of radiation damage events are difficult to quantify, particularly because the influence of the local environment and of the geometry of the specimen are difficult to grasp. Apart from ionization damage and the damage of the specimen due to heating, the elastic interaction between the electrons and the atoms in the specimen can lead to the creation of defects. This third type of radiation damage is knock-on radiation damage. In principle, knock-on radiation damage is controllable by adjusting the high tension of the microscope.

The central point about knock-on damage is that the momentum of the electrons is changed by elastic scattering with the atoms in the specimen. The distribution of the momentum of the electrons after interacting with the specimen is reflected in the diffraction pattern. However, the change of an electron's momentum simply means that an atom must experience a force. The force leads to a kinetic energy, i.e. the atom is accelerated. In the case where the kinetic energy is large enough, the atom can be permanently displaced from its original position in the specimen. This is similar to a billiard game: one just has to imagine that the incident ball has a mass which is roughly 1,000 times smaller than the stationary balls. Hence, the kinetic energy of the incident electron needs to be sufficiently large in order to cause the 'stationary' atoms to move. However, once the electron's initial energy is sufficiently large, either a simple point defect is created or a defect cascade is initiated. The figure of merit is the recoil energy $T_{\max}$

$$T_{\max} = 4\frac{m_0}{m_{\text{atom}}} E_0 \left(1 + \frac{E_0}{2m_0c^2}\right), \tag{9.29}$$

which is the amount of kinetic energy an electron can transfer to an atom (Corbett, 1966). As can be expected from the analogy of a billiard game, $T_{\max}$ decreases with the mass $m_{\text{atom}}$ of the atom. The rest mass of the electron is $m_0$ and $c$ is the speed of light in vacuum. The billiard balls are not freely movable, though. If the kinetic energy transferred to an atom is below a certain critical value, the probability for permanent atomic displacement vanishes. The atom stays in its place and transfers its kinetic energy into thermal energy. However, if the recoil energy exceeds this critical value, the atom is permanently displaced from its position. The critical recoil energy required to initiate displacement is called the threshold recoil energy $T_d$ (Corbett, 1966). This value is a material-specific quantity, which in particular depends on the coordination of the atom and the atomic bonds. Furthermore, for anisotropic materials, $T_d$ depends on the direction. Hence, the other important measure which is used to assess the probability of radiation damage is the minimum threshold recoil energy $T_{d,min}$, which is the value of $T_d$ corresponding to the weakest direction. From this we can conclude that if $T_{max}$ exceeds $T_{d,min}$, knock-on radiation damage becomes feasible in an electron microscope. While $T_{max}$ can be derived from Eq. (9.29), values for the (minimum) threshold recoil energy can be found in the literature (see, e.g. Jung, 1991).

Apart from knowing whether radiation damage is possible, it is also important to know whether knock-on damage is likely to occur. Therefore, the binary picture revealing whether radiation damage can occur at all or not needs to be rendered to a picture which tells us whether radiation is likely. Hence, the central question is: how many damage events occur during the acquisition of an image? The number of damage events $n_d$ can be estimated by (see, e.g. Jouffrey, 1983):

$$n_d = n_{in}\, \sigma_{McF}\, n_{atoms}\, t, \tag{9.30}$$

where $n_{in}$ is the flux of electrons, i.e., the number of incident electrons per time and area, $\sigma_{McF}$ is the total scattering cross-section for atom displacement according to McKinley and Feshbach (1948), $n_{atoms}$ is the mean number of atoms illuminated by the electron beam and $t$ is the illumination time. In principle, all these parameters can be derived for a particular experimental setup, except $\sigma_{McF}$. The total scattering cross-section for atom displacement $\sigma_{McF}$ is given by (see, e.g. Corbett, 1966)

$$\sigma_{McF} = A\, Z^2\, \frac{(1-\beta^2)}{\beta^4} \times \left[(\kappa_T - 1) - \beta^2 \ln \kappa_T + \pi \frac{Z}{137}\, \beta\, (2\sqrt{\kappa_T} - 2 - \ln \kappa_T)\right], \tag{9.31}$$

where we employed the following abbreviations:

$$A = 10^{-14} \pi \frac{\text{kg}^2\text{m}^2}{\text{C}^4} \frac{e^4}{m_e^2}, \qquad \beta = \frac{v}{c} \quad \text{and} \quad \kappa_T = \frac{T_{max}}{T_{d,min}}.$$

Equation (9.31) provides the scattering cross-section in square meters. The displacement scattering cross-section $\sigma_{McF}$ thus provides a mean to evaluate the likelihood of radiation damage for a certain primary electron energy $E_0$ and a certain material characterized by $T_{d,min}$.

Figure 9.11 plots the scattering cross-sections for three relatively light elements: lithium, boron (in boron nitride) and aluminum. First of all, one notices that the scattering cross-sections $\sigma_{McF}$ have a starting energy where they become finite positive. This value reflects the threshold energy of the incident electron. Below this threshold energy, the incident electron does not possess enough energy to cause knock-on radiation damage. According to Fig. 9.11, this threshold energy for Li is around 25 keV, for B around 70 keV and for Al around 170 keV. Hence, if these

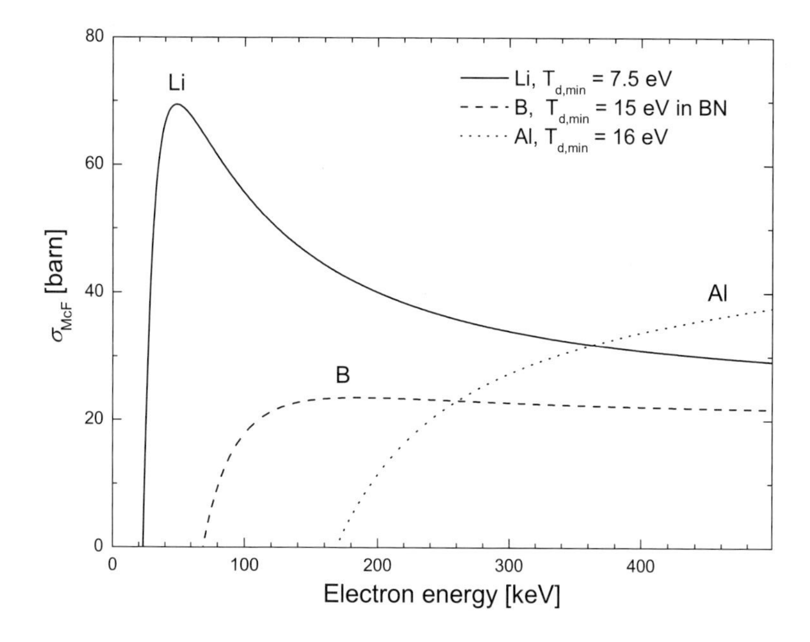


Fig. 9.11 Displacement scattering cross-sections $\sigma_{\mathrm{McF}}$ according to McKinley and Feshbach (1948) for Li ($T_{\mathrm{d,min}} \approx 7.5$ eV, see Rossell *et al.*, 2009), for boron in boron nitride ($T_{\mathrm{d,min}} \approx 15$ eV, see Zobelli *et al.*, 2007) and for aluminum ($T_{\mathrm{d,min}} \approx 16$ eV, see, e.g. Erni *et al.*, 2003a). The cross-sections are given in barn: $1\,\mathrm{barn} = 10^{-28}\mathrm{m}^2$.

materials are imaged with an electron microscope operated below this threshold acceleration voltage, knock-on radiation damage can be avoided. However, though one would expect that the knock-on radiation damage increases with increasing electron energy, this is not the case for Li and B. The maximum displacement scattering cross-section for Li is around 50 keV and for boron in boron nitride it is around 160 keV. Hence, if lithium is investigated at 50 keV primary electron energy (or boron nitride around 160 keV), the expected knock-on radiation damage is maximal. Indeed, for aluminum this maximum occurs at a primary electron energy of about 10 MeV. The point is that for light elements of low $T_{\mathrm{d,min}}$, there is a maximal probability for knock-on damage in the range in which high-resolution electron microscopes are operated (Rossell *et al.*, 2009). For this reason, one cannot simply assume that the knock-on radiation damage decreases with decreasing high tension of the microscope. However, once the high tension of the microscope is below the threshold energy, i.e. where $\sigma_{\mathrm{McF}}$ crosses the $x$-axis, no knock-on radiation damage has to be expected.

The fact that radiation damage needs to be considered in modern electron microscopes, which aim at achieving novel levels of resolution and sensitivity, is not *a priori* a negative point. Indeed, this is the point where aberration correction becomes crucially important. With an aberration-corrected electron microscope, whose chromatic blurring can be reduced by employing, for example, an electron monochromator, the high tension of the microscope can largely be adjusted according to the requirements of the specimen. Sufficient resolution is maintained because of the aberration corrector and the monochromator minimizes the information loss due to partial temporal coherence. Of course, it is not always necessary to adjust the primary electron energy of the microscope below the threshold energy, but the high tension can be adjusted such that knock-on radiation damage is kept small, i.e. such that the observed structure reflects the pristine structure of the specimen.

The flexibility of adjusting the high tension of the electron microscope according to the requirements of the specimen without losing the feasibility of accessing atomic-structure information — this is one of the main advantages of aberration-corrected electron microscopy. For a spherical aberration-corrected electron microscope, the wavelength of the electrons is not as crucial for atomic-resolution imaging as it is for conventional electron microscopes. This flexibility opens the way to analyze novel nanomaterials with a resolution and sensitivity that makes it possible to access the very basic atomic skeleton of these delicate materials. Most importantly, this can be realized by employing an experimental setup which warrants the integrity of the specimen.

Considering radiation damage as one of the limits in electron microscopy, one has to admit that this certainly is not a new insight. However, its importance is amplified by the level of resolution and sensitivity that can be achieved with aberration-corrected microscopes (see, e.g. Scherzer, 1970). Another limitation in modern electron microscopy which is also imposed by the specimen is the finite size of the object. Atoms are not infinitely small — they have a finite scattering cross-section. As such, it can indeed happen that the observable resolution is not determined by the optical performance of the microscope but by the finite size of the object, i.e. by the size of the atoms. Object-defined resolution can be important once an optical resolution of clearly better than one Ångström is achieved (see, e.g. Nellist *et al.*, 2004). This is the type of resolution that modern microscopes routinely achieve. Still, one might think that the resolution limit imposed by the object makes further advancements of electron optical instrumentation unnecessary. However, as already pointed out above, it is not primary the resolution which is beneficial in aberration-corrected electron microscopy — it is also the enhanced sensitivity. Though the resolution might be limited by the object, an improved optical performance still enables increasing of the sensitivity. A high sensitivity is crucial to directly access quantitative information about the specimen, like chemical information or local distortions. Moreover, the further development of electron microscopes will enlarge the flexibility by which the experimental conditions can be adjusted according to the requirements of the specimen (see, e.g. Rose, 2010), expanding, for instance, the application regime to the study of dynamic processes as this has already been initiated (see, e.g. Barwick *et al.*, 2008; Kim *et al.*, 2008).

Aberration-corrected electron microscopy moves our point of observation significantly closer to a level at which we can say where which atom is and what it does, where less interpretation is needed to read micrographs.

# Appendix A

# Physical Constants, Abbreviations, Acronyms and Symbols

## A.1 Physical Constants

Table A.1 List of physical constants.[a]

| Quantity | Symbol | Value |
|---|---|---|
| Elementary charge | $e$ | $1.602177 \times 10^{-19}\mathrm{C}$ |
| Electron rest mass | $m_0$ | $9.1094 \times 10^{-31}\mathrm{kg}$ |
| Imaginary unit | $\mathrm{i}$ | $\sqrt{-1}$ |
| Speed of light | $c$ | $2.9979246 \times 10^{8}\,\frac{\mathrm{m}}{\mathrm{s}^2}$ |
| Planck's constant | $\hbar = \frac{h}{2\pi}$ | $1.05457 \times 10^{-34}\,\mathrm{J\,s}$ |

[a]SI units are employed throughout the book.

## A.2 Abbreviations and Acronyms

Table A.2 List of abbreviations and acronyms used in the text.

| Abbreviation/Acronym | |
|---|---|
| ADF | Annular dark field |
| BF | Bright field |
| BFP | Back focal plane |
| CCD (camera) | Charge-Coupled Device (camera) |
| FFT | Fast Fourier Transform |
| HAADF | High angle annular dark field |
| HP | Hexapole (lens) |
| HRTEM | High-Resolution Transmission Electron Microscopy |
| MTF | Modulation Transfer Function |
| NCSI | Negative spherical aberration (CS) Imaging |
| OL | Objective lens |
| OP | Octupole lens |
| QP | Quadrupole (lens) |
| TCC | Transmission Cross Coefficient |
| TEM | Transmission Electron Microscopy |
| SEM | Scanning Electron Microscopy |
| STEM | Scanning Transmission Electron Microscopy |

# Bibliography

Archard, B.D. (1955). Two new simplified systems for the correction of spherical aberration in electron lenses, *Proceedings of the Physical Society. Section B* **68**, pp. 156–164.

Barthel, J. and Thust, A. (2008). Quantification of the information limit of transmission electron microscopes, *Physical Review Letters* **101**, p. 200801.

Barwick, B., Park, H.S., Kwon, O.H., Baskin, J.S. and Zewail, A.H. (2008). 4D imaging of transient structures and morphologies in ultrafast electron microscopy, *Science* **322**, pp. 1227–1231.

Batson, P.E. (2009). Control of parasitic aberrations in multipole optics, *Journal of Electron Microscopy* **58**, pp. 123–130.

Batson, P.E., Dellby, N. and Krivanek, O.L. (2002). Sub-Ångstrom resolution using aberration corrected electron optics, *Nature* **418**, pp. 617–620.

Beck, V.D. (1979). A hexapole spherical aberration corrector, *Optik* **53**, pp. 241–255.

Benner, G., Essers, E., Matijevic, M., Orchowski, A., Schlossmacher, P., Thesen, A., Haider, M. and Hartel, P. (2004). Performance of monochromized and aberration-corrected TEMs, *Microscopy and Microanalysis* (Suppl. 2) **10**, pp. 108–109.

Bethe, H. (1928). Theory der Beugung von Elektronen an Kristallen, *Annalen der Physik* **87**, pp. 55–129.

Bonevich, J.E. and Marks, L.D. (1988). Contrast transfer theory for non-linear imaging, *Ultramicroscopy* **26**, pp. 313–320.

Born, M. and Wolf, E. (2001). *Principles of Optics*, 7th edn. (Cambridge University Press).

Busch, H. (1922). Eine neue Methode zur e/m-Bestimmung (Vorläuffige Mitteilung.), *Physikalishce Zeitschrift* **23**, pp. 438–441.

Busch, H. (1926). Berechnung der Bahn von Kathodenstrahlen im axialsymmetrischen elektromagnetischen Felde, *Annalen der Physik* **81**, pp. 974–933.

Buseck, P.R., Cowley, J.M. and Eyering, L. (1992). *High-Resolution Transmission Electron Microscopy and Associated Techniques* (Oxford University Press, New York and Oxford).

Campbell, G.H., King, W.E. and Cohen, D. (1997). Analysis of experimental error in high resolution electron micrographs, *Microscopy and Microanalysis* **3**, pp. 451–457.

Chang, L.Y., Kirkland, A.I. and Titchmarsh, J.M. (2006). On the importance of fifth-order spherical aberration for a fully corrected electron microscope, *Ultramicroscopy* **106**, pp. 301–306.

Chen, J.H., Zandbergen, H.W. and Van Dyck, D. (2004). Atomic imaging in aberration-corrected high-resolution transmission electron microscopy, *Ultramicroscopy* **98**, pp. 81–97.

Coene, W. and Jansen, A.J.E.M. (1992). Image delocalization and high resolution transmission electron microscopic imaging with a field emission gun, *Scanning Microscopy Supplement* **6**, pp. 379–403.

Coene, W.M.J., Thust, A., Op de Beeck, M. and Van Dyck, D. (1996). Maximum-likelihood method for focus-variation image reconstruction in high resolution transmission electron microscopy, *Ultramicroscopy* **64**, pp. 109–135.

Corbett, J.W. (1966). Electron radiation damage in semiconductors and metals, in F. Seitz and D. Turnbull (eds.), *Solid State Physics, Suppl. 7* (Academic Press, New York).

Cosgriff, E., D'Alfonso, A., Allen, L., Findlay, S., Kirkland, A. and Nellist, P. (2008). Three-dimensional imaging in double aberration-corrected scanning confocal electron microscopy, Part I: Elastic scattering, *Ultramicroscopy* **108**, pp. 1558–1566.

Cowley, J.M. (1969). Image contrast in a transmission scanning electron microscope, *Applied Physics Letters* **15**, pp. 58–59.

Cowley, J.M., Merkulov, V.I. and Lannin, J.S. (1996). Imaging of light-atom nanocrystals with a thin annular detector in STEM, *Ultramicroscopy* **65**, pp. 61–70.

Cowley, J.M. and Moodie, A.F. (1957). The scattering of electrons by atoms and crystals. I. A new theoretical approach, *Acta Crystallography* **10**, pp. 609–619.

Cowley, J.M. and Moodie, A.F. (1959a). The scattering of electrons by atoms and crystals. II. The effect of finite source size, *Acta Crystallography* **12**, pp. 353–359.

Cowley, J.M. and Moodie, A.F. (1959b). The scattering of electrons by atoms and crystals. III. Single-crystal diffraction patterns, *Acta Crystallography* **12**, pp. 360–367.

Crewe, A.V. (1966). Scanning electron microscopes: Is high resolutuion possible? *Science* **154**, pp. 729–738.

Crewe, A.V. (1982). A system for the correction of axial aperture aberrations in electron lenses, *Optik* **60**, pp. 271–281.

Crewe, A.V. (1984). The sextupole corrector. 1. Algebraic calculations, *Optik* **69**, pp. 24–29.

Crewe, A.V. (1987). Optimization of small electron probes, *Ultramicroscopy* **23**, pp. 159–168.

Crewe, A.V. (1997). The scanning transmission electron microscope, in J. Orloff (ed.), *Handbook of Charged Particle Optics* (CRC Press), pp. 401–427.

Crewe, A.V. and Kopf, D. (1980). Sextupole system for the correction of spherical-aberration, *Optik* **55**, pp. 1–10.

Crewe, A.V. and Salzman, D.B. (1982). On the optimum resolution of a corrected STEM, *Ultramicroscopy* **9**, pp. 373–378.

D'Alfonso, A., Cosgriff, E., Findlay, S., Behan, G., Kirkland, A., Nellist, P. and Allen, L. (2008). Three-dimensional imaging in double aberration-corrected scanning confocal electron microscopy, Part II: Inelastic scattering, *Ultramicroscopy* **108**, pp. 1567–1578.

Danev, R., Glaeser, R.M. and Nagayama, K. (2009). Practical factors affecting the performance of a thin-film phase plate for transmission electron microscopy, *Ultramicroscopy* **109**, pp. 312–325.

De Broglie, L. (1950). *Optique Électronique et Corpusculaire* (Hermann, Paris).

De Jong, A.F. and Van Dyck, D. (1992). Ultimate resolution and information in electron microscopy: general principles, *Ultramicroscopy* **47**, pp. 266–281.

De Jong, A.F. and Van Dyck, D. (1993). Ultimate resolution and information in electron microscopy II. The information limit of transmission electron microscopes, *Ultramicroscopy* **49**, pp. 66–80.

De Ruijter, W.J. (1995). Imaging properties and applications of slow-scan charge-coupled device cameras suitable for electron microscopy, *Micron* **26**, pp. 247–275.

Dellby, N., Krivanek, O.L. and Murfitt, M.F. (2009). Optimized quadrupole–octupole $C_3/C_5$ aberration corrector for STEM, *Physics Procedia* **1**, pp. 179–183.

Dellby, N., Krivanek, O.L., Nellist, P.D., Batson, P.E. and Lupini, A.R. (2001). Progress in aberration-corrected scanning transmission electron microscopy, *Journal of Electron Microscopy* **50**, pp. 177–185.

Deltrap, J. H.M. (1964). Correction of spherical aberration with combined quadrupole–octupole units, in *Proceedings of the 3th European Region Conference on Electron Microscopy, Rome*, pp. 45–46.

Dwyer, C., Erni, R. and Etheridge, J. (2008). Method to measure spatial coherence of subangstrom electron beams, *Applied Physics Letters* **93**, p. 021115.

Egerton, R.F. (1996). *Electron Energy-Loss Spectroscopy in the Electron Microscope*, 2nd edn. (Plenum Press, New York and London).

Erni, R. and Browning, N.D. (2005). Valence electron energy-loss spectroscopy in monochromated scanning transmission electron microscopy, *Ultramicroscopy* **104**, pp. 176–192.

Erni, R., Freitag, B., Hartel, P., Müller, H., Tiemeijer, P., van der Stam, M., Stekelenburg, M., Hubert, D., Specht, P. and Garibay-Febles, V. (2006). Atomic scale analysis of planar defects in polycrystalline diamond, *Microscopy and Microanalysis* **12**, pp. 492–497.

Erni, R., Heinrich, H. and Kostorz, G. (2003a). On the internal structure of GuinierPreston zones in Al3 at.% Ag, *Philosophical Magazine Letters* **83**, pp. 599–609.

Erni, R., Heinrich, H. and Kostorz, G. (2003b). Quantitative characterization of chemical inhomogeneities in Al-Ag using high-resolution Z-contrast STEM, *Ultramicroscopy* **94**, pp. 125–133.

Erni, R., Rossell, M.D., Kisielowski, C. and Dahmen, U. (2009). Atomic-resolution imaging with a sub-50-pm electron probe, *Physical Review Letters* **102**, p. 096101.

Erni, R., Rossell, M.D. and Nakashima, P.N.H. (2010). Optimization of exit-plane waves restored from HRTEM through-focal series, *Ultramicroscopy* **110**, pp. 151–161.

Fertig, J. and Rose, H. (1979). On the theory of image formation in the electron microscope II, *Optik* **54**, pp. 165–191.

Fertig, J. and Rose, H. (1981). Resolution and contrast of crystalline objects in high-resolution scanning transmission electron microscopy, *Optik* **59**, pp. 407–429.

Frank, J. (1976). Determination of source size and energy spread from electron-micrographs using the method of Young's fringes, *Optik* **44**, pp. 379–391.

Freitag, B., Kujawa, S., Mul, P.M., Ringnald, J. and Tiemeijer, P. (2005). Breaking the spherical and chromatic aberration barrier in transmission electron microscopy, *Ultramicroscopy* **102**, pp. 209–214.

Freitag, B., Kujawa, S., Mul, P.M., Tiemeijer, P. and Snoeck, E. (2004). First experimental proof of spatial resolution improvement in a monochromized and $C_S$-corrected TEM, *Microscopy and Microanalysis (*Suppl. 2) **10**, pp. 978–979.

Gaj, M. (1971). Fifth-order field aberration coefficients for an optical surface of rotational symmetry, *Applied Optics* **10**, pp. 1642–1647.

Gamm, B., Schultheiss, K., Gerthsen, D. and Schröder, R.R. (2008). Effect of a physical phase plate on contrast transfer in an aberration-corrected transmission electron microscope, *Ultramicroscopy* **108**, pp. 878–884.

Girit, Ç.Ö., Meyer, J.C., Erni, R., Rossell, M.D., Kisielowski, C., Yang, L., Park, C.-H., Crommie, M.F., Cohen, M.L., Louie, S.G. and Zettl, A. (2009). Graphene at the edge: stability and dynamics, *Science* **323**, pp. 1705–1708.

Glaser, W. (1952). *Grundlagen der Elektronenoptik* (Springer Verlag, Vienna).

Grivet, P. (1972). *Electron Optics*, 2nd edn. (Pergamon Press, Oxford).

Haider, M., Bernhardt, W. and Rose, H. (1982). Design and test of an electric and magnetic dodecapole lens, *Optik* **63**, pp. 9–23.

Haider, M., Braunshausen, G. and Schwan, E. (1995). Correction of the spherical aberration of 200 kV TEM by means of a hexapole-corrector, *Optik* **99**, pp. 167–179.

Haider, M., Müller, H. and Uhlemann, S. (2008a). Present and future hexapole aberration correctors for high-resolution electron microscopy, in P. W. Hawkes (ed.), *Advances in Electronics and Electron Physics*, Vol. 153 (Academic Press), pp. 44–119.

Haider, M., Müller, H., Uhlemann, S., Zach, J., Loebau, U. and Hoeschen, R. (2008b). Prerequisites for a $C_C/C_S$-corrected ultrahigh-resolution TEM, *Ultramicroscopy* **108**, pp. 167–178.

Haider, M., Uhlemann, S. and Zach, J. (2000). Upper limits for the residual aberrations of high-resolution aberration-corrected STEM, *Ultramicroscopy* **81**, pp. 163–175.

Haider, M., Uhlmann, S., Rose, H., Kabius, B. and Urban, K. (1998). Electron microscopy image enhanced, *Nature* **392**, pp. 768–769.

Haigh, S.J., Sawada, H. and Kirkland, A.I. (2009a). Atomic structure imaging beyond conventional resolution limits in the transmission electron microscope, *Physical Review Letters* **103**, p. 126101.

Haigh, S.J., Sawada, H. and Kirkland, A.I. (2009b). Optimal tilt magnitude determination for aberration-corrected super resolution exit wave function reconstruction, *Philosophical Transaction of the Royal Society A* **367**, pp. 3755–3771.

Hartel, P., Rose, H. and Dinges, C. (1996). Conditions and reasons for incoherent imaging in STEM, *Ultramicroscopy* **63**, pp. 93–114.

Hawkes, P.W. (2004). Recent advances in electron optics and electron microscopy, *Annales de la Fondations Louis de Broglie* **29**, pp. 837–855.

Hawkes, P.W. (2007). Aberration correction, in P.W. Hawkes and J.C.H. Spence (eds.), *Science of Microscopy*, Vol. 1 (Springer), pp. 696–747.

Hawkes, P.W. (2009a). Aberration correction past and present, *Philosophical Transaction of the Royal Society A* **367**, pp. 3637–3664.

Hawkes, P.W. (2009b). Aberrations, in J. Orloff (ed.), *Handbook of Charged Particle Optics* (CRC Press), pp. 209–340.

Hawkes, P.W. and Kasper, E. (1989a). *Principles of Electron Optics*, Vol. 1, Basic Geometrical Optics (Academic Press, London).

Hawkes, P.W. and Kasper, E. (1989b). *Principles of Electron Optics*, Vol. 2, Applied Geometrical Optics (Academic Press, London).

Hawkes, P.W. and Kasper, E. (1994). *Principles of Electron Optics*, Vol. 3, Wave Optics (Academic Press, London).

Hetherington, C.J.D., Chang, L.-Y.S., Haigh, S., Nellist, P.D., Gontard, L.C., Dunin-Borkowski, R.E. and Kirkland, A.I. (2008). High-resolution TEM and the application of direct and indirect aberration correction, *Microscopy and Microanalysis* **14**, pp. 60–67.

Hillier, J. (1946). Further improvement of the resolving power of the electron microscope, *Journal of Applied Physics* **17**, pp. 307–309.

Hillyard, S. and Silcox, J. (1995). Detector geometry, thermal diffuse scattering and strain effects in ADF STEM imaging, *Ultramicroscopy* **58**, pp. 6–17.

Huang, W.J., Sun, R., Tao, J., Menard, L.D., Nuzzo, R.G. and Zuo, J.M. (2008). Coordination-dependent surface atomic contraction in nanocrystals revealed by coherent diffraction, *Nature Materials* **7**, pp. 308–313.

Hutchison, J.L., Titchmarsh, J.M., Cockayne, D.J.H., Doole, R.C., Hetherington, C.J., Kirkland, A.I. and Sawada, H. (2005). A versatile double aberration-corrected, energy filtered HREM/STEM for materials science, *Ultramicroscopy* **103**, pp. 7–15.

Intaraprasonk, V., Xin, H.L. and Muller, D.A. (2008). Analytic derivation of optimal imaging conditions for incoherent imaging inaberration-corrected electron microscopes, *Ultramicroscopy* **108**, pp. 1454–1466.

Ishizuka, K. (1980). Contrast transfer of crystal images in TEM, *Ultramicroscopy* **5**, pp. 55–65.

Jackson, J.D. (1998). *Classical Electrodynamics*, 3rd edn. (John Wiley).

James, E.M. and Browning, N.D. (1999). Practical aspects of atomic resolution imaging and analysis in STEM, *Ultramicroscopy* **78**, pp. 125–139.

Jia, C.L., Lentzen, M. and Urban, K. (2003). Atomic-resolution imaging of oxygen in perovskite ceramics, *Science* **299**, pp. 870–873.

Jia, C.L., Mi, S.-B., Urban, K., Vrejoiu, I., Alexe, M. and Hesse, D. (2008). Atomic-scale study of electric dipoles near charged and uncharged domain walls in ferroelectric films, *Nature Materials* **7**, pp. 57–61.

Jiye, X. and Crewe, A.V. (1985). Correction of spherical and coma aberrations with a sextupole-round lens-sextupole system, *Optik* **69**, pp. 141–146.

Jouffrey, B. (1983). Sur quelques aspects des collisions électron-atome: cas élastique et inélastique, in B. Jouffrey, A. Bourret and C. Colliex (eds.), *Cours de l'école de microscopie électronique en science des matériaux, Bombannes 1981* (CNRS), pp. 85–184.

Jouffrey, B. and Karlik, M. (1992). First attempt towards the direct determination of the Guinier-Preston zones (GP1) copper content in Al-1.7% Cu alloy, *Microscopy Microanalysis Microstructure* **3**, pp. 243–257.

Jung, P. (1991). Atomic defects in metals, in H. Ullmaier (ed.), *Landolt-Börnstein, New Series*, Vol. 25 (Springer Verlag).

Kabius, B., Hartel, P., Haider, M., Müller, H., Uhlemann, S., Loebau, U., Zach, J. and Rose, H. (2009). First application of $C_C$-corrected imaging for high-resolution and energy-filtered TEM, *Journal of Electron Microscopy* **58**, pp. 147–155.

Kabius, B. and Rose, H. (2008). Novel aberration correction concepts, in P.W. Hawkes (ed.), *Advances in Electronics and Electron Physics*, Vol. 153 (Academic Press), pp. 261–281.

Kim, J.S., LaGrange, T., Reed, B.W., Taheri, M.L., Armstrong, M.R., King, W.E., Browning, N.D. and Campbell, G.H. (2008). Imaging of transient structures using nanosecond in situ TEM, *Science* **321**, pp. 1472–1475.

Kirkland, A.I., Saxton, W.O., Chau, K.-L., Tsuno, K. and Kawasaki, M. (1995). Super-resolution by aperture synthesis: tilt series reconstruction in CTEM, *Ultramicroscopy* **57**, pp. 355–374.

Kirkland, E.J. (1982). Non-linear high resolution image processing of conventional transmission electron micrographs. I. Theory. *Ultramicroscopy* **9**, pp. 45–64.

Kirkland, E.J. (1984). Improved high resolution image processing of bright field electron micrographs. I. Theory, *Ultramicroscopy* **15**, pp. 151–172.

Kirkland, E.J. (1998). *Advanced Computing in Electron Microscopy* (Plenum Press, New York).

Kirkland, E.J., Loane, R.F. and Silcox, J. (1987). Simulation of annular dark field STEM images using a modified multislice method, *Ultramicroscopy* **23**, pp. 77–96.

Kirkland, E.J., Siegeland, B.M., Uyeda, N. and Fujiyoshi, Y. (1985). Improved high resolution image processing of bright field electron micrographs. II. Experiment, *Ultramicroscopy* **17**, pp. 87–104.

Kisielowski, C., Freitag, B., Bischoff, M., van Lin, H., Lazar, S., Knippels, G., Tiemeijer, P., van der Stam, M., von Harrach, S., Stekelenburg, M., Haider, M., Uhlemann, S., Müller, H., Hartel, P., Kabius, B., Miller, D., Petrov, I., Olson, E.A., Donchev, T., Kenik, E.A., Lupini, A.R., Bentley, J., Pennycook, S.J., Anderson, I.M., Minor, A.M., Schmid, A.K., Duden, T., Radmilovic, V., Ramasse, Q.M., Watanabe, M., Erni, R., Stach, E.A., Denes, P. and Dahmen, U. (2008). Detection of single atoms and buried defects in three dimensions by aberration-corrected electron microscope with 0.5-Å information limit, *Microscopy and Microanalysis* **14**, pp. 469–477.

Kisielowski, C., Hetherington, C.J.D., Wang, Y.C., Kilaas, R., O'Keefe, M.A. and Thust, A. (2001). Imaging columns of the light elements carbon, nitrogen and oxygen with sub-Ångstrom resolution, *Ultramicroscopy* **89**, pp. 243–263.

Kleber, W., Bautsch, H.-J. and Bohm, J. (1990). *Einfürung in die Kristallographie*, 17th edn. (Verlag Technik, Berlin).

Knoll, M. and Ruska, E. (1932). Das Elektronenmikroskop, *Zeitschrift für Physik* **78**, pp. 318–339.

Kohl, H. and Rose, H. (1985). Theory of image formation by inelastically scattered electrons in the electron microscope, in P.W. Hawkes (ed.), *Advances in Electronics and Electron Physics*, Vol. 65 (Academic Press), pp. 173–227.

Koster, A.J. and de Jong, A.F. (1991). Measurements of the spherical aberration coefficient of transmission electron microscopes by beam-tilt-induced image displacements, *Ultramicroscopy* **38**, pp. 235–240.

Krivanek, O., Corbin, G., Dellby, N., Elston, B., Keyse, R., Murfitt, M., Own, C., Szilagyi, Z. and Woodruff, J. (2008a). An electron microscope for the aberration-corrected era, *Ultramicroscopy* **108**, pp. 179–195.

Krivanek, O., Nellist, P., Dellby, N., Murfitt, M. and Szilagyi, Z. (2003). Towards sub-0.5 Å electron beams, *Ultramicroscopy* **96**, pp. 229–237.

Krivanek, O.L. (1994). Three-fold astigmatism in high-resolution transmission electron microscopy, *Ultramicroscopy* **55**, pp. 419–433.

Krivanek, O.L., Dellby, N., Keyse, R.J., Murfitt, M.F., Own, C.S. and Scilagyi, Z.S. (2008b). Advances in aberration-corrected scanning transmission electron microscopy and electron energy-loss spectroscopy, in P.W. Hawkes (ed.), *Advances in Electronics and Electron Physics*, Vol. 153 (Academic Press), pp. 121–160.

Krivanek, O.L., Dellby, N. and Lupini, A.R. (1999). Towards sub-Å electron beams, *Ultramicroscopy* **78**, pp. 1–11.

Krivanek, O.L., Dellby, N. and Murfitt, M.F. (2009a). Aberration correction in electron microscopy, in J. Orloff (ed.), *Handbook of Charged Particle Optics* (CRC Press), pp. 601–640.

Krivanek, O.L., Dellby, N., Spence, A.J., Camps, R.A. and Brown, L.M. (1997). Aberration correction in the STEM, in J.M. Rodenburg (ed.), *Electron Microscopy and Analysis Group Conf. 1997 (EMAG97), Institute of Physics Conference Series*, Vol. 153 (IOP Publishing), pp. 35–40.

Krivanek, O.L., Gubbens, A.J., Dellby, N. and Meyer, C.E. (1992). Design and first applications of a post-column imaging filter, *Microscopy Microanalysis Microstructure* **3**, pp. 187–199.

Krivanek, O.L., Kundmann, M.K. and Kimoto, K. (1995). Spatial resolution in EFTEM elemental maps, *Journal of Microscopy* **180**, pp. 277–287.

Krivanek, O.L. and Mooney, P.E. (1993). Applications of slow-scan CCD cameras in transmission electron microscopy, *Ultramicroscopy* **49**, pp. 95–108.

Krivanek, O.L., Ursin, J.P., Bacon, N.J., Corbin, G.J., Dellby, N., Hrncirik, P., Murfitt, M.F., Own, C.S. and Szilagyi, Z.S. (2009b). High-energy-resolution monochromator for aberration-corrected scanning transmission electron microscopy/electron energy-loss spectroscopy, *Philosophical Transaction of the Royal Society A* **367**, pp. 3683–3697.

LeBeau, J.M., D'Alfonso, A.J., Findlay, S.D., Stemmer, S. and Allen, L.J. (2009). Quantitative comparisons of contrast in experimental and simulated bright-field scanning transmission electron microscopy images, *Physical Review Letters* **80**, p. 174106.

Lenková, B. (2009). Electrostatic lenses, in J. Orloff (ed.), *Handbook of Charged Particle Optics* (CRC Press), pp. 161–207.

Lentzen, M. (2008). Contrast transfer and resolution limits for sub-Angstrom high-resolution transmission electron microscopy, *Microscopy and Microanalysis* **14**, pp. 16–26.

Li, Z.Y., Young, N.P., Di Vece, M., Palomba, S., Palmer, R.E., Bleloch, A.L., Curley, B.C., Johnston, R.L., Jiang, J. and Yuan, J. (2008). Three-dimensional atomic-scale structure of size-selected gold nanoclusters, *Nature* **451**, pp. 46–48.

Lichte, H. (1991). Optimum focus for taking electron holograms, *Ultramicroscopy* **38**, pp. 13–22.

Loane, R.F., Xu, P. and Silcox, J. (1992). Incoherent imaging of zone axis crystals with ADF STEM, *Ultramicroscopy* **40**, pp. 121–138.

Lupini, A., Borisevich, A., Idrobo, J., Christen, H., Biegalski, M. and Pennycook, S. (2009). Characterizing the two- and three-dimensional resolution of an improved aberration-corrected STEM, *Microscopy and Microanalysis* **15**, pp. 441–453.

Lupini, A.R. (2001). *Aberration Correction in STEM* (PhD thesis, University of Cambridge).

Majorovits, E., Barton, B., Schultheiss, K., Gerthsen, F.P.-W.D. and Schröder, R.R. (2007). Optimizing phase contrast in transmission electron microscopy with an electrostatic (Boersch) phase plate, *Ultramicroscopy* **107**, pp. 213–226.

Malm, J. and O'Keefe, M.A. (1993). Using convergence and spread-of-focus parameters to model spatial and temporal coherence in HRTEM image simulations, in G.W. Bailey and C.L. Rieder (eds.), *Proceedings of the 51st Annual Meeting of the Microscopy Society of America* (San Francisco Press), pp. 974–975.

Matsumura, S., Toyohara, M. and Tomokiyo, Y. (1990). Strain contrast of coherent precipitates in bright-field images under zone axis incidence, *Philosophical Magazine A* **62**, pp. 653–670.

McKinley, W.A. and Feshbach, H. (1948). The Coulomb scattering of relativistic electrons by nuclei, *Physical Review* **74**, pp. 1759–1763.

Metherell, A.J.F. (1975). Diffraction of Electrons by Perfect Crystals, in U. Valdre and E. Ruedl (eds.), *Electron Microscopy in Materials Science*, Vol. 2 (Commission of the European Communities), pp. 401–552.

Meyer, J.C., Chuvilin, A., Algara-Siller, G., Biskupek, J. and Kaiser, U. (2009). Selective sputtering and atomic resolution imaging of atomically thin boron nitride membranes, *Nano Letters* **9**, pp. 2683–2689.

Meyer, R.R., Kirkland, A.I., Dunin-Borkowski, R.E. and Hutchison, J.L. (2000). Experimental characterisation of CCD cameras for HREM at 300 kV, *Ultramicroscopy* **85**, pp. 9–13.

Mitsuishi, K., Takeguchi, M., Kondo, Y., Hosokawa, F., Okamoto, K., Sannomiya, T., Hori, M., Iwama, T., Kawazoe, M. and Furuya, K. (2006). Ultrahigh-vacuum third-order spherical aberration $C_S$ corrector for a scanning transmission electron microscope, *Microscopy and Microanalysis* **12**, pp. 456–460.

Möllenstedt, G. (1956). Elektronenmikroskopische Bilder mit einem nach O. Scherzer korrigiertem Objectiv, *Optik* **13**, pp. 209–215.

Müller, H., Uhlemann, S., Hartel, P. and Haider, M. (2006). Advancing the hexapole $C_S$-corrector for the scanning transmission electron microscope, *Microscopy and Microanalysis* **12**, pp. 442–455.

Müller, H., Uhlemann, S., Hartel, P. and Haider, M. (2008). Aberration-corrected optics: from an idea to a device, *Physics Proceedia* **1**, pp. 167–178.

Muller, D.A., Edwards, B., Kirkland, E.J. and Silcox, J. (2001). Simulation of thermal diffuse scattering including a detailed phonon dispersion curve, *Ultramicroscopy* **86**, pp. 371–380.

Muller, D.A., Kirkland, E.J., Thomas, M.G., Grazu., J.L., Fitting, L. and Weyland, M. (2006). Room design for high-performance electron microscopy, *Ultramicroscopy* **106**, pp. 1033–1040.

Nakashima, P. and Johnson, A. (2003). Measuring the PSF from aperture images of arbitrary shape an algorithm, *Ultramicroscopy* **94**, pp. 135–148.

Nellist, P. and Pennycook, S.J. (1998). Subangstrom resolution by underfocused incoherent transmission electron microscopy, *Physical Review Letters* **81**, pp. 4156–4159.

Nellist, P.D., Chisholm, M.F., Dellby, N., Krivanek, O.L., Murfitt, M.F., Szilagyi, Z.S., Lupini, A.R., Borisevich, A., Jr., W. H.S. and Pennycook, S.J. (2004). Direct subangstrom imaging of a crystal lattice, *Science* **305**, p. 1741.

Nellist, P.D. and Pennycook, S.J. (1999). Incoherent imaging using dynamically scattered coherent electrons, *Ultramicroscopy* **78**, pp. 111–124.

Nellist, P.D. and Pennycook, S.J. (2000). The principle and interpretation of annular dark-field Z-contrast imaging, in *Advances in Imaging and Electron Physics*, Vol. 113, pp. 147–203.

O'Keefe, M.A. (2008). Seeing atoms with aberration-corrected sub-Ångström electron microscopy, *Ultramicroscopy* **108**, pp. 196–209.

O'Keefe, M.A., Allard, L.F. and Blom, D.A. (2008). Young's fringes are not evidence of HRTEM resolution, *Microscopy and Microanalysis* (Suppl. 2) **14**, pp. 834–835.

O'Keefe, M.A., Nelson, E.C., Wang, Y.C. and Thust, A. (2001a). Sub-Ångström resolution of atomistic structures below 0.8 Å, *Philosophical Magazine B* **81**, pp. 1861–1878.

O'Keefe, M.A., Nelson, E.C., Wang, Y.C. and Thust, A. (2001b). Sub-Ångström high-resolution transmission electron microscopy at 300 kev, *Ultramicroscopy* **89**, pp. 215–241.

Op de Beeck, M., Van Dyck, D. and Coene, W. (1996). Wave function reconstruction in HRTEM: the parabola method, *Ultramicroscopy* **64**, pp. 167–183.

Otten, M.T. and Coene, W.M.J. (1993). High-resolution imaging on a field-emission tem, *Ultramicroscopy* **48**, pp. 77–91.

Overwijk, M.H.F., Bleeker, A.J. and Thust, A. (1997). Correction of three-fold astigmatism for ultra-high-resolution TEM, *Ultramicroscopy* **67**, pp. 163–170.

Pennycook, S.J. and Jesson, D.E. (1991). High-resolution Z-contrast imaging of crystals, *Ultramicroscopy* **37**, pp. 14–38.

Plies, E. and Rose, H. (1971). Über die axialen Bildfehler magnetischer Ablenksysteme mit krummer Achse, *Optik* **34**, pp. 171–190.

Pogany, A.P. and Turner, P.S. (1968). Reciprocity in electron diffraction and microscopy, *Acta Crytallographica* **A 24**, pp. 103–109.

Pöhner, W. and Rose, H. (1974). Die Auflösungsgrenze korrigierter elektronenmikroskope bei verbeulter Wellenfläche, *Optik* **41**, pp. 69–89.

Pulvermacher, H. (1981). Der Transmissions-Kreuz-Koeffizient für die elektronenmikroskopische Abbildung bei partiell kohärenter Beleuchtung und elektrischer Instabilität, *Optik* **60**, pp. 45–60.

Rafferty, B., Nellist, P.D. and Pennycook, S.J. (2001). On the origin of transverse incoherence in Z-contrast stem, *Journal of Electron Microscopy* **50**, pp. 227–233.

Ramasse, Q.M. and Bleloch, A.L. (2005). Diagnosis of aberrations from crystalline samples in scanning transmission electron microscopy, *Ultramicroscopy* **106**, pp. 37–56.

Rose, H. (1967). Über den sphärischen und chromatischen Fehler unrunder Elektronenlinsen, *Optik* **25**, pp. 587–597.

Rose, H. (1968a). Über die Berechnung der Bildfehler elektronenoptischer Systeme mit gerader Achse, Teil I, *Optik* **27**, pp. 466–474.

Rose, H. (1968b). Über die Berechnung der Bildfehler elektronenoptischer Systeme mit gerader Achse, Teil II, *Optik* **27**, pp. 497–514.

Rose, H. (1971a). Abbildungseigenschaften sphärisch korrigierter elektronenoptischer Achromate, *Optik* **33**, pp. 1–24.

Rose, H. (1971b). Elektronenoptische Aplanate, *Optik* **34**, pp. 285–311.

Rose, H. (1972). Zur Gaußian Dioptrik elektrisch-magnetischer Zylinderlinsen, *Optik* **36**, pp. 19–36.

Rose, H. (1974). Phase contrast in scanning transmission electron microscopy, *Optik* **39**, pp. 416–434.

Rose, H. (1975). Zur Theory der Bildentstehung im Elektronen-Mikroskop I, *Optik* **42**, pp. 217–244.

Rose, H. (1976a). Image formation by inelastically scattered electrons in electron microscopy, *Optik* **45**, pp. 139–158.

Rose, H. (1976b). Image formation by inelastically scattered electrons in electron microscopy. II. *Optik* **45**, pp. 187–208.

Rose, H. (1981). Correction of aperture aberrations in magnetic systems with threefold symmetry, *Nuclear Instruments and Methods* **187**, pp. 187–199.

Rose, H. (1990). Outline of a spherically corrected semiaplanatic medium-voltage transmission electron microscope, *Optik* **85**, pp. 19–24.

Rose, H. (1994). Correction of aberrations, a promising means for improving the spatial and energy resolution of energy-filtering electron microscopes, *Ultramicroscopy* **56**, pp. 11–25.

Rose, H. (1999). Prospects for realizing a sub-Å sub-eV resolution EFTEM, *Ultramicroscopy* **78**, pp. 13–25.

Rose, H. (2008). History of direct aberration correction, in P. W. Hawkes (ed.), *Handbook of Charged Particle Optics*, Vol. 153 (Academic Press), pp. 3–39.

Rose, H. (2009a). *Geometrical Charged-Particle Optics* (Springer-Verlag, Berlin Heidelberg).

Rose, H. (2009b). Historical aspects of aberration correction, *Journal of Electron Microscopy* **58**, pp. 77–85.

Rose, H. (2010). Theoretical aspects of image formation in the aberration-corrected electron microscope, *Ultramicroscopy*, doi:10.1016/j.ultramic.2009.10.003.

Rose, H. and Krahl, D. (1995). Electron optics of imaging energy filters, in L. Reimer (ed.), *Energy-Filtering Transmission Electron Microscopy*, Vol. 71, Springer Series in Optical Sciences (Springer-Verlag, Berlin Heidelberg New York), pp. 43–149.

Rossell, M.D., Erni, R., Asta, M., Radmilovic, V. and Dahmen, U. (2009). Atomic-resolution imaging of lithium in $Al_3Li$ precipitates, *Physical Review B* **80**, p. 024110.

Ruska, E. (1987). The development of the electron microscope and of electron microscopy, *Reviews of Modern Physics* **59**, pp. 627–638.

Sawada, H., Sasaki, T., Hosokawa, F., Yuasa, S., Terao, M., Kawazoe, M., Nakamichi, T., Kaneyama, T., Kondo, Y., Kimoto, K. and Suenaga, K. (2009). Correction of higher order geometrical aberration by triple 3-fold astigmatism field, *Journal of Electron Microscopy* **58**, pp. 341–347.

Saxton, W.O. (1978). Computer techniques for image processing in electron microscopy, in *Advances in Electronics and Electron Physics,* Suppl. 10 (Academic Press, New York).

Saxton, W.O. (1994). What is the focus variation method? Is it new? Is it direct? *Ultramicroscopy* **55**, pp. 171–181.

Saxton, W.O. (1995). Observation of lens aberrations for very high-resolution electron microscopy, I. Theory, *Journal of Microscopy* **179**, pp. 201–213.

Saxton, W.O. and Smith, D.J. (1985). The determination of atomic positions in high-resolution electron micrographs, *Ultramicroscopy* **18**, pp. 39–48.

Scherzer, O. (1933). Zur Theorie der elektronenoptischen Linsenfehler, *Zeitschrift für Physik* **80**, pp. 193–202.

Scherzer, O. (1936a). Die schwache elektrische Einzellinse gerinster sphärischerer Aberration, *Zeitschrift für Physik* **101**, pp. 23–26.

Scherzer, O. (1936b). Über einige Fehler von Elektronenlinsen, *Zeitschrift für Physik* **101**, pp. 593–606.

Scherzer, O. (1939). Das theoretisch erreichbare Auflösungsvermögen des Elektronenmikroskops, *Zeitschrift für Physik* **114**, pp. 427–434.

Scherzer, O. (1941). Die unteren Grenzen der Brennweite und des chromatischen Fehlers von magnetischen Elektronenlinsen, *Zeitschrift für Physik* **118**, pp. 461–466.

Scherzer, O. (1947). Sphärische und chromatische Korrektur von Elektronen-Linsen, *Optik* **2**, pp. 114–132.

Scherzer, O. (1949). The theoretical resolution limit of the electron microscope, *Journal of Applied Physics* **20**, pp. 20–27.

Scherzer, O. (1970). Die Strahlenschädigung der Objekte als Grenze für die hochauflösende Elektronenmikroskopie, *Berichte der Bunsen-Gesellschaft* **74**, pp. 1154–1167.

Scherzer, O. (1982). How not to correct an electron lens, *Ultramicroscopy* **9**, p. 385.

Scherzer, O. and Typke, D. (1967/1968). Die Auflösungsgrenze eines in zwei Schnitten sphärisch korrigierten Objektives, *Optik* **26**, pp. 564–573.

Schiske, P. (1968). Zur Frage der Bildrekonstruktion durch Fokusreihen, in *Proceedings of the 4th European Region Conference on Electron Microscopy, Rome*, pp. 145–146.

Schiske, P. (2002). Image reconstruction by means of focus series, *Journal of Microscopy* **207**, p. 154.

Schwartz, L.H. and Cohen, J.B. (1987). *Diffraction from Materials*, 2nd edn. (Springer Verlag, Berlin).

Seeliger, P. (1949). Versuche der sphärische Korrektur von Elektronenlinsen mittels nicht rotationssymmetrischer Abbildungselemente, *Optik* **5**, pp. 490–496.

Seeliger, P. (1951). Die sphärische Korrektur von Elektronenlinsen mittels nicht-rotationssymmetrischer Abbildungselemente, *Optik* **8**, pp. 311–317.

Seeliger, P. (1953). über die Justierung sphärisch korrigierter elektronenoptischer Systeme, *Optik* **10**, pp. 29–41.

Shao, Z. (1988). On the fifth order aberration in a sextupole corrected probe forming system, *Review of Scientific Instruments* **59**, pp. 2429–2437.

Shao, Z. and Crewe, A.V. (1987). Chromatic aberration effects in small electron probes, *Ultramicroscopy* **23**, pp. 169–174.

Smith, D.J. (2009). Development of aberration-corrected electron microscopy, *Microscopy and Microanalysis* **14**, pp. 2–15.

Spence, J.C.H. (1981). *Experimental High-Resolution Electron Microscopy* (Oxford University Press, New York).

Stenkamp, D. (1998). Detection and quantitative assessment of image aberrations from single HRTEM lattice images, *Journal of Microscopy* **190**, pp. 194–203.

Swanson, L.W. and Schwind, G.A. (1997). A review of the Zro/W Schottky cathode, in J. Orloff (ed.), *Handbook of Charged Particle Optics* (CRC Press), pp. 77–102.

Tang, D., Zandbergen, H.W., Jansen, J., Op de Beeck, M. and Van Dyck, D. (1996). Fine-tuning of the focal residue in exit-wave reconstruction, *Ultramicroscopy* **64**, pp. 265–276.

Thust, A., Coene, W.M.J., Op de Beeck, M. and Van Dyck, D. (1996). Focal-series reconstruction in HRTEM: simulation studies on non-periodic objects, *Ultramicroscopy* **64**, pp. 211–230.

Tiemeijer, P.C. (1999a). Measurement of Coulomb interactions in an electron beam monochromator, *Ultramicroscopy* **78**, pp. 53–62.

Tiemeijer, P.C. (1999b). Operation modes of a TEM monochromator, in *Electron Microscopy and Analysis Group Conf. 1999 (EMAG99), Institute of Physics Conference Series*, Vol. 161 (IOP Publishing), pp. 191–194.

Tiemeijer, P.C., Bischoff, M., Freitag, B. and Kisielowski, C. (2008). Using a monochromator to improve the resolution in focal-series reconstructed TEM down to 0.5 Å, in M. Luysberg, K. Tillmann and T. Weirich (eds.), *Proceedings of the 14th European Microscopy Congress (EMC), Aachen 2008, Instrumentation and Methods*, Vol. 1 (Springer-Verlag), pp. 53–54.

Tonomura, A., Allard, L.F., Pozzi, G., Joy, D.C. and Ono, Y.A. (1995). *Electron Holography* (Elsevier Science B.V., Amsterdam).

Tsuno, K. (2009). Magnetic lenses for electron microscopy, in J. Orloff (ed.), *Handbook of Charged Particle Optics* (CRC Press), pp. 129–159.

Typke, D. and Dierksen, K. (1995). Determination of image aberrations in high-resoluion electron microscopy using diffractogram and cross-correlation methods, *Optik* **99**, pp. 155–166.

Uhlemann, S. and Haider, M. (1998). Residual wave aberrations in the first spherical aberration corrected transmission electron microscope, *Ultramicroscopy* **72**, pp. 109–119.

Uno, S., Honda, K., Nakamura, N., Matsuya, M. and Zach, J. (2005). Aberration correction and its automatic control in scanning electron microscopes, *Optik* **116**, pp. 438–448.

Urban, K.W., Jia, C.-L., Houben, L., Lentzen, M., Mi, S.-B. and Tillmann, K. (2009). Negative spherical aberration ultrahigh-resolution imaging in corrected transmission electron microscopy, *Philosophical Transaction of the Royal Society A* **367**, pp. 3735–3753.

Van Aert, S. and Van Dyck, D. (2006). Resolution of coherent and incoherent imaging systems reconsidered — classical criteria and a statistical alternative, *Optics Express* **14**, pp. 3830–3839.

van Benthem, K., Lupini, A.R., Kim, M., Baik, H.S., Doh, S.-J., Lee, J.-H., Oxley, M.P., Findlay, S.D., Allen, L.J., Luck, J.T. and Pennycook, S.J. (2005). Three-dimensional imaging of individual hafnium atoms inside a semiconductor device, *Applied Physics Letters* **87**, p. 034104.

van Benthem, K., Lupini, A.R., Oxley, M.P., Findlay, S.D., Allen, L.J. and Pennycook, S.J. (2006). Three-dimensional ADF imaging of individual atoms by through-focal series scanning transmission electron microscopy, *Ultramicroscopy* **106**, pp. 1062–1068.

Van Dyck, D. (1999). A simple theory for dynamic electron diffraction in crystals, *Solid State Communications* **109**, pp. 501–505.

Van Dyck, D., Lichte, H. and van der Mast, K.D. (1996). Sub-Ångström structure characterisation: the Brite-Euram route towards one Ångström, *Ultramicroscopy* **64**, pp. 1–15.

Van Dyck, D., Van Aert, S., den Dekker, A.J. and van den Bos, A. (2003). Is atomic resolution transmission electron microscopy able to resolve and refine amorphous structures? *Ultramicroscopy* **98**, pp. 27–42.

von Ardenne, M. (1938a). Das Elektronen-Rastermikroskop; Praktische Ausführung, *Zeitschrift für technische Physik* **19(11)**, pp. 407–416.

von Ardenne, M. (1938b). Das Elektronen-Rastermikroskop; Theoretische Grundlagen, *Zeitschrift für Physik* **109**, pp. 553–572.

von Harrach, H.S. (1995). Instrumental factors in high-resolution FEG STEM, *Ultramicroscopy* **58**, pp. 1–5.

Wade, R.H. and Frank, J. (1977). Electron microscope transfer function for partially coherent axial illumination and chromatic defocus spread, *Optik* **49**, pp. 81–92.

Walther, T., Quandt, E., Stegmann, H., Thesen, A. and Benner, G. (2006). First experimental test of a new monochromated and aberration-corrected 200 kV field-emission scanning transmission electron microscope, *Ultramicroscopy* **106**, pp. 963–969.

Wang, Z.L. (1995). *Elastic and Inelastic Scattering in Electron Diffraction and Imaging* (Plenum Press, New York).

Watanabe, K., Nakanishi, N., Yamazaki, T., Kawasaki, M., Hashimoto, I. and Shiojiri, M. (2003). Effect of the incident probe on HAADF STEM images, *Physica Status Solidi (b)* **235**, pp. 179–188.

Watanabe, K., Yamazaki, T., Kikuchi, Y., Kotaka, Y., Kawasaki, M., Hashimoto, I. and Shiojiri, M. (2001). Atomic-resolution incoherent high-angle annular dark field STEM images of si(011), *Physical Review B* **63**, p. 085316.

Williams, D.B. and Carter, C.B. (1996). *Transmission Electron Microscopy* (Plenum Press, New York).

Xiu, K. and Gibson, J.M. (2001). Study of quadrupole–octopole $C_C$ corrector for the large gap HREM, *Optik* **112**, pp. 521–530.

Yamazaki, T., Kawasaki, M., Watanabe, K., Hashimoto, I. and Shiojiri, M. (2001). Artificial bright spots in atomic-resolution high-angle annular dark field STEM images, *Journal of Electron Microscopy* **50**, pp. 517–521.

Zach, J. (2009). Chromatic correction: a revolution in electron microscopy? *Philosophical Transaction of the Royal Society A* **367**, pp. 3699–3707.

Zach, J. and Haider, M. (1995). Correction of spherical and chromatic aberration in a low-voltage SEM, *Optik* **98**, pp. 112–118.

Zandbergen, H.W. and Van Dyck, D. (2000). Exit wave reconstructions using through focus series of HREM images, *Microscopy Research and Technique* **49**, pp. 301–323.

Zemlin, F. and Schiske, P. (2000). Measurement of the phase contrast transfer function and the cross-correlation peak using Young's interference fringes, *Microscopy Research and Technique* **49**, pp. 301–323.

Zemlin, F., Weiss, K., Schiske, P., Kunath, W. and Herrmann, K.-H. (1978). Coma-free alignment of high resolution electron microscopes with the aid of optical diffractograms, *Ultramicroscopy* **3**, pp. 49–60.

Zobelli, A., Gloter, A., Ewels, C.P., Seifert, G., and Colliex, C. (2007). Electron knock-on cross section of carbon and boron nitride nanotubes, *Physical Review B* **75**, p. 245402.

Zuo, J.M., Vartanyants, I., Gao, M., Zhang, R. and Nagahara, L.A. (2003). Atomic resolution imaging of a carbon nanotube from diffraction intensities, *Science* **300**, pp. 1419–1421.

[illegible]

# Index

# About the Author

Rolf Erni received his doctoral degree from the Swiss Federal Institute of Technology Zurich (ETH Zurich) where he was employed as a research assistant at the (formerly) Institute of Applied Physics. He was awarded the ETH medal for his thesis. Thereafter he carried out postdoctoral studies at the University of California at Davis and at the National Center for Electron Microscopy (NCEM), Lawrence Berkeley National Laboratory. He then joined FEI Company as a high-end application specialist and was later in the role of a senior system engineer. Before returning to NCEM as a staff scientist, he spent a period of time as a faculty member at the EMAT institute (Electron Microscopy for Materials Science) of the University of Antwerp. Rolf Erni is now head of the Electron Microscopy Center of the Swiss Federal Laboratories for Materials Science and Technology (Empa). His research interests cover various topics in electron microscopy, such as ultra-high resolution and low-voltage electron microscopy imaging as well as valence electron energy-loss spectroscopy. In collaboration with various research groups, he has published several articles in numerous fields ranging from atomic-scale studies of precipitates in binary alloys, valence electron energy-loss spectroscopy, interplanetary dust particles, carbon-based nanomaterials of reduced dimensionality, semiconductor nanostructures and in particular in the field of aberration-corrected ultra-high resolution electron microscopy imaging.